AF361792

Advanced Eco-Friendly Wood-Based Composites II

Advanced Eco-Friendly Wood-Based Composites II

Editors

Roman Reh
Lubos Kristak
Muhammad Adly Rahandi Lubis
Seng Hua Lee
Petar Antov

MDPI • Basel • Beijing • Wuhan • Barcelona • Belgrade • Manchester • Tokyo • Cluj • Tianjin

Editors

Roman Reh
Faculty of Wood Sciences and
Technology
Technical University in
Zvolen
Zvolen
Slovakia

Lubos Kristak
Faculty of Wood Sciences and
Technology
Technical University in
Zvolen
Zvolen
Slovakia

Muhammad Adly Rahandi
Lubis
Research Center for Biomass
and Bioproducts
National Research and
Innovation Agency
Cibinong
Indonesia

Seng Hua Lee
Department of Wood
Industry, Faculty of Applied
Sciences
Universiti Teknologi MARA
Pahang Branch Campus
Jengka
Bandar Tun Razak
Malaysia

Petar Antov
Department of Mechanical
Wood Technology, Faculty of
Forest Industry
University of Forestry
Sofia
Bulgaria

Editorial Office
MDPI
St. Alban-Anlage 66
4052 Basel, Switzerland

This is a reprint of articles from the Special Issue published online in the open access journal *Forests* (ISSN 1999-4907) (available at: www.mdpi.com/journal/forests/special_issues/friendly_Wood_Composites).

For citation purposes, cite each article independently as indicated on the article page online and as indicated below:

LastName, A.A.; LastName, B.B.; LastName, C.C. Article Title. *Journal Name* **Year**, *Volume Number*, Page Range.

ISBN 978-3-0365-7665-7 (Hbk)
ISBN 978-3-0365-7664-0 (PDF)

Contents

Petar Antov, Seng Hua Lee, Muhammad Adly Rahandi Lubis, Lubos Kristak and Roman Réh
Advanced Eco-Friendly Wood-Based Composites II
Reprinted from: *Forests* 2023, 14, 826, doi:10.3390/f14040826 . . . . **1**

Marius Cătălin Barbu, Eugenia Mariana Tudor, Katharina Buresova and Alexander Petutschnigg
Assessment of Physical and Mechanical Properties Considering the Stem Height and Cross-Section of *Paulownia tomentosa* (Thunb.) Steud. *x elongata* (S.Y.Hu) Wood
Reprinted from: *Forests* 2023, 14, 589, doi:10.3390/f14030589 . . . . **9**

Xiu Hao, Yanglun Yu, Chunmei Yang and Wenji Yu
In Situ Detection of the Flexural Fracture Behaviors of Inner and Outer Bamboo-Based Composites
Reprinted from: *Forests* 2023, 14, 515, doi:10.3390/f14030515 . . . . **29**

Norwahyuni Mohd Yusof, Paridah Md Tahir, Seng Hua Lee, Mohd Khairun Anwar Uyup, Redzuan Mohammad Suffian James and Syeed Saifulazry Osman Al-Edrus et al.
Effects of Adhesive Types and Structural Configurations on Shear Performance of Laminated Board from Two Gigantochloa Bamboos
Reprinted from: *Forests* 2023, 14, 460, doi:10.3390/f14030460 . . . . **39**

Apri Heri Iswanto, Jajang Sutiawan, Atmawi Darwis, Muhammad Adly Rahandi Lubis, Marta Pedzik and Tomasz Rogoziński et al.
Influence of Isocyanate Content and Hot-Pressing Temperatures on the Physical–Mechanical Properties of Particleboard Bonded with a Hybrid Urea–Formaldehyde/Isocyanate Adhesive
Reprinted from: *Forests* 2023, 14, 320, doi:10.3390/f14020320 . . . . **55**

Norwahyuni Mohd Yusof, Lee Seng Hua, Paridah Md Tahir, Redzuan Mohammad Suffian James, Syeed Saifulazry Osman Al-Edrus and Rasdianah Dahali et al.
Effects of Boric Acid Pretreatment on the Properties of Four Selected Malaysian Bamboo Strips
Reprinted from: *Forests* 2023, 14, 196, doi:10.3390/f14020196 . . . . **69**

Richard Kminiak, Miroslav Němec, Rastislav Igaz and Miloš Gejdoš
Advisability-Selected Parameters of Woodworking with a CNC Machine as a Tool for Adaptive Control of the Cutting Process
Reprinted from: *Forests* 2023, 14, 173, doi:10.3390/f14020173 . . . . **85**

Jozef Mitterpach, Roman Fojtík, Eva Machovčáková and Lenka Kubíncová
Life Cycle Assessment of a Road Transverse Prestressed Wooden–Concrete Bridge
Reprinted from: *Forests* 2022, 14, 16, doi:10.3390/f14010016 . . . . **97**

Marta Pedzik, Karol Tomczak, Dominika Janiszewska-Latterini, Arkadiusz Tomczak and Tomasz Rogoziński
Management of Forest Residues as a Raw Material for the Production of Particleboards
Reprinted from: *Forests* 2022, 13, 1933, doi:10.3390/f13111933 . . . . **113**

Iveta Marková, Martina Ivaničová, Linda Makovická Osvaldová, Jozef Harangózo and Ivana Tureková
Ignition of Wood-Based Boards by Radiant Heat
Reprinted from: *Forests* 2022, 13, 1738, doi:10.3390/f13101738 . . . . **125**

Marius Cătălin Barbu, Katharina Buresova, Eugenia Mariana Tudor and Alexander Petutschnigg
Physical and Mechanical Properties of *Paulownia tomentosa x elongata* Sawn Wood from Spanish, Bulgarian and Serbian Plantations
Reprinted from: *Forests* **2022**, *13*, 1543, doi:10.3390/f13101543 . . . . **143**

Muhammad Iqbal Maulana, Muhammad Adly Rahandi Lubis, Fauzi Febrianto, Lee Seng Hua, Apri Heri Iswanto and Petar Antov et al.
Environmentally Friendly Starch-Based Adhesives for Bonding High-Performance Wood Composites: A Review
Reprinted from: *Forests* **2022**, *13*, 1614, doi:10.3390/f13101614 . . . . **155**

forests

Editorial

Advanced Eco-Friendly Wood-Based Composites II

Petar Antov [1,*], Seng Hua Lee [2], Muhammad Adly Rahandi Lubis [3,4], Lubos Kristak [5] and Roman Réh [5]

1 Faculty of Forest Industry, University of Forestry, 1797 Sofia, Bulgaria
2 Department of Wood Industry, Faculty of Applied Sciences, Universiti Teknologi MARA (UiTM), Cawangan Pahang Kampus Jengka, Bandar Tun Razak 26400, Pahang, Malaysia; leesenghua@hotmail.com
3 Research Center for Biomass and Bioproducts, National Research and Innovation Agency, Cibinong 16911, Indonesia; muha142@brin.go.id
4 Research Collaboration Center for Biomass and Biorefinery between BRIN and Universitas Padjajaran, National Research and Innovation Agency, Cibinong 16911, Indonesia
5 Faculty of Wood Sciences and Technology, Technical University in Zvolen, 96001 Zvolen, Slovakia; kristak@tuzvo.sk (L.K.); reh@tuzvo.sk (R.R.)
* Correspondence: p.antov@ltu.bg

Citation: Antov, P.; Lee, S.H.; Lubis, M.A.R.; Kristak, L.; Réh, R. Advanced Eco-Friendly Wood-Based Composites II. *Forests* **2023**, *14*, 826. https://doi.org/10.3390/f14040826

Received: 13 April 2023
Accepted: 17 April 2023
Published: 18 April 2023

The ongoing twin transition of the wood-based panel industry towards a green, digital, and more resilient bioeconomy is essential for a successful transformation, with the aim of decarbonising the sector and implementing a circular development model, transforming linear industrial value chains to minimize pollution and waste generation, and providing more sustainable growth and jobs. This green transition represents an opportunity to place the wood-based panel industry on a new path of more sustainable and inclusive growth, tackling climate change and reducing our dependence on fossil-derived raw materials, thus improving the industry's resource efficiency and security.

A crucial circular economy principle is exploiting natural resources more effectively to produce various value-added wood-based products, as the demand for wood and wood-based components is anticipated to triple between 2010 and 2050. In efforts to promote effective recycling and reuse, the upcycling of wood and wood-based materials and the search for substitute raw materials, recent legislative regulations and increased awareness of social environments have posed new challenges to both industry and academia. These regulations and laws are related to enhancing the "cascading use" of wood or prioritising the value-added, non-fuel applications of wood and other lignocellulosic resources.

Wood composites are manufactured from different wood and non-wood lignocellulosic raw materials, bonded together with synthetic or bio-based adhesives and used for particular value-added applications and service requirements [1–9]. Conventional wood-based composites are manufactured with synthetic, formaldehyde-based resins, commonly produced from petroleum-based components, such as urea, phenol and melamine [10–13]. The use of these thermoset adhesives has several drawbacks related to the release of harmful volatile organic compounds, such as formaldehyde emissions from the created wood-based composites. Free formaldehyde emissions from the created wood-based composites has been linked to seriously detrimental human health effects, including irritation of the eyes, nose, throat and skin; nausea (short-term exposure); as well as respiratory problems and cancer (long-term exposure) [14–18]. The transition towards a circular, low-carbon wood-based panel industry, increased environmental concerns related to the use of unsustainable petroleum-based resources and the strict legislative requirements of free formaldehyde release from engineered wood composites have tremendously increased the research and development of 'green', eco-friendly wood-based composites [19–25], optimal valorisation of available lignocellulosic resources [26–30], and use of alternative raw materials [31–39]. The adverse free formaldehyde emission from wood-based composites can be mitigated by coating the surfaces of finished composites, by adding various organic or inorganic formaldehyde scavengers to synthetic wood adhesives, or by using bio-based, environmentally friendly wood adhesives [40–49]. The manufacture of binderless wood-based composites is another viable option since wood as a natural raw material is composed of biopolymeric constituents, i.e., cellulose, lignin, and hemicelluloses [50–54].

In this Special Issue, 11 well-written, authentic pieces of research and critical analysis are collated to show instances of recent technological advances in the design, manufacture, characteristics, and future uses of environmentally friendly wood and wood-based composites.

Barbu et al. [55] investigated and evaluated the physical and mechanical characteristics of *Paulownia tomentosa × elongata* plantation wood. These characteristics were determined taking into consideration the effects of cross section position and stem height. This study was conducted due to the increased interest in Paulownia as a fast-growing tree species. The authors came to the conclusion that, in terms of wood density and dimensional stability, Paulownia plantation wood becomes stable after the fifth year of growth, and they recommended harvesting trees older than seven years in order to maximise the yield of sawn wood. This recommendation was made in light of the fact that the authors harvested trees younger than five years in order to obtain optimal results.

Bamboo is another sustainable and eco-friendly material that has attracted significant study interest in recent years due to its multiple advantages and abundance, as well as ability to be recycled and reused. Although bamboo is seen as a promising substitute of wood, its poor stiffness and culm diameter are the key reasons restricting its widespread use. To address these issues, bamboo culms can be disassembled into flat thin lamellae and bonded together with an adhesive to create a certifiable structural material known as laminated bamboo [56–58].

The components of bamboo-based composites were investigated by Hao et al. [59] for their bending performance, fracture toughness, and enhancement mechanism. According to the authors' reports, the bamboo composites exhibited greater fracture toughness, compared to bamboo itself. Additionally, the composites exhibited longer deformation and less damage to fibre and parenchymal cell walls. The mechanical strength of cell walls, particularly parenchymal cell walls, was found to be enhanced by phenol–formaldehyde resin, as evidenced by an increase in indented modulus and hardness. According to the authors, the main factor affecting the fracture toughness of bamboo-based composites was the crosslinking effects of phenol–formaldehyde resin with the cell wall and fibres.

The shear performance of laminated boards fabricated from two Malaysian bamboo species was studied by Mohd Yusof et al. [60]. The two species studied were semantan (*Gigantochloa scortechinii*) and beting (*Gigantochloa levis*). Using phenol–resorcinol-formaldehyde (PRF) and polyurethane (PUR) adhesive systems, three-layer laminated bamboo panels with two lay-up patterns, perpendicular and parallel, and three strip arrangements (vertical, horizontal, and mixed) were fabricated. Board delamination, bamboo failure, and shear strength were all measured. It was determined that the lay-up pattern and adhesive type were the primary determinants of shear performance. The authors reported higher values for shear strength and bamboo failure for laminated bamboo boards bonded with PRF compared to those bonded with PUR resin. PUR-bonded bamboo, on the other hand, had a significantly lower rate of delamination, indicating a more durable bond. Overall, PRF was found to be the superior adhesive for bonding laminated bamboo boards due to its superior shear performance.

Particleboard made of sengon (*Paraserianthes falcataria*) wood was fabricated by Iswanto et al. [61]. In their study, single-layer particleboard with a density of 750 kg.m^3 was produced. Urea–formaldehyde (UF) resin added with methylene diphenyl diisocyanate (MDI) was used as a binder for the particleboard. The physico-mechanical properties of the resultant particleboards were explored. Four different hot-pressing temperatures (130, 140, 150, and 160 °C) were used to produce the particleboard. Based on a total adhesive content of 12%, the used UF/MDI mixtures were composed of 100% UF and 0% MDI, 85% UF and 15% MDI, 70% UF and 30% MDI, and 55 UF and 45% MDI, respectively. Hot pressing at 140 °C with an adhesive system consisting of 85UF/15MDI produced particleboard with physical and mechanical properties meeting the requirements for type 8 boards, as specified in JIS A5908-2003. Additionally, the particleboard fulfilled the requirements for type 2 boards according to EN 312 standards.

Another interesting study was carried out by Yusof et al. [62] on the influence of boric acid pretreatment on bamboo strips. The physical and mechanical performance of the pre-treated bamboo strips was assessed after boric acid treatment. Adhesion properties were also studied, as well as the morphological characteristics of the bamboo strips. These bamboo strips were derived from four widely distributed bamboo species in Malaysia: *Gigantochloa scortechinii*, *Gigantochloa levis*, *Bambusa vulgaris* and *Dendrocalamus asper*. According to the authors' findings, treating bamboo strips with boric acid improved their wettability, dimensional stability, and mechanical properties, resulting in a greater potential for use in composite applications. Most importantly, treatment with boric acid may improve the biological durability of the bamboo strips and broaden the range of potential applications for these laminated panels in the exterior environment.

The ongoing digitalization of the wood-based panel industry via the adoption of Industry 4.0 principles and technological advances, referring to enhanced automation and use of smart, data-driven manufacturing systems, is a prerequisite for the green and digital transformation of sector, enhancing its competitiveness and sustainability.

Kminiak et al. [63] used a computer numerical control machine for the adaptive control of cutting processes to examine the impact of various input parameters on processing wood-based composites (particleboards). The authors conducted experiments to determine the relationship between feed speed, revolutions, and radial depth of cut, as well as the equivalence of sound pressure level and milling tool temperature. The obtained results show that the noise level and temperature of the milling tool were affected by all of the investigated parameters, with the rate of radial depth of cut having the greatest influence on the rise in temperature, and the number of revolutions having the greatest influence on the sound pressure level.

Buildings are designed and constructed with careful consideration given to the selection and application of structural materials that are renewable and friendly to the environment. When compared to a reinforced concrete road bridge of the same span and load, the performance of a cross-prestressed timber-reinforced concrete bridge is superior. Mitterpach et al. [64] used the LCA principle to investigate and evaluate the environmental performance of each structure. The results show that the timber-reinforced concrete bridge was more eco-friendly than the steel-concrete road bridges. The findings have important implications for evaluating the ecological effectiveness of building components and structures.

Adhering to circular economy practices, which include the upcycling of raw materials and the increased utilization of by-products to manufacture new products with added value, Pędzik et al. [65] studied the possibilities of using forest residues generated from Scots pine harvesting as a substitutional material for manufacturing particleboards. Markedly, the composites, fabricated from forest biomass residues, exhibited satisfactory mechanical properties, fulfilling the requirements for type P5 particleboards, suitable for load-bearing applications for use in humid conditions in accordance with EN 312 standard. However, the lower dimensional stability of the produced composites allowed their classification as type P2 particleboards, suitable for internal use (including furniture) in dry conditions.

Particleboard and oriented strand boards (OSB) are two types of wood-based composite that, if burned, could create a dangerous environment in homes and public buildings. Marková et al. subjected unfinished particleboards and OSB panels were to radiant heat testing and evaluation [66]. Mass loss and time-to-ignition of the composites were reported to be significantly affected by heat flux. The experiment findings reveal that the ignition time and the temperature at which thermal decomposition occurred were both significantly higher for OSB panels than for particleboards.

Paulownia (*Paulownia tomentosa* (Tunb.) × *elongata* (S.Y. Hu)) sawn wood from three European plantation sites was studied for its physical and mechanical properties by Barbu et al. [67]. The results conclusively show that Paulownia wood's physical and mechanical qualities were significantly affected by the growing conditions. Paulownia wood was found to have a significant promise as an alternative natural feedstock to be used in specialised applications, such as non-load-bearing structural components and thermal insulation, despite having inferior physical and mechanical properties compared to traditional tree species.

Finally, Maulana et al. [68] conducted a comprehensive overview on the latest advancements in the field of "green," eco-friendly, starch-based wood adhesives. These can be used to produce wood-based composites that are non-toxic, have low emissions, superior properties and a reduced negative impact on the environment. The authors described and analyzed the vast potential of starch as a cheap and abundant natural feedstock for use in wood adhesives. New methods of starch modification were also discussed, with the goal of enhancing the effectiveness of starch-based wood adhesives.

A significant precondition for the ongoing movement toward the production of environmentally friendly, high-performance wood-based composites is the industry's ongoing transformation from a linear to a circular bioeconomy. This transition is a strong prerequisite for this production trend. This Special Issue provides a detailed summary of potential developments in the design, production and applications of sustainable, environmentally friendly wood-based composites with enhanced properties and a reduced carbon footprint, which form the focus of this discussion.

Author Contributions: Conceptualization, P.A., S.H.L., M.A.R.L., L.K. and R.R.; Writing and editing, P.A., S.H.L., M.A.R.L., L.K. and R.R. All authors have read and agreed to the published version of the manuscript.

Acknowledgments: This research was supported by the project "Properties and application of innovative biocomposite materials in furniture manufacturing", no. H C--1215/04.2022, carried out at the University of Forestry, Sofia, Bulgaria. This research was also supported by the Slovak Research and Development Agency under contracts No. APVV-20-004, APVV-19-0269 and No. SK-CZ-RD-21-0100.

Conflicts of Interest: The authors declare no conflict of interest.

References

1. Pizzi, A.; Papadopoulos, A.N.; Policardi, F. Wood composites and their polymer binders. *Polymers* **2020**, *12*, 1115. [CrossRef]
2. Barbu, M.C.; Reh, R.; Irle, M. Wood-based composites. In *Research Developments in Wood Engineering and Technology*; Aguilera, A., Davim, J.P., Eds.; IGI Global: Hershey, PA, USA, 2014; Chapter 1; pp. 1–45.
3. Zanuttini, R.; Negro, F. Wood-Based Composites: Innovation towards a Sustainable Future. *Forests* **2021**, *12*, 1717. [CrossRef]
4. Papadopoulos, A.N. Advances in Wood Composites III. *Polymers* **2021**, *13*, 163. [CrossRef] [PubMed]
5. Mirski, R.; Derkowski, A.; Kawalerczyk, J.; Dziurka, D.; Walkiewicz, J. The Possibility of Using Pine Bark Particles in the Chipboard Manufacturing Process. *Materials* **2022**, *15*, 5731. [CrossRef]
6. Savov, V. Nanomaterials to Improve Properties in Wood-Based Composite Panels. In *Emerging Nanomaterials*; Taghiyari, H.R., Morrell, J.J., Husen, A., Eds.; Springer: Cham, Switzerland, 2023; pp. 135–153.
7. Lee, S.H.; Lum, W.C.; Boon, J.G.; Kristak, L.; Antov, P.; Rogoziński, T.; Pędzik, M.; Taghiyari, H.R.; Lubis, M.A.R.; Fatriasari, W.; et al. Particleboard from Agricultural Biomass and Recycled Wood Waste: A Review. *J. Mater. Res. Technol.* **2022**, *20*, 4630–4658. [CrossRef]
8. Taghiyari, H.R.; Morrell, J.J.; Husen, A. Emerging Nanomaterials for Forestry and Associated Sectors: An Overview. In *Emerging Nanomaterials*; Taghiyari, H.R., Morrell, J.J., Husen, A., Eds.; Springer: Cham, Switzerland, 2023; pp. 1–24.
9. Sandberg, D.; Gorbacheva, G.; Lichtenegger, H.; Niemz, P.; Teischinger, A. Advanced Engineered Wood-Material Concepts. In *Springer Handbook of Wood Science and Technology*; Niemz, P., Teischinger, A., Sandberg, D., Eds.; Springer Handbooks; Springer: Cham, Switzerland, 2023; pp. 1835–1888.
10. Mantanis, G.I.; Athanassiadou, E.T.; Barbu, M.C.; Wijnendaele, K. Adhesive systems used in the European particleboard, MDF and OSB industries. *Wood Mater. Sci. Eng.* **2018**, *13*, 104–116. [CrossRef]
11. Dorieh, A.; Ayrilmis, N.; Pour, M.F.; Movahed, S.G.; Kiamahalleh, M.V.; Shahavi, M.H.; Hatefnia, H.; Mehdinia, M. Phenol formaldehyde resin modified by cellulose and lignin nanomaterials: Review and recent progress. *Int. J. Biol. Macromol.* **2022**, *222*, 1888–1907. [CrossRef] [PubMed]
12. Valyova, M.; Ivanova, Y.; Koynov, D. Investigation of free formaldehyde quantity in the production of plywood with modified urea-formaldehyde resin. *Int. J. Wood Des. Technol.* **2017**, *6*, 72–76.

13. Dorieh, A.; Pour, M.F.; Movahed, S.G.; Pizzi, A.; Selakjani, P.P.; Kiamahalleh, M.V.; Hatefnia, H.; Shahavi, M.H.; Aghaei, R. A review of recent progress in melamine-formaldehyde resin based nanocomposites as coating materials. *Prog. Org. Coat.* **2022**, *165*, 106768. [CrossRef]

14. Kumar, R.N.; Pizzi, A. Environmental Aspects of Adhesives–Emission of Formaldehyde. In *Adhesives for Wood and Lignocellulosic Materials*; Wiley-Scrivener Publishing: Hoboken, NJ, USA, 2019; pp. 293–312.

15. Walkiewicz, J.; Kawalerczyk, J.; Mirski, R.; Dziurka, D.; Wieruszewski, M. The Application of Various Bark Species as a Fillers for UF Resin in Plywood Manufacturing. *Materials* **2022**, *15*, 7201. [CrossRef] [PubMed]

16. Bekhta, P.; Sedliačik, J.; Noshchenko, G.; Kačík, F.; Bekhta, N. Characteristics of Beech Bark and its Effect on Properties of UF Adhesive and on Bonding Strength and Formaldehyde Emission of Plywood Panels. *Eur. J. Wood Prod.* **2021**, *79*, 423–433. [CrossRef]

17. Savov, V.; Antov, P.; Zhou, Y.; Bekhta, P. Eco-Friendly Wood Composites: Design, Characterization and Applications. *Polymers* **2023**, *15*, 892. [CrossRef] [PubMed]

18. International Agency for Research on Cancer. *IARC Classifies Formaldehyde as Carcinogenic to Humans*; International Agency for Research on Cancer: Lyon, France, 2004.

19. Ninikas, K.; Mitani, A.; Koutsianitis, D.; Ntalos, G.; Taghiyari, H.R.; Papadopoulos, A.N. Thermal and Mechanical Properties ofGreen Insulation Composites Made from Cannabis and Bark Residues. *J. Compos. Sci.* **2021**, *5*, 132. [CrossRef]

20. Antov, P.; Savov, V.; Trichkov, N.; Krišťák, Ľ.; Réh, R.; Papadopoulos, A.N.; Taghiyari, H.R.; Pizzi, A.; Kunecová, D.; Pachikova, M. Properties of High-Density Fiberboard Bonded with Urea–Formaldehyde Resin and Ammonium Lignosulfonate as a Bio-Based Additive. *Polymers* **2021**, *13*, 2775. [CrossRef] [PubMed]

21. Savov, V.; Valchev, I.; Antov, P.; Yordanov, I.; Popski, Z. Effect of the Adhesive System on the Properties of Fiberboard Panels Bonded with Hydrolysis Lignin and Phenol-Formaldehyde Resin. *Polymers* **2022**, *14*, 1768. [CrossRef]

22. Reh, R.; Kristak, L.; Antov, P. Advanced Eco-Friendly Wood-Based Composites. *Materials* **2022**, *15*, 8651. [CrossRef] [PubMed]

23. Valchev, I.; Yordanov, Y.; Savov, V.; Antov, P. Optimization of the Hot-Pressing Regime in the Production of Eco-Friendly Fibreboards Bonded with Hydrolysis Lignin. *Period. Polytech. Chem. Eng.* **2022**, *66*, 125–134. [CrossRef]

24. Hidayati, S.; Budiyanto, E.F.; Saputra, H.; Hadi, S.; Iswanto, A.H.; Solihat, N.N.; Antov, P.; Lee, S.H.; Fatriasari, W.; Salit, M.S. Characterization of Formacell Lignin Derived from Black Liquor as a Potential Green Additive for Advanced Biocomposites. *J. Renew. Mater.* **2023**, *11*, 2861–2875. [CrossRef]

25. Bekhta, P.; Noshchenko, G.; Réh, R.; Kristak, L.; Sedliačik, J.; Antov, P.; Mirski, R.; Savov, V. Properties of Eco-Friendly Particleboards Bonded with Lignosulfonate-Urea-Formaldehyde Adhesives and pMDI as a Crosslinker. *Materials* **2021**, *14*, 4875. [CrossRef] [PubMed]

26. Reinprecht, L.; Iždinský, J. Composites from Recycled and Modified Woods—Technology, Properties, Application. *Forests* **2022**, *13*, 6. [CrossRef]

27. Pędzik, M.; Kwidziński, Z.; Rogoziński, T. Particles from Residue Wood-Based Materials from Door Production as an Alternative Raw Material for Production of Particleboard. *Drv. Ind.* **2022**, *73*, 351–357. [CrossRef]

28. Koynov, D.; Valyova, M.; Parzhov, E.; Lee, S.H. Utilization of Scots Pine (*Pinus sylvestris* L.) Timber with Defects in Production of Engineered Wood Products. *Drv. Ind.* **2023**, *74*, 71–79. [CrossRef]

29. Ahamad, W.N.; Salim, S.; Lee, S.H.; Abdul Ghani, M.A.; Mohd Ali, R.A.; Md Tahir, P.; Fatriasari, W.; Antov, P.; Lubis, M.A.R. Effects of Compression Ratio and Phenolic Resin Concentration on the Properties of Laminated Compreg Inner Oil Palm and Sesenduk Wood Composites. *Forests* **2023**, *14*, 83. [CrossRef]

30. Pędzik, M.; Auriga, R.; Kristak, L.; Antov, P.; Rogoziński, T. Physical and Mechanical Properties of Particleboard Produced with Addition of Walnut (*Juglans regia* L.) Wood Residues. *Materials* **2022**, *15*, 1280. [CrossRef] [PubMed]

31. Barbu, M.C.; Sepperer, T.; Tudor, E.M.; Petutschnigg, A. Walnut and Hazelnut Shells: Untapped Industrial Resources and Their Suitability in Lignocellulosic Composites. *Appl. Sci.* **2020**, *10*, 6340. [CrossRef]

32. Kain, G.; Morandini, M.; Stamminger, A.; Granig, T.; Tudor, E.M.; Schnabel, T.; Petutschnigg, A. Production and Physical–Mechanical Characterization of Peat Moss (*Sphagnum*) Insulation Panels. *Materials* **2021**, *14*, 6601. [CrossRef]

33. Barbu, M.C.; Montecuccoli, Z.; Förg, J.; Barbeck, U.; Klímek, P.; Petutschnigg, A.; Tudor, E.M. Potential of Brewer's Spent Grain as a Potential Replacement of Wood in pMDI, UF or MUF Bonded Particleboard. *Polymers* **2021**, *13*, 319. [CrossRef]

34. Rammou, E.; Mitani, A.; Ntalos, G.; Koutsianitis, D.; Taghiyari, H.R.; Papadopoulos, A.N. The Potential Use of Seaweed (*Posidonia oceanica*) as an Alternative Lignocellulosic Raw Material for Wood Composites Manufacture. *Coatings* **2021**, *11*, 69. [CrossRef]

35. Pędzik, M.; Janiszewska, D.; Rogoziński, T. Alternative Lignocellulosic Raw Materials in Particleboard Production: A Review. *Ind. Crop. Prod.* **2021**, *174*, 114162. [CrossRef]

36. Chaydarreh, K.C.; Lin, X.; Guan, L.; Yun, H.; Gu, J.; Hu, C. Utilization of tea oil camellia (*Camellia oleifera* Abel.) shells as alternative raw materials for manufacturing particleboard. *Ind. Crop. Prod.* **2021**, *161*, 113221. [CrossRef]

37. Khalaf, Y.; El Hage, P.; Mihajlova, J.; Bergeret, A.; Lacroix, P.; El Hage, R. Influence of agricultural fibers size on mechanical and insulating properties of innovative chitosan-based insulators. *Constr. Build. Mater.* **2021**, *287*, 123071. [CrossRef]

38. Grigorov, R.; Mihajlova, J.; Savov, V. Physical and Mechanical Properties of Combined Wood-Bases Panels with Participation of Particles from Vine Sticks in Core Layer. *Innov. Wood. Ind. Eng. Des.* **2020**, *1*, 42–52.

39. Santos, J.; Pereira, J.; Ferreira, N.; Paiva, N.; Ferra, J.; Magalhães, F.D.; Martins, J.M.; Dulyanska, Y.; Carvalho, L.H. Valorisation of non-timber by-products from maritime pine (*Pinus pinaster, Ait*) for particleboard production. *Ind. Crop. Prod.* **2021**, *168*, 113581. [CrossRef]
40. Lykidis, C. Formaldehyde Emissions from Wood-Based Composites: Effects of Nanomaterials. In *Emerging Nanomaterials*; Taghiyari, H.R., Morrell, J.J., Husen, A., Eds.; Springer: Cham, Switzerland, 2023; pp. 337–360.
41. Kawalerczyk, J.; Walkiewicz, J.; Dziurka, D.; Mirski, R.; Brózdowski, J. APTES-Modified Nanocellulose as the Formaldehyde Scavenger for UF Adhesive-Bonded Particleboard and Strawboard. *Polymers* **2022**, *14*, 5037. [CrossRef] [PubMed]
42. Kristak, L.; Antov, P.; Bekhta, P.; Lubis, M.A.R.; Iswanto, A.H.; Reh, R.; Sedliacik, J.; Savov, V.; Taghiayri, H.; Papadopoulos, A.N.; et al. Recent Progress in Ultra-Low Formaldehyde Emitting Adhesive Systems and Formaldehyde Scavengers in Wood-Based Panels: A Review. *Wood Mater. Sci. Eng.* **2022**, *18*, 1–20. [CrossRef]
43. Kawalerczyk, J.; Walkiewicz, J.; Woźniak, M.; Dziurka, D.; Mirski, R. The effect of urea-formaldehyde adhesive modification with propylamine on the properties of manufactured plywood. *J. Adhes.* **2022**. [CrossRef]
44. Medved, S.; Gajsek, U.; Tudor, E.M.; Barbu, M.C.; Antonovic, A. Efficiency of bark for reduction of formaldehyde emission fromparticleboards. *Wood Res.* **2019**, *64*, 307–315.
45. Arias, A.; González-Rodríguez, S.; Vetroni Barros, M.; Salvador, R.; de Francisco, A.C.; Piekarski, C.M.; Moreira, M.T. Recent developments in bio-based adhesives from renewable natural resources. *J. Clean. Prod.* **2021**, *314*, 127892. [CrossRef]
46. Hussin, M.H.; Abd Latif, N.H.; Hamidon, T.S.; Idris, N.N.; Hashim, R.; Appaturi, J.N.; Brosse, N.; Ziegler-Devin, I.; Chrusiel, L.; Fatriasari, W.; et al. Latest advancements in high-performance bio-based wood adhesives: A critical review. *J. Mater. Res. Technol.* **2022**, *21*, 3909–3946. [CrossRef]
47. Gonçalves, S.; Ferra, J.; Paiva, N.; Martins, J.; Carvalho, L.H.; Magalhães, F.D. Lignosulphonates as an Alternative to Non-Renewable Binders in Wood-Based Materials. *Polymers* **2021**, *13*, 4196. [CrossRef]
48. Sari, R.A.L.; Lubis, M.A.R.; Sari, R.K.; Kristak, L.; Iswanto, A.H.; Mardawati, E.; Fatriasari, W.; Lee, S.H.; Reh, R.; Sedliacik, J.; et al. Properties of Plywood Bonded with Formaldehyde-Free Adhesive Based on Poly(vinyl alcohol)–Tannin–Hexamine at Different Formulations and Cold-Pressing Times. *J. Compos. Sci.* **2023**, *7*, 113. [CrossRef]
49. Gumowska, A.; Kowaluk, G. Physical and Mechanical Properties of High-Density Fiberboard Bonded with Bio-Based Adhesives. *Forests* **2023**, *14*, 84. [CrossRef]
50. Puspaningrum, T.; Haris, Y.H.; Sailah, I.; Yani, M.; Indrasti, N.S. Physical and mechanical properties of binderless medium density fiberboard (MDF) from coconut fiber. *IOP Conf. Ser. Earth Environ. Sci.* **2020**, *472*, 012011. [CrossRef]
51. Jerman, M.; Böhm, M.; Dušek, J.; Černý, R. Effect of steaming temperature on microstructure and mechanical, hygric, and thermal properties of binderless rape straw fiberboards. *Build. Environ.* **2022**, *223*, 109474. [CrossRef]
52. Kaffashsaei, E.; Yousefi, H.; Nishino, T.; Matsumoto, T.; Mashkour, M.; Madhoushi, M. Binderless Self-densified 3 mm-Thick Board Fully Made from (Ligno)cellulose Nanofibers of Paulownia Sawdust. *Waste Biomass Valor.* **2023**. [CrossRef]
53. Ferrandez-Garcia, A.; Ferrandez-Garcia, M.T.; Garcia-Ortuño, T.; Ferrandez-Villena, M. Influence of the Density in Binderless Particleboards Made from Sorghum. *Agronomy* **2022**, *12*, 1387. [CrossRef]
54. Zhang, D.; Zhang, A.; Xue, L. A review of preparation of binderless fiberboards and its self-bonding mechanism. *Wood Sci. Technol.* **2015**, *49*, 661–679. [CrossRef]
55. Barbu, M.C.; Tudor, E.M.; Buresova, K.; Petutschnigg, A. Assessment of Physical and Mechanical Properties Considering the Stem Height and Cross-Section of *Paulownia tomentosa* (Thunb.) Steud. x *elongata* (S.Y.Hu) Wood. *Forests* **2023**, *14*, 589. [CrossRef]
56. Lee, S.H.; Md Tahir, P.; Osman Al-Edrus, S.S.; Uyup, M.K.A. (Eds.) Bamboo Resources, Trade, and Utilisation. In *Multifaceted Bamboo*; Springer: Singapore, 2023; pp. 1–14.
57. Dwianto, W.; Darmawan, T.; Nugroho, N.; Lubis, M.A.R.; Bahanawan, A.; Adi, D.S.; Triwibowo, D. Development of Compressed Bamboo Lamination from Curved Cross-Section Slats. In *Multifaceted Bamboo*; Md Tahir, P., Lee, S.H., Osman Al-Edrus, S.S., Uyup, M.K.A., Eds. Springer: Singapore, 2023; pp. 193–216.
58. Abidin, W.N.S.N.Z.; Al-Edrus, S.S.O.; Hua, L.S.; Ghani, M.A.A.; Bakar, B.F.A.; Ishak, R.; Qayyum Ahmad Faisal, F.; Sabaruddin, F.A.; Kristak, L.; Lubis, M.A.R.; et al. Properties of Phenol Formaldehyde-Bonded Layered Laminated Woven Bamboo Mat Boards Made from *Gigantochloa scortechinii*. *Appl. Sci.* **2023**, *13*, 47. [CrossRef]
59. Hao, X.; Yu, Y.; Yang, C.; Yu, W. In Situ Detection of the Flexural Fracture Behaviors of Inner and Outer Bamboo-Based Composites. *Forests* **2023**, *14*, 515. [CrossRef]
60. Mohd Yusof, N.; Md Tahir, P.; Lee, S.H.; Anwar Uyup, M.K.; James, R.M.S.; Osman Al-Edrus, S.S.; Kristak, L.; Reh, R.; Lubis, M.A.R. Effects of Adhesive Types and Structural Configurations on Shear Performance of Laminated Board from Two Gigantochloa Bamboos. *Forests* **2023**, *14*, 460. [CrossRef]
61. Iswanto, A.H.; Sutiawan, J.; Darwis, A.; Lubis, M.A.R.; Pędzik, M.; Rogoziński, T.; Fatriasari, W. Influence of Isocyanate Content and Hot-Pressing Temperatures on the Physical–Mechanical Properties of Particleboard Bonded with a Hybrid Urea–Formaldehyde/Isocyanate Adhesive. *Forests* **2023**, *14*, 320. [CrossRef]
62. Yusof, N.M.; Hua, L.S.; Tahir, P.M.; James, R.M.S.; Al-Edrus, S.S.O.; Dahali, R.; Roseley, A.S.M.; Fatriasari, W.; Kristak, L.; Lubis, M.A.R.; et al. Effects of Boric Acid Pretreatment on the Properties of Four Selected Malaysian Bamboo Strips. *Forests* **2023**, *14*, 196. [CrossRef]
63. Kminiak, R.; Němec, M.; Igaz, R.; Gejdoš, M. Advisability-Selected Parameters of Woodworking with a CNC Machine as a Tool for Adaptive Control of the Cutting Process. *Forests* **2023**, *14*, 173. [CrossRef]

64. Mitterpach, J.; Fojtík, R.; Machovčáková, E.; Kubíncová, L. Life Cycle Assessment of a Road Transverse Prestressed Wooden–Concrete Bridge. *Forests* **2023**, *14*, 16. [CrossRef]
65. Pędzik, M.; Tomczak, K.; Janiszewska-Latterini, D.; Tomczak, A.; Rogoziński, T. Management of Forest Residues as a Raw Material for the Production of Particleboards. *Forests* **2022**, *13*, 1933. [CrossRef]
66. Marková, I.; Ivaničová, M.; Osvaldová, L.M.; Harangózo, J.; Tureková, I. Ignition of Wood-Based Boards by Radiant Heat. *Forests* **2022**, *13*, 1738. [CrossRef]
67. Barbu, M.C.; Buresova, K.; Tudor, E.M.; Petutschnigg, A. Physical and Mechanical Properties of *Paulownia tomentosa x elongata* Sawn Wood from Spanish, Bulgarian and Serbian Plantations. *Forests* **2022**, *13*, 1543. [CrossRef]
68. Maulana, M.I.; Lubis, M.A.R.; Febrianto, F.; Hua, L.S.; Iswanto, A.H.; Antov, P.; Kristak, L.; Mardawati, E.; Sari, R.K.; Zaini, L.H.; et al. Environmentally Friendly Starch-Based Adhesives for Bonding High-Performance Wood Composites: A Review. *Forests* **2022**, *13*, 1614. [CrossRef]

Article

Assessment of Physical and Mechanical Properties Considering the Stem Height and Cross-Section of *Paulownia tomentosa* (Thunb.) Steud. *x elongata* (S.Y.Hu) Wood

Marius Cătălin Barbu [1,2], Eugenia Mariana Tudor [1,2,*], Katharina Buresova [1] and Alexander Petutschnigg [1,3]

1   Green Engineering and Circular Design Department, Salzburg University of Applied Sciences, Markt 136a, 5431 Kuchl, Austria; cmbarbu@unitbv.ro (M.C.B.); kburesova.htw-m2021@fh-salzburg.ac.at (K.B.); alexander.petutschnigg@fh-salzburg.ac.at (A.P.)
2   Faculty of Furniture Design and Wood Engineering, Transilvania University of Brasov, B-dul. Eroilor nr. 29, 500036 Brasov, Romania
3   Institute of Wood Technology and Renewable Materials, University of Natural Resources and Life Sciences (BOKU), Konrad Lorenz-Straße 24, 3340 Tulln an der Donau, Austria
*   Correspondence: eugenia.tudor@fh-salzburg.ac.at

Abstract: The aim of this study is to analyze the properties of *Paulownia tomentosa x elongata* plantation wood from Serbia, considering the influence of the stem height (0 to 1 m and 4.5 to 6 m above soil level—height spot) and radial position from the pith to bark (in the core, near the bark, and in between these zones—cross-section spot). The results show that most properties are improved when the samples were taken from upper parts of the tree (height spot) and from the near bark spot (cross-section spot). The mean density measured 275 kg/m^3 at the stem height between 4.5–6 m and 245 kg/m^3 for the samples collected from 0–1 m trunk height. The density had the highest value on the spot near bark (290 kg/m^3), for the mature wood at a height of 4.5–6 m, and near pith had a mean density of 230 kg/m^3. The Brinell hardness exhibited highest values in the axial direction (23 N/mm^2) and near bark (28 N/mm^2). The bending strength was 41 N/mm^2 for the trunk's height range of 4.5–6 m and 45 N/mm^2 in the cross-section, close to cambium. The three-point modulus of elasticity (MOR) of the samples taken at a stem height of 4.5 to 6 m was up to 5000 N/mm^2, and on the spot near bark, the MOR measured 5250 N/mm^2. Regarding compressive strength, in the cross-section, near the pith, the mean value was the highest with 23 N/mm^2 (4.5–6 m), whilst it was 19 N/mm^2 near bark. The tensile strength was, on average, 40 N/mm^2 for both 0–1 m and 4.5–6 m trunk height levels and 49 N/mm^2 between bark and pith. The screw withdrawal resistance measured 58 N/mm for the samples extracted at a stem height of 4.5 to 6 m and 92 N/mm for the specimens collected near pith. This study stresses the influence, in short-rotation Paulownia timber, of indicators, such as juvenile and mature wood (difference emphasized after the fifth year of growth) and height variation, on the physical and mechanical properties of sawn wood. This study will help utilize more efficient sustainable resources, such as Paulownia plantation wood. This fast-growing hardwood species from Europe is adequate as a core material in sandwich applications for furniture, transport, sport articles, and lightweight composites, being considered the European Balsa.

Keywords: Paulownia; wood properties; position in stem; plantation wood; Balsa

Citation: Barbu, M.C.; Tudor, E.M.; Buresova, K.; Petutschnigg, A. Assessment of Physical and Mechanical Properties Considering the Stem Height and Cross-Section of *Paulownia tomentosa* (Thunb.) Steud. *x elongata* (S.Y.Hu) Wood. *Forests* **2023**, *14*, 589. https://doi.org/10.3390/f14030589

Academic Editor: Xinzhou Wang

Received: 3 February 2023
Revised: 9 March 2023
Accepted: 13 March 2023
Published: 16 March 2023

1. Introduction

Paulownia is a fast growing tree that originates in Asia, having at least nine species within the family [1]; *Paulownia tomentosa* (Thunb.), *Paulownia elongata* (S. Y. Hu), and *Paulownia fortunei* (Seem.) Hemsl. are intensely used and studied [2–4]. Paulownia hybrids can combine, for example, *Paulownia tomentosa x elongata* [5] or *Paulownia elongata x Paulownia fortunei* [6,7]. In Europe, the most studied clones are Paulownia in vitro 112—Oxytree [1,8,9], BIO 125 [10], or Cotevisa 2 [11].

In the last decades, interest for agroforestry plantations of Paulownia for industrial use has risen in Europe. As a result, the ecological footprint of Paulownia lumber from Asia is diminished due to elimination of oversea transport costs. Paulownia plantations contribute to the scaling down of soil hazards by tree intercropping in farmlands [7] and crop fields. For example, when intercropped with wheat, the production rate was increased, similar to the increase in ginger production [12]. It is recommended to intercrop Paulownia with winter crops and vegetables, considering that the trees in the dormant season will not compete for water and nutrients during the cold season [13]. Paulownia protects systems against erosion, flooding, or wind damages [14], reduces soil degradation (by absorbing heavy metals), diminishes air pollution (by cleaning the air of harmful gases due to its huge leaves), and improves the microclimate [15].

Paulownia's bark is commonly processed with its wood because the bark is thin, difficult to remove, and accounts for less than 1% of the overall volume [16].

This species arrived in Europe over a century ago as a decorative tree. Paulownia and its hybrids were naturalized in Europe after an adjustment process dependent on climatic zones and soils. Literature on the properties of Paulownia wood sourced from European plantations is in continuous development, yet still scarce [1], considering that studies on its properties commenced a decade ago on the old continent [7].

This tree is renowned due to it being fast growing with a low density of 260 to 300 kg/m^3 [5], gaining in the last period the nickname of "European Balsa" [3] or the aluminum of wood. The density correlates sharply with other physical and mechanical properties and influences the woodworking and drying processes. The maximum moisture content of this wood species measures 350% [17]; therefore, kiln-drying schedules should be attentively selected and controlled. Paulownia air-dried wood is normally without drying defects [18].

The fiber saturation point is 31.15% for *Paulownia fortune* and 29% for *Paulownia tomentosa*. The chemical composition of the cell wall is as follows: 51% cellulose, 30% hemicellulose, 23.5% lignin, and 11.8% extractives [17]. Depending on the species, the porosity of Paulownia ranges between 75 and 88% [19].

Since the end of the last century, several countries, such as Austria, Italy, France, Serbia, Spain, Romania, Bulgaria, Hungary, Israel, and Ukraine, are experimenting with Paulownia plantations [4,7,20]. Paulownia was gradually introduced in South America (Argentina, Brazil, Paraguay) and Australia and is used for timber production [2]. Paulownia exhibits facile processability [21], acceptable fire resistance, and high rate of carbon absorption [22]. Magar et al. [12] calculated a rate of 33 t CO_2 per hectare per year for Paulownia. Paulownia implemented in agroforestry systems can help to reduce the greenhouse emissions in cities and neighborhoods of farmhouses, highlighting the carbon storage potential, which is about 50 fold, 30 fold, and 20 fold higher compared to oak, beech, and lime trees, respectively [23], and at least 10 times more than any other tree species [24].

An adult Paulownia tree has a trunk of 10 to 20 m, with an increment of about 3 m/year and a 40 cm diameter [25]. Harvesting of 15-year-old Paulownia trees from plantations with about 2000 trees per hectare results in valuable timber [26]. Paulownia wood is used for furniture construction [26], packaging, plywood, solid wood products (single-layered), and pulp production [27]. The increased content of holo-cellulose (81%) determines a higher pulp yield and acceptable strength properties [17]. Paulownia wood is also interesting because it does not crack or warp and is not susceptible to decay [28]. It is light-colored, soft, easy to carve, dimensionally stable, generally knot free and straight-grained [18], uniform texture, and clear after plaining [4]. The absence of knots and fibers that run parallel to the longitudinal axes are due to appropriate Paulownia plantation management, considering the orientation (southern facing exposure with wind protection) and well-drained soil (pH-value of 5–8). The tree should be irrigated until the root system is robust [29]. The Paulownia is a light-loving tree and even slight shade can cause deformation in saplings. At 70% shade, the younger trees can be completely damaged. Paulownia trees are coppiced (the sapling is cut back to ground level, to determine the formation of a new shoot) during the second year of life (technical cut, maintaining 2 cm of plant) [30]. Only the best buds will be selected to fully grow [29]. When the saplings are 2 or 3 years old, the pruning should be done during the growing season, when new branches develop. The purpose of the pruning is to raise the value of the timber by gaining much cleaner roundwood (without knots). Unnecessary lateral branches should be removed; the branches of the crown, however, should not be cut during the year of their emergence, since they will constitute the sympodial monochasium of the trunk [29]. The condition is to keep the first 6–7 m of the stem clear of branches (75%) for a stem height of 7–8 m; after this height, the canopy can be allowed to reach its natural shape [30].

Lower quality Paulownia trunks can serve as raw materials for biomass and biofuels [5], with potential for second generation bioethanol production [31]. The phenolic compounds in Paulownia have antioxidant properties; therefore, can be used for medical purposes [12]. In Asia, different elements as leaves, flowers, fruits, and barks were used for centuries and served as traditional medicines [31].

Paulownia could be considered invasive according to [3,32,33]. Invasive tree species are able to survive, to reproduce, and to spread over the landscape, sometimes at disturbing quotes [34], aggressively competing with native plants. As reported by [35], the spreading of this tree species can be considered as high or wide-ranging. Paulownia is not included in the updated list of invasive alien species of the European Union [36].

To analyze the potential of Paulownia timber, is important to scrutinize the properties of the length and cross-section of the tree stem. In this way, the influence of juvenile and mature wood on the physical and mechanical properties of Paulownia plantation wood can be studies. In most species, juvenile wood has a low density [37], not only in plantation, but also in naturally grown trees. This can be attributed to the shorter tracheids (for softwood) and fibers (for hardwood) with thinner cell walls. Transverse shrinkage is lower in a vast majority of wood species. Juvenile wood has a significantly lower modulus of rupture, modulus of elasticity, and reduced compressive and tensile strength [37]. Wood properties of short-rotation plantation Paulownia trees can be controlled competently through forest management. In general, the longer the rotation age, the more mature wood is harvested; therefore, the better the mechanical properties for productive purposes. Estevez et al. [6] estimated a minimum of 5 years for felling of Paulownia for its use as solid fuel (pellets) and as solid wood

This research is a follow-up study of [5]. The aim of this study is to assess the properties of *Paulownia tomentosa x elongata* plantation wood from Serbia, cut from different trunk heights, namely 0–1 m and 4.5–6 m, and from three cross-section sites (near to bark, near to the pith, and in between.

2. Materials and Methods

The raw material was provided by Glendor Holding GmbH Company (Kilb, Austria) from a Paulownia plantation in Serbia (Figure 1). It consisted of two *Paulownia tomentosa x elongata* trees (7 years old) cut at different stem heights (0–1 m and 4.5–6 m), resulting in 4 logs (Figure 2).

The logs were labelled 1–4, with log 1 and 2 belonging to the first tree and log 3 and 4 to the second tree (Figure 2).

The logs 1 to 4 were processed to sawn wood (Figure 3). Stem 1 and 3 were extracted, at 0–1 m height, arising from the ground. Stem 2 and 4 were cut from the tree trunk at a height of 4.5–6 m (Figure 4). These two stem height levels were chosen due to log cleanliness and knots-free areas without growth defects.

Figure 1. Paulownia plantation managed by Glendor GmbH with 7-year-old trees.

Figure 2. Four Paulownia logs extracted from two trees at different heights (0–1 m, 4.5–6 m).

Figure 3. Marking of Paulownia sawn wood in the cross-section prior to the preparation of testing specimens.

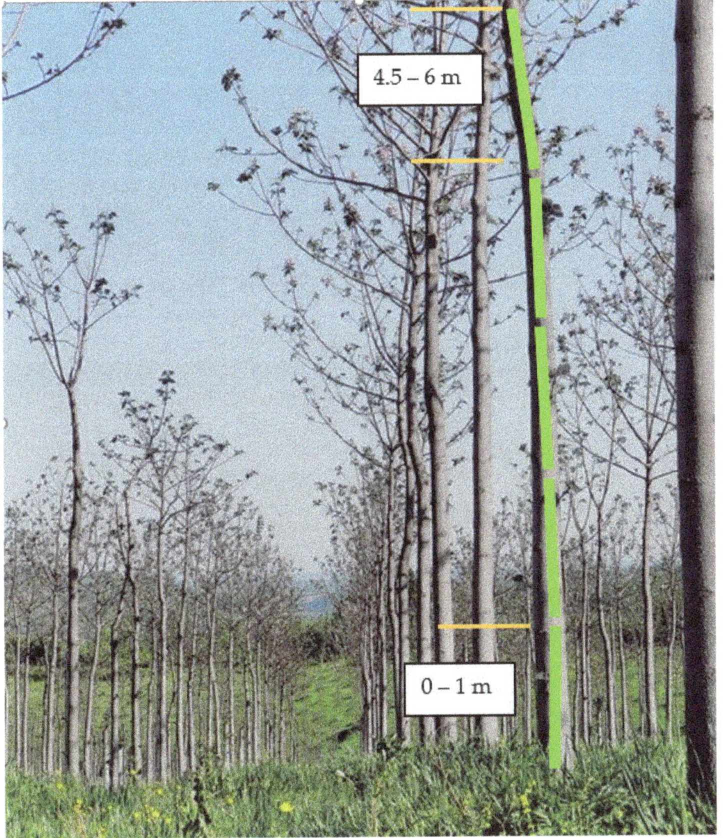

Figure 4. Two spots on the Paulownia stem (0–1 m and 4.5–6 m) selected after harvesting for the research investigation.

The diameters of the four logs and cutting direction (parallel and perpendicular to the grain) are presented in Table 1.

Table 1. Diameters of Paulownia logs depending on stem height.

Log (Extraction Position)	Diameter (cm)
1 (0–1 m) (cut parallel to the grain)	32
2 (4.5–6 m) (cut parallel to the grain)	27.5
3 (0–1 m) (perpendicular to the grain)	26
4 (4.5–6 m) (perpendicular to the grain)	21

The sawn wood extracted from stems was labelled using letters from A to F (Figure 3). The codification used to characterize the position in the cross-section was K—assigned for the core area, A—assigned for the outer area, and M—assigned for the middle area between the core and outer area (Figures 3–5). Determination of the properties within the cross-section was carried out at the low and high height levels and the mean value was calculated. The samples were extracted from the cross-section from sawn boards, eliminating the spots with pith, and were divided equally for the spots representing the core, middle, and edge. The results are shown in Figure 5. The samples were straight grained and generally knot free.

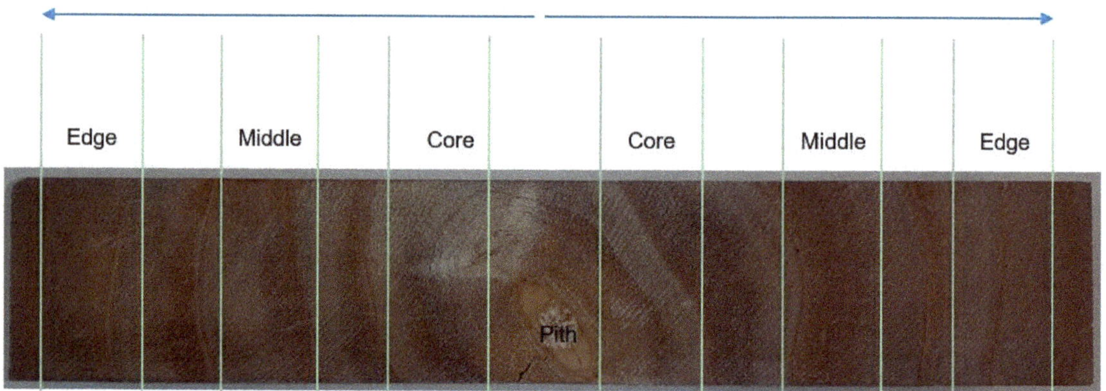

Figure 5. Radial areas from the pith to margins (core, middle, and edge) for the extraction of the samples.

Prior to testing, the raw material was conditioned at 20 °C and 65% relative air humidity for at least 30 days, until a constant mass was achieved. The moisture content after conditioning was 12%.

Several tests were carried out according to EN, ISO, and DIN norms to determine the physical and mechanical properties of Paulownia plantation wood from Serbia (Table 2), differentiated through the position in trunk and in the cross-section. The mechanical properties were tested with the Zwick Roell 250 universal testing machine (Ulm, Germany) and Emco Test Automatic (Hallein, Austria).

Table 2. Test preparation for Paulownia wood samples: norms, dimensions, and number of testing specimens.

Test	Norm	Number of Samples	Sample Dimension [mm]
Swelling and shrinkage	DIN 52184:1979-05	12	20 × 20 × 10
Bulk density (kg/m^3)	ISO 3131:1996	12	
Brinell hardness (N/mm^2)	EN 1534:2011-01	10	50 × 50 × 10
3-point modulus of elasticity (MOE) (N/mm^2)	DIN 52186:1978-06	12	20 × 20 × 360
3-point modulus of rupture (MOR) (N/mm^2)	DIN 52186:1978-06		
Compressive strength (N/mm^2)	DIN 52185:1976-09	12	20 × 20 × 50
Tensile shear strength (N/mm^2)	DIN 52188:1979-05	16	20 × 6 at predetermined breaking point
Screw withdrawal resistance (N/mm)	EN 320:2011-07	12	50 × 50

3. Results and Discussion

This section presents the results of the tests regarding density, sorption behavior, Brinell hardness, three-point modulus of rupture, three-point modulus of elasticity, compressive strength, tensile shear strength, and screw withdrawal resistance, measured at 0–1 m and at 4.5–6 m in the tree trunk height and also depending on the position in the cross-section (near to bark, in the core near the pith, and in between these two areas).

3.1. Density

Within the same tree, significant variations in density occurred from the bark to the pith, and up the trunk from level ground [38]. Variations in density in the cross-section of the stem are correlated with the amount of juvenile and mature wood [37]. Compared to mature wood, juvenile wood is characterized by a lower density, that increases significantly with new growth rings [4].

The density of Paulownia samples, measured according to ISO 3131:1996 [39], increases at 4.5–6 m (Figure 6). The average value of the bulk density at the basal portion of the stem (0–1 m) measured 245 kg/m^3, and at a height of 4.5–6 m, it was 274 kg/m^3 (30 kg/m^3 difference in density between these height spots). The wood extracted from the upper part of Paulownia stem (4.5–6 m) has a higher bulk density, which can be explained by the longer fiber length and lower microfibril angle. This tendency can be considered atypical, because for most of tree species, the samples extracted from the top part of the tree have lower density and strength properties [38,40].

In the cross-section, the bulk density increases from the core (near pith) to the edge (Figure 7). Near the bark (edge), the density was measured at 289 kg/m^3, in the middle it was 257 kg/m^3, and in the core it was 232 kg/m^3. There is a clear difference between juvenile (pith) and mature wood (near bark), with the former exhibiting a higher density due to longer fibers and considerably thicker cell walls [37]. This progressive increase in bulk density, from 202 kg/m^3 at the first ring (juvenile wood) towards 270 kg/m^3 for the sixth ring (mature wood), was determined by [4], who demonstrated that there is no significant difference in densities between the fifth and sixth rings. This suggests that the wood from the first to fourth rings can be considered juvenile wood, and from the fifth ring it can be considered mature wood. Besides a lower density, juvenile wood has a larger microfibril angle than mature wood [38].

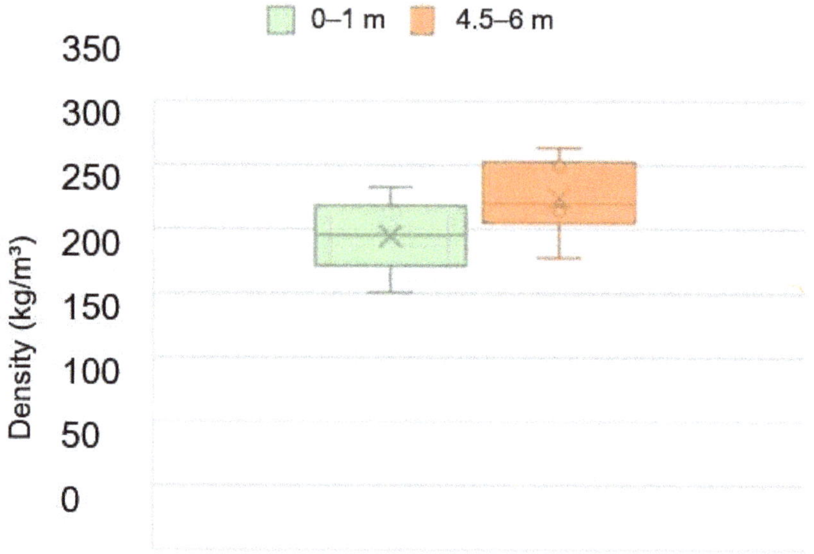

Figure 6. Density of Paulownia samples depending on the position in the tree trunk (0–1 m, 4.5–6 m).

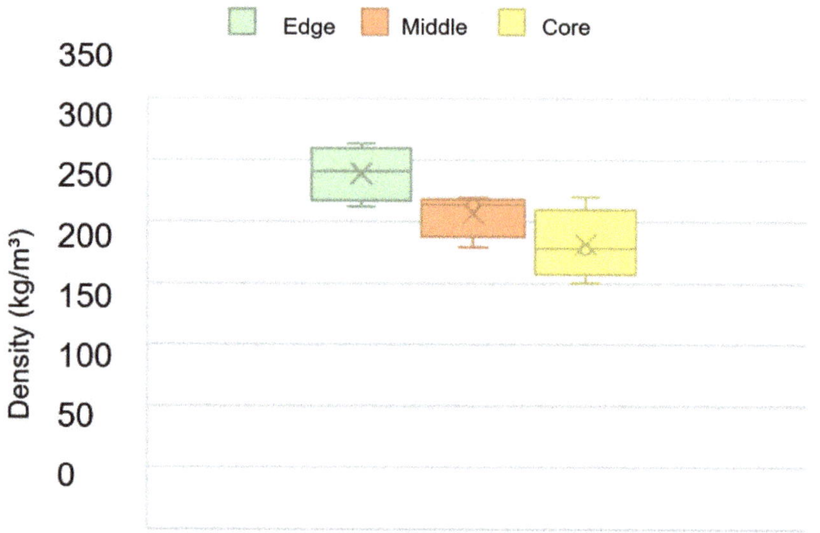

Figure 7. Density of Paulownia samples depending on the position in the cross-section (edge, middle, core).

Similar density ranges, but with no other details about the sample's spot and origin in tree, were reported in the study of Akyldiz and Kol [25], with a mean value of 272 kg/m^3 for the basic species *Paulownia tomentosa* from Turkey. Esteves et al. [6] measured a density of 460 kg/m^3 for Paulownia wood sourced from Portugal, which is way higher than the density of spruce with 430 kg/m^3 according to [41]. The lowest density of 216 kg/m^3 for a Paulownia plantation wood in Spain was reported by [19]. Jakubowski [7] showed that, at a 12% moisture content, Paulownia wood density varies from 220 to 350 kg/m^3, with an average value of 270 kg/m^3. This variability in density is justified by the growth conditions (soil, temperature, climate). Densities higher than 400 kg/m^3 were determined for *Paulownia tomentosa* in Turkey [25] and Portugal [6], and for *Paulownia* Siebold and Zucc. (from Bulgaria) [42].

3.2. Sorption Behavior

Figure 8 shows that the sorption behavior of wood, determined according to DIN 52184:1979 [43], was minimally influenced by the position in the tree stem. The mostly increased swelling was measured in the tangential section. In the axial and radial direction, wood swelling is less, as described by [38].

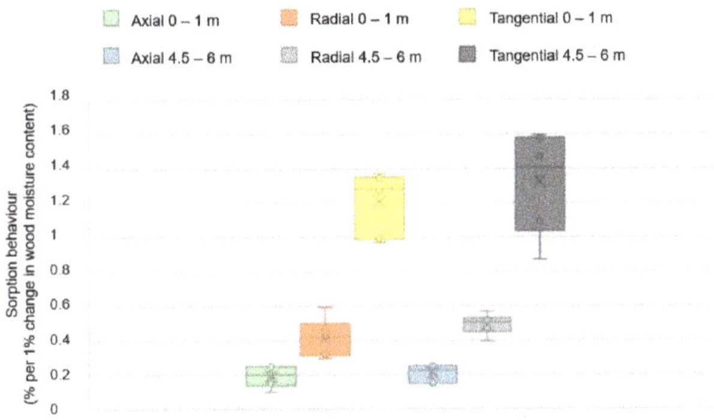

Figure 8. Sorption behavior in the axial, radial, and tangential direction of Paulownia samples extracted at 0–1 m and 4.5–6 m from the tree stem.

From Figure 9, the results show that in the axial direction, the wood swelling was the lowest (0.199%). In the radial direction, the shrinking was higher (0.456%). The differential swelling and shrinkage in the tangential direction was the highest (1.266%), which was also highlighted by [38].

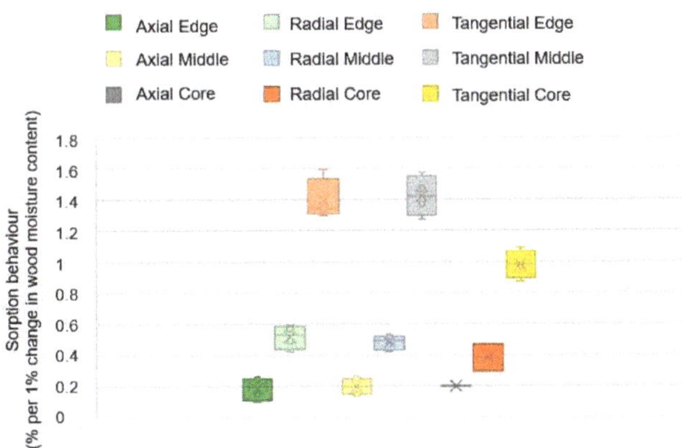

Figure 9. Sorption behavior in the axial, radial, and tangential direction of Paulownia samples extracted from the edge, middle, and from the core of the tree.

The samples seized from the edge of the stem showed very similar shrinkage compared to the samples extracted from the middle area of the cross-section. For example, the samples tested in the axial direction extracted from the edge area measured a shrinkage of 0.186%, whereas the shrinkage of the testing specimens extracted from the middle area was 0.199%. The shrinkage of the samples tested in the tangential direction showed only minimal change.

In the cross-section, an average shrinkage of 1.394% in the edge area, 1.422% in the middle area, and 0.981% in the core area were measured.

Similar values of shrinkage, in all cutting directions, were determined by [4]. For the first and second year rings, the longitudinal shrinkage was under 0.25%, decreasing towards areas with mature wood (fifth and sixth annual ring) to 0.1%. The radial shrinkage had similar mean values in the stem's cross-section. The tangential shrinkage exhibited higher values for the first ring (6%), with a 25% decrease towards mature wood in the sixth ring [4]. These values are consistent with the findings of [17,44,45]. The high microfibril angle of juvenile wood is one of the main parameters that influences shrinkage and shrinkage anisotropy [46].

It is important to emphasize the lower swelling ratios of Paulownia wood, which can be associated to narrower core rays [7]. The medullar rays control the wood in radial direction and determine swelling up to 4% [47]. Moreover, the narrow core rays did not influence higher swelling rates in the tangential direction [5]. It was observed that juvenile wood had the lowest transverse shrinkage, which is consistent to the study of [37]. The swelling and shrinkage behavior of wood is correlated with the bulk density, the proportion of latewood, the anatomical structure, and the lignin amount. The low longitudinal swelling or shrinkage is due to the orientation of the fibrils in the longitudinal direction and the relatively low proportion of cell walls lying transverse to the direction of the fibers [38].

3.3. Brinell Hardness

Brinell hardness (HB) [48] increases with height in all main sections (axial, radial, and tangential) (Figure 10). The HB values were, on average, one third higher for the samples originating from a height of 4.5–6 m of the tree stem, especially in the axial direction. For the samples collected from a height of 0 to 1 m from the stem, the Brinell hardness was 4.16 N/mm^2 in the radial direction, 5.21 N/mm^2 in the tangential direction, and 17.87 N/mm^2 in the axial direction. For the samples extracted from log at the height of 4.5 to 6 m, these values of Brinell hardness increased in the radial direction to 6.98 N/mm^2, in the tangential direction to 6.24 N/mm^2, and in the axial direction to 23.23 N/mm^2. There is a direct correlation between Brinell hardness and density [38].

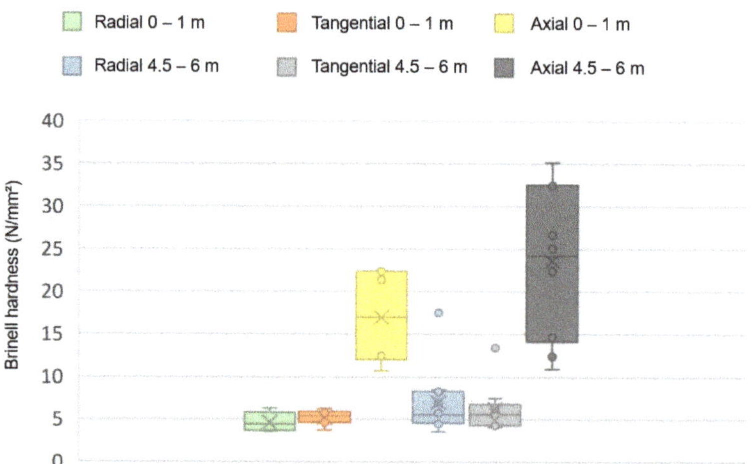

Figure 10. Brinell hardness in the axial, radial, and tangential sections of Paulownia samples extracted at 0–1 m and 4.5–6 m from the tree stem.

By analyzing the tree stem cross-section (Figure 11), Brinell hardness (HB) in the radial direction tended to increase from the seventh to the first growth ring. Near bark, HB in the radial direction was 5.19 N/mm^2 and increased to 7.65 N/mm^2 near the pith.

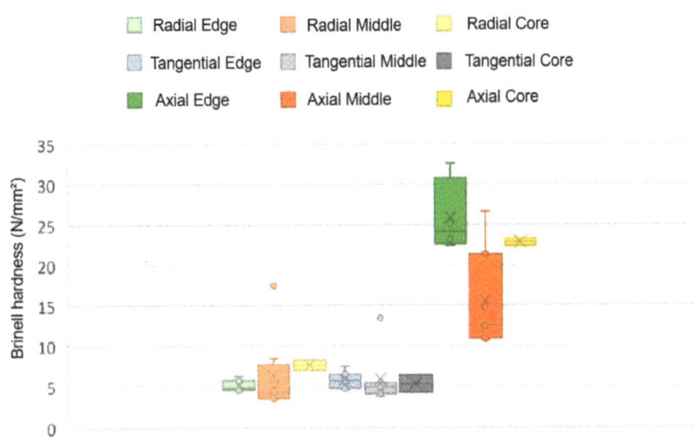

Figure 11. Brinell hardness in the axial, radial, and tangential section of Paulownia samples extracted from 10%, 50%, and 90% from the cross-section of Paulownia's tree stem.

At the seventh growth ring, near the bark, the tangential HB value was 5.87 N/mm^2, and near the pith (first ring), it reached 5.32 N/mm^2, which showed only a slight downward tendency.

The values of the HB in the transversal direction measured a maximum of 27.61 N/mm^2 on the seventh ring. In the middle area (between the third and fourth rings), HB was the lowest at 15.23 N/mm^2. Near the first growth ring, HB was 22.85 N/mm^2.

The low values of Brinell hardness were influenced by the low density of Paulownia wood, determining an increased indentation. Under a load parallel to the direction of the fibers, HB is at least 2.5 fold greater than the hardness in the radial or tangential direction [38]. Brinell hardness of *Paulownia tomentosa x elongata* plantation wood increases from the first to the sixth rings and measures the highest values in the longitudinal direction, as suggested by [4]. In the transversal direction, Paulownia hardness increases at the upper part of the tree stem.

In the axial direction, at a trunk height of 0–1 m, the HB value was similar with the findings of [1,5,25,42], but at a height of 4.5 to 6 m, the maximum of 35 N/mm^2 calculated in the present study had no equivalent in the studied literature. Akyldiz and Kol [25] determined a maximum of longitudinal HB of 19 N/mm^2, 9 N/mm^2 in the tangential, and 8 N/mm^2 in the radial direction. Lower values of HB were determined by [49] as follows: 10.61 N/mm^2 in the longitudinal direction, 5.63 N/mm^3 in the tangential direction, and 5.46 N/mm^2 in the radial direction. A comparison with other studies regarding HB of Paulownia samples extracted from different positions of the stem height is not yet possible considering state of the art of Paulownia short-rotation plantation wood.

3.4. Modulus of Rupture (MOR) and Modulus of Elasticity (MOE)

The three-point MOR, determined according to DIN 52186:1978 [50], exhibits the same tendency as bulk density (3.1): it increases with height (Figure 12) and it is at its highest near the bark (Figure 13). MOR was 33.9 N/mm^2 at a height of 0–1 m and 41.18 N/mm^2 at a height of 4.5–6 m. On the edge (Figure 10), MOR had an average value of 44.78 N/mm^2. In the central area, MOR was 33.44 N/mm^2, and in the core area, it was 31.27 N/mm^2.

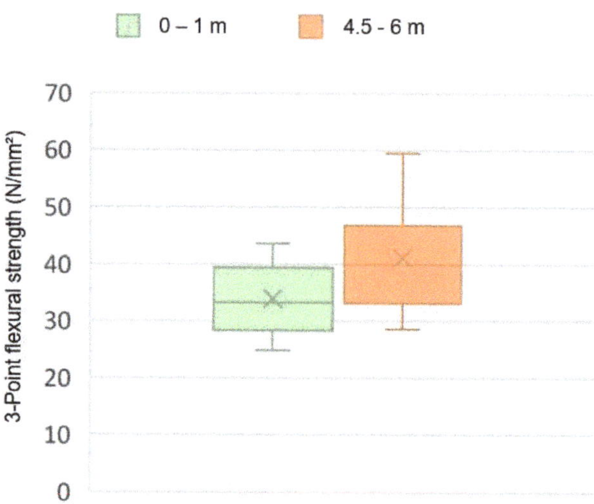

Figure 12. Modulus of rupture of Paulownia samples extracted at 0−1 m and 4.5−6 m from the tree stem.

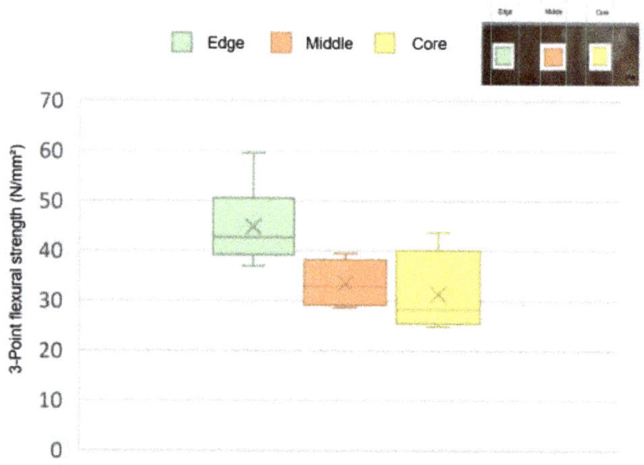

Figure 13. Modulus of rupture in the cross-section of the Paulownia's tree stem.

These MOR values are consistent with the interval of 23.98–43.56 N/mm^2 presented in the review study of [7]. Lachowicz and Giedrowicz [19] reported values from similar up to double at 23.89–53.17 N/mm^2, with an average of 38.63 N/mm^2. Esteves et al. [6] measured a higher three-point flexural strength of 53.5 N/mm^2 for Paulownia plantation wood from Portugal.

In the trunk area of 0–1 m, the MOE was 4123.75 N/mm^2, and at a height of 4.5–6 m, it was 4941.22 N/mm^2 (Figure 14).

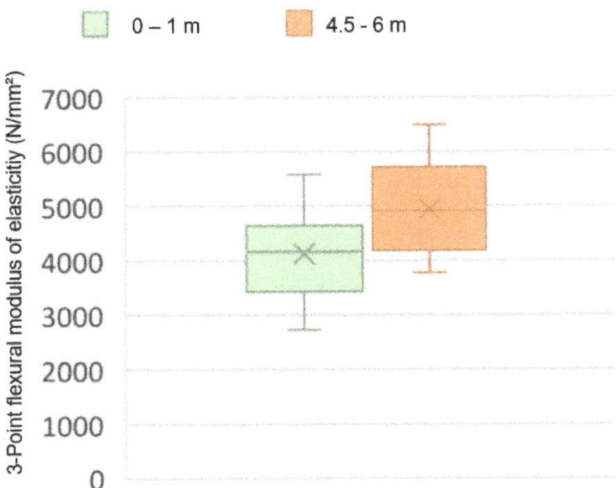

Figure 14. Modulus of elasticity of Paulownia samples extracted at 0−1 m and 4.5−6 m from the tree stem.

For the trunk cross-section, a decreasing tendency of MOE can be observed from the edge towards the core (Figure 15). On the edge, an average bending modulus of elasticity was measured as 5249.37 N/mm^2. In the core area, MOE was significantly lower, namely 3842.75 N/mm^2.

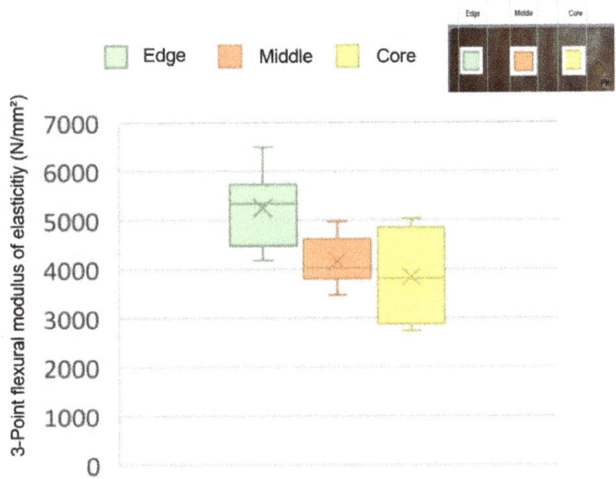

Figure 15. Modulus of elasticity in the cross-section of the Paulownia's tree stem.

Relatively similar values were reported by [5], from 4500 to 4900 N/m^2. Almost half of these values, namely 1900 N/mm^2, was MOE analyzed by [19]. The bending properties of Paulownia were significantly influenced by the density and porosity of the wood. Kiaei [17] measured a high porosity rate of 83% for Paulownia species sourced from Iran. In this study, MOR was, on average, 41 N/mm^2 and MOE was 3740 N/mm^2, which are consistent with the finding of this study considering a trunk height of 0–1 m and the area positioned between the pith and near the bark of the samples of Paulownia sourced from Serbia. The high microfibril angle (mostly of juvenile wood) significantly influenced the elastic behavior of Paulownia plantation wood. A comparison with other studies regarding elastic properties of Paulownia samples extracted from different height positions from the stem or cross-sections cannot be made considering insufficient data on Paulownia plantation wood.

3.5. Compressive Strength

The compressive strength, according to DIN 52185:1976 [51], of Paulownia wood is highest at a height of 4.5–6 m, with a value of 23.41 N/mm^2, and also at the edge of the log cross-section, with 25.11 N/mm^2 (Figures 16 and 17). At the height of 0–1 m, the compressive strength reached 19.41 N/mm^2. By scrutinizing the compressive strength of the samples extracted from the trunk's cross-section, the compressive strength in the middle area was 18.65 N/mm^2 and in the core area 19.25 N/mm^2.

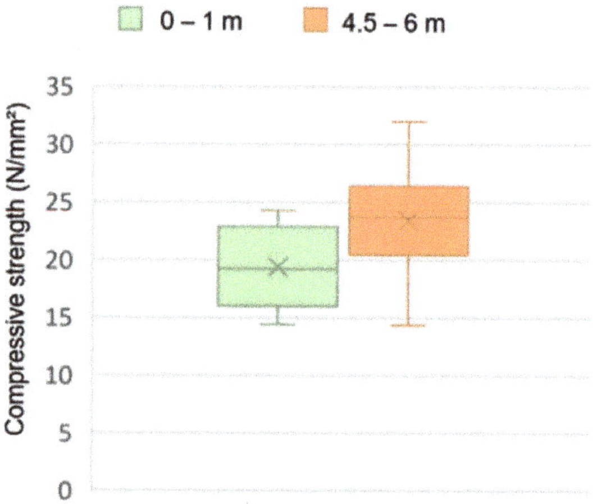

Figure 16. Compressive strength of Paulownia samples extracted at 0−1 m and 4.5−6 m from the tree stem.

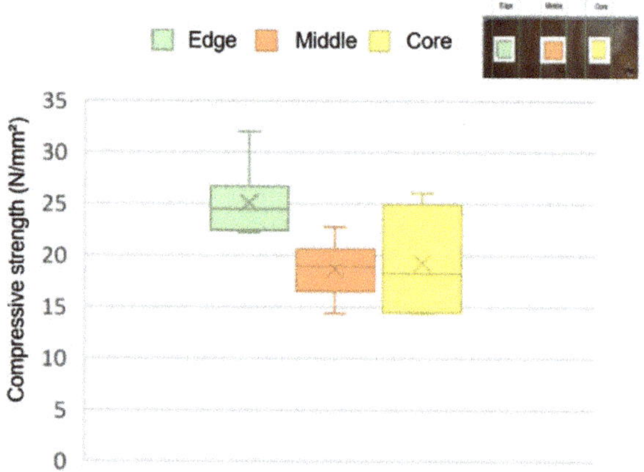

Figure 17. Compressive strength of Paulownia samples extracted from the cross-section.

The compressive strength of Paulownia wood analyzed in the present study is similar to the findings of [5], where compressive strength ranged from 19 to 23 N/mm^2. Higher values were reported by [25], of 26 N/mm^2, and significantly higher values (36 N/mm^2) were measured by [52]; however, much lower values, 14 N/mm^2, have been reported in the study of [19]. A comparison with other studies regarding the compressive strength of

Paulownia samples extracted from different positions from stem height or cross-section cannot be made yet considering state of the art of Paulownia short-rotation plantation wood.

3.6. Tensile Strength

Tensile strength, according to DIN 52188:1979 [53], decreases slightly when the samples were extracted from the top part of the tree (4.5 to 6 m) (Figure 18). The average value of the tensile strength was 40.44 N/mm^2 from 0–1 m height and 39.85 N/mm^2 from 4.5–6 m height.

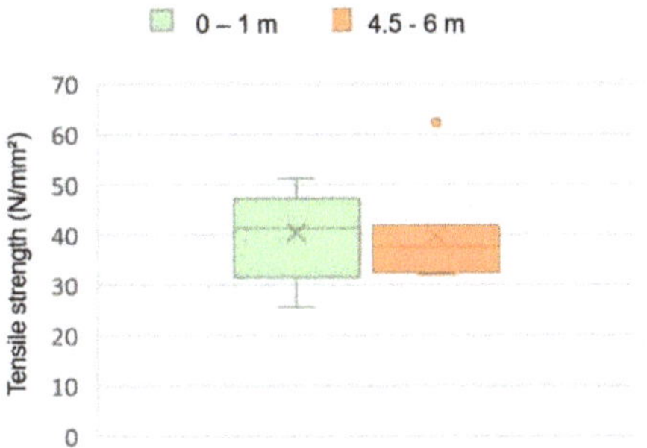

Figure 18. Tensile strength of Paulownia samples extracted at 0−1 m and 4.5−6 m from the tree stem.

It can be observed that in the core area of the tree trunk, the tensile strength decreased significantly (31.07 N/mm^2), because the density in the core is very low and, therefore, the wood can undergo little tensile stress. Between the bark and pith, the tensile strength reached a maximum average value of 49 N/mm^2. The tensile strength near the bark was 40.94 N/mm^2 (Figure 19).

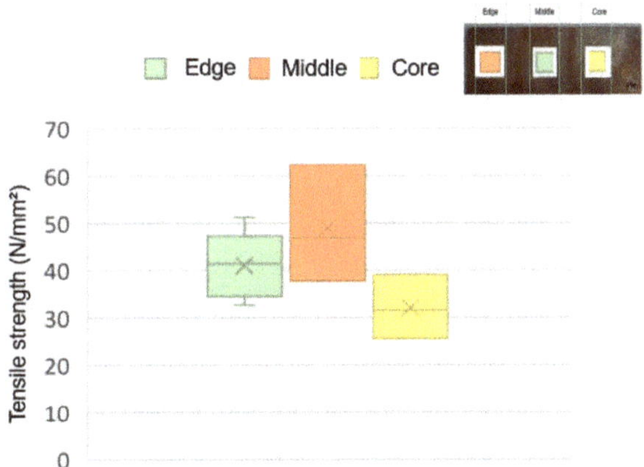

Figure 19. Tensile strength of Paulownia samples extracted from the cross-section.

The results of tensile strength are consistent with the findings presented by [5], from 36 to 44 N/mm^2 and 33 N/mm^2, and as reported by [44]. A comparison with other studies regarding the tensile strength of samples extracted from different positions of the stem height or cross-section are not yet available.

3.7. Screw Withdrawal Resistance (SWR)

The higher the position of the extracted samples in the tree trunk, the higher the screw withdrawal resistance (SWR). SWR was determined in concordance to EN 320:2011 [54]. In the area close to the ground (0–1 m), the mean value of SWR was 54.77 N/mm, and at a height of 4.5–6 m, it reached 57.82 N/mm (Figure 20).

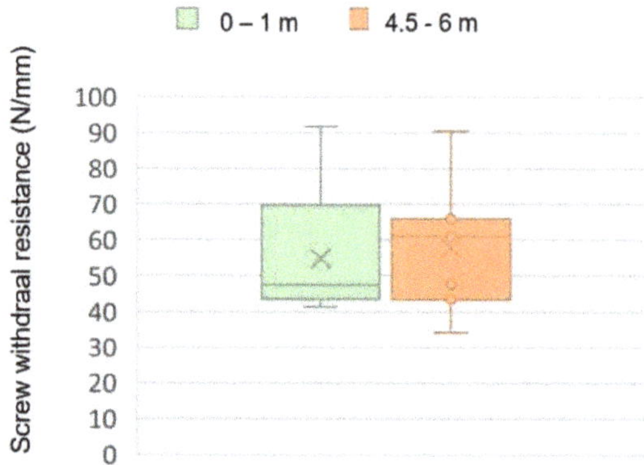

Figure 20. Screw withdrawal resistance of Paulownia samples extracted at 0–1 m and 4.5–6 m from the tree stem.

In the stem cross-section (Figure 21), SWR measured 55.39 N/mm between the fifth and seventh annual ring, 47.93 N/mm in the middle area (third to fourth growth ring), and near the second and first annual ring, it was 41 N/mm. Typical values of SWR range from 31 to 57 N/mm, per the results from the studies of [5,25]. Akyldiz [55] determined significantly lower SWR in the tangential, radial, and transversal direction for *Paulownia tomentosa* Steud., namely 19 N/mm, 18 N/mm, and 16 N/mm for the samples with a moisture content of 12%.

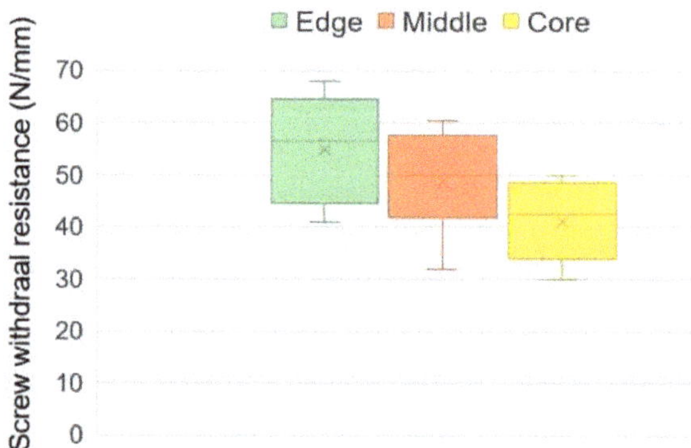

Figure 21. Screw withdrawal resistance of Paulownia samples extracted from the cross-section.

4. Conclusions

The present study assessed the physical and mechanical properties of Paulownia plantation wood, which were determined considering the tree stem height and position in the cross-section (near the bark, in the core, and in between these areas). The sampling height and cross-section site are of importance to any studies and in the processing of Paulownia plantation wood.

Fast-growing sawn wood from short-rotation plantations consists mostly of juvenile wood, with a lower density and larger microfibril angle than adult wood. There are clear differences between mature and juvenile wood, with higher strength properties for the former. Juvenile wood has a considerably lower modulus of rupture and elasticity, and maximum crushing strength in compression to that of mature wood. The relatively lower values of overall mechanical properties are dictated by density, with an average of 270 kg/m^3. Moreover, the results of this study indicate that *Paulownia tomentosa x elongata* plantation wood stabilizes from the fifth year of growth in terms of material density and shrinkage. For practical purposes, it should be considered the transition from juvenile to mature wood and the fact that high quality Paulownia timber can be harvested when the trees are older than 7 years. In this way, the yield of quality Paulownia sawn wood increases substantially.

Paulownia sawn wood extracted from the upper part of tree trunk (4.5–6 m) has a longer fiber length and lower microfibril angle, hence a higher bulk density that significantly influences the mechanical properties. This tendency can be considered atypical for Paulownia, because for most tree species, the samples extracted from the top part of the tree have lower density and strength properties.

There are large fluctuations in strength within a log and, therefore, large variations in mechanical properties. This could be important in raw material assessing, processing, targeting the added value and sustainable new products.

Due to its physical and mechanical properties Paulownia wood cannot be used in for the manufacture of structural components, where strong criteria are imposed for CE certification.

For further investigations, it is important to pay closer attention to the properties of Paulownia lumber depending on the position in the stem. A precise determination of wood characteristics should involve collecting samples from the whole tree height, at different tree ages, with a maximum of 15 years, that deliver the best yield for Paulownia short-rotation plantation timber.

Author Contributions: Conceptualization, E.M.T. and K.B.; methodology, E.M.T.; validation, A.P. and E.M.T.; formal analysis, K.B.; investigation, K.B.; resources, K.B. and E.M.T.; data curation, A.P.; writing—original draft preparation, K.B. and E.M.T.; writing—review and editing, E.M.T.; visualization, M.C.B. and A.P.; supervision, A.P. and E.M.T.; project administration, M.C.B. All authors have read and agreed to the published version of the manuscript.

Funding: This research received no external funding.

Institutional Review Board Statement: Not applicable.

Informed Consent Statement: Not applicable.

Data Availability Statement: Not applicable.

Acknowledgments: The authors would like to thank to Arien Crul, for providing all Paulownia roundwood from Serbia, Thomas Schnabel (FH Salzburg) for the support with design of experiment, Ing. Thomas Wimmer, for the measurements conducted in the facilities of FH Salzburg and Helmut Radauer, BSc. for the support with sample preparing.

Conflicts of Interest: The authors declare no conflict of interest.

References

1. Koman, S.; Feher, S. Physical and mechanical properties of Paulownia clone in vitro 112. *Eur. J. Wood Prod.* **2020**, *78*, 421–423. [CrossRef]
2. Pasiecznik, N. *Paulownia tomentosa* (paulownia). *CABI Compend.* **2022**. [CrossRef]
3. Huber, C.; Moog, D.; Stingl, R.; Pramreiter, M.; Stadlmann, A.; Baumann, G.; Praxmarer, G.; Gutmann, R.; Eisler, H.; Müller, U. Paulownia (*Paulownia elongata* S.Y.Hu)—Importance for forestry and a general screening of technological and material properties. *Wood Mater. Sci. Eng.* **2023**, *18*, 1–13. [CrossRef]
4. Fos, M.; Oliver-Villanueva, J.-V.; Vazquez, M. Radial variation in anatomical wood characteristics and physical properties of *Paulownia elongata* x *Paulownia fortunei* hybrid Cotevisa 2 from fast-growing plantations. *Eur. J. Wood Prod.* **2023**, *81*. [CrossRef]
5. Barbu, M.C.; Buresova, K.; Tudor, E.M.; Petutschnigg, A. Physical and Mechanical Properties of *Paulownia tomentosa* x elongata Sawn Wood from Spanish, Bulgarian and Serbian Plantations. *Forests* **2022**, *13*, 1543. [CrossRef]
6. Esteves, B.; Cruz-Lopes, L.; Viana, H.; Ferreira, J.; Domingos, I.; Nunes, L.J.R. The Influence of Age on the Wood Properties of *Paulownia tomentosa* (Thunb.) Steud. *Forests* **2022**, *13*, 700. [CrossRef]
7. Jakubowski, M. Cultivation Potential and Uses of Paulownia Wood: A Review. *Forests* **2022**, *13*, 668. [CrossRef]
8. Stochmal, A.; Moniuszko-Szajwaj, B.; Szumacher-Strabel, M.; Cieślak, A. *Paulownia Clon In Vitro 112*®*: The Tree of the Future: 21st World Congress on Nutrition and Food Science*; Journal of Nutrition & Food Sciences: Sydney, Australia, 2018.
9. Kadlec, J.; Novosadová, K.; Pokorný, R. Impact of Different Pruning Practices on Height Growth of Paulownia Clon in Vitro 112®. *Forests* **2022**, *13*, 317. [CrossRef]

10. Criscuoli, I.; Brunetti, M.; Goli, G. Characterization of *Paulownia elongata* x fortunei (BIO 125 clone) Roundwood from Plantations in Northern Italy. *Forests* **2022**, *13*, 1841. [CrossRef]

11. Dżugan, M.; Miłek, M.; Grabek-Lejko, D.; Hęclik, J.; Jacek, B.; Litwińczuk, W. Antioxidant Activity, Polyphenolic Profiles and Antibacterial Properties of Leaf Extract of Various *Paulownia* spp. Clones. *Agronomy* **2021**, *11*, 2001. [CrossRef]

12. Magar, L.B.; Khadka, S.; Joshi, J.R.R.; Pokharel, U.; Rana, N.; Thapa, P.; Sharma, K.R.S.R.; Khadka, U.; Marasini, B.P.; Parajuli, N. Total Biomass Carbon Sequestration Ability Under the Changing Climatic Condition by *Paulownia tomentosa* Steud. *Int. J. Appl. Sci. Biotechnol.* **2018**, *6*, 220–226. [CrossRef]

13. Wang, Q.; Shogren, J.F. Characteristics of the crop-paulownia system in China. *Agric. Ecosyst. Environ.* **1992**, *39*, 145–152. [CrossRef]

14. Feng, Y.; Cui, L.; Zhao, Y.; Qiao, J.; Wang, B.; Yang, C.; Zhou, H.; Chang, D. Comprehensive selection of the wood properties of Paulownia clones grown in the hilly region of southern China. *BioResources* **2020**, *15*, 1098–1111. [CrossRef]

15. Abbasi, M.; Pishvaee, M.S.; Bairamzadeh, S. Land suitability assessment for *Paulownia cultivation* using combined GIS and Z-number DEA: A case study. *Comput. Electron. Agric.* **2020**, *176*, 105666. [CrossRef]

16. López, F.; Pérez, A.; Zamudio, M.A.; de Alva, H.E.; García, J.C. Paulownia as raw material for solid biofuel and cellulose pulp. *Biomass Bioenergy* **2012**, *45*, 77–86. [CrossRef]

17. Kiaei, M. Technological properties of iranian cultivated paulownia wood (*Paulownia fortunei*). *Cellul. Chem. Technol.* **2013**, *47*, 735–743.

18. Dogu, D.; Tuncer, F.D.; Bakir, D.; Candan, Z. Characterizing microscopic changes of paulownia wood under thermal compression. *BioResources* **2017**, *12*, 5279–5295. [CrossRef]

19. Lachowicz, H.; Giedrowicz, A. Charakterystyka jakości technicznej drewna paulowni COTE−2. *Sylwan* **2020**, *164*, 414–423. [CrossRef]

20. Moreno, J.L.; Bastida, F.; Ondoño, S.; García, C.; Andrés-Abellán, M.; López-Serrano, F.R. Agro-forestry management of *Paulownia* plantations and their impact on soil biological quality: The effects of fertilization and irrigation treatments. *Appl. Soil Ecol.* **2017**, *117*, 46–56. [CrossRef]

21. Lee, S.H.; Lum, W.C.; Boon, J.G.; Kristak, L.; Antov, P.; Pędzik, M.; Rogoziński, T.; Taghiyari, H.R.; Lubis, M.A.R.; Fatriasari, W.; et al. Particleboard from agricultural biomass and recycled wood waste: A review. *J. Mater. Res. Technol.* **2022**, *20*, 4630–4658. [CrossRef]

22. Yorgun, S.; Yıldız, D.; Şimşek, Y.E. Activated carbon from paulownia wood: Yields of chemical activation stages. *Energy Sources Part A Recovery Util. Environ. Eff.* **2016**, *38*, 2035–2042. [CrossRef]

23. Icka, P.; Damo, R.; Icka, E. Paulownia Tomentosa, a Fast Growing Timber. *Ann. Valahia Univ. Agric.* **2016**, *10*, 14–19. [CrossRef]

24. BioTree. Paulownia Environment. Available online: https://paulowniatrees.eu/learn-more/paulownia-environment/ (accessed on 20 February 2023).

25. Akyildiz, M.H.; Kol, H.S. Some technological properties and uses of paulownia (*Paulownia tomentosa* Steud.) wood. *J. Environ. Biol.* **2010**, *31*, 351–355. [PubMed]

26. Ab Latib, H.; Choon Liat, L.; Ratnasingam, J.; Law, E.L.; Abdul Azim, A.A.; Mariapan, M.; Natkuncaran, J. Suitability of paulownia wood from Malaysia for furniture application. *BioResources* **2020**, *15*, 4727–4737. [CrossRef]

27. Yadav, N.K.; Vaidya, B.N.; Henderson, K.; Lee, J.F.; Stewart, W.M.; Dhekney, S.A.; Joshee, N. A Review of Paulownia Biotechnology: A Short Rotation, Fast Growing Multipurpose Bioenergy Tree. *AJPS* **2013**, *4*, 2070–2082. [CrossRef]

28. Ayrilmis, N.; Kaymakci, A. Fast growing biomass as reinforcing filler in thermoplastic composites: *Paulownia elongata* wood. *Ind. Crops Prod.* **2013**, *43*, 457–464. [CrossRef]

29. El-Showk, N.; El-Showk, S. The Paulownia Tree: An Alternative for Sustainable Forestry. 2003, pp. 1–10. Available online: http://www.cropdevelopment.org/docs/PaulowniaBrochure_print.pdf (accessed on 25 January 2023).

30. Crul, A. Paulownia Plantages Management in Central Europe. Coppicing and Pruning. Expert interview. 2023. Available online: https://treevest.de/wir-ueber-uns/ (accessed on 25 January 2023).

31. Rodríguez-Seoane, P.; Díaz-Reinoso, B.; Moure, A.; Domínguez, H. Potential of *Paulownia* sp. for biorefinery. *Ind. Crops Prod.* **2020**, *155*, 112739. [CrossRef]

32. Chongpinitchai, A.R.; Williams, R.A. The response of the invasive princess tree (*Paulownia tomentosa*) to wildland fire and other disturbances in an Appalachian hardwood forest. *Glob. Ecol. Conserv.* **2021**, *29*, e01734. [CrossRef]

33. Snow, W.A. Ornamental, crop, or invasive? The history of the Empress tree (*Paulownia*) in the USA. *For. Trees Livelihoods* **2015**, *24*, 85–96. [CrossRef]

34. van Wilgen, B.W.; Zengeya, T.A.; Richardson, D.M. A review of the impacts of biological invasions in South Africa. *Biol. Invasions* **2022**, *24*, 27–50. [CrossRef]

35. Remaley, T. Non-Native Plants—Rock Creek Park. Available online: https://www.nps.gov/rocr/learn/nature/non-native-plants.htm (accessed on 7 February 2023).

36. European Comission. List of Invasive Alien Species of Union Concern. Available online: https://ec.europa.eu/environment/nature/invasivealien/list/index_en.htm (accessed on 25 January 2023).

37. Bao, F.C.; Jiang, Z.H.; Jiang, X.M.; Lu, X.X.; Luo, X.Q.; Zhang, S.Y. Differences in wood properties between juvenile wood and mature wood in 10 species grown in China. *Wood Sci. Technol.* **2001**, *35*, 363–375. [CrossRef]

38. Niemz, P.; Sonderegger, W.U. *Holzphysik: Eigenschaften, Prüfung und Kennwerte*; 2., aktualisierte Auflage; Hanser: München, Germany, 2021; ISBN 978-3-446-46749-1.
39. *ISO 3131:1996-06-01*; Wood-Determination of Density for Physical and Mechanical Tests. International Organization for Standardization: Brussels, Belgium, 1996.
40. Ayarkawa, J. The influence of site and axial position in the tree on the density and strength properties of the wood of Pterygota Marcoarpa K. Schum. *Ghana J. For.* **1998**, *6*, 34–41.
41. Grosser, D. *Die Hölzer Mitteleuropas: Ein Mikrophotographischer Lehratlas*; Reprint der 1. Aufl. von 1977; Kessel: Remagen, Germany, 2007; ISBN 3935638221.
42. Bardarov, N.; Popovska, T. Examination of the properties of local origin Paulownia wood (*Paulownia* sp. Siebold & Zucc.). *Manag. Sustain. Dev.* **2017**, *63*, 75–78.
43. *DIN 52184:1979-05*; Testing of Wood; Determination of Swelling and Shrinkage. Deutsches Institut für Normung: Berlin, Germany, 1979.
44. Komán, S.; Vityi, A. Physical and mechanical properties of *Paulownia tomentosa* wood planted in Hungaria. *Wood Res.* **2017**, *62*, 335–340.
45. Sedlar, T.; Šefc, B.; Drvodelić, D.; Jambreković, V.; Kučinić, M.; Ištok, I. Physical Properties of Juvenile Wood of Two Paulownia Hybrids. *Drv. Ind.* **2020**, *71*, 179–184. [CrossRef]
46. Donaldson, L. Microfibril angle: Measurement, variation and relationships—A review. *IAWA J.* **2008**, *29*, 345–386. [CrossRef]
47. Martínez-Martínez, V.; del Alamo-Sanza, M.; Menéndez-Miguélez, M.; Nevares, I. Method to estimate the medullar rays angle in pieces of wood based on tree-ring structure: Application to planks of Quercus petraea. *Wood Sci. Technol.* **2018**, *52*, 519–539. [CrossRef]
48. *EN 1534:2011-01*; Wood Flooring-Determination of Resistance to Indentation-Test Method. European Committee for Standardization: Brussels, Belgium, 2011.
49. Mania, P.; Hartlieb, K.; Mruk, G.; Roszyk, E. Selected Properties of Densified Hornbeam and Paulownia Wood Plasticised in Ammonia Solution. *Materials* **2022**, *15*, 4984. [CrossRef]
50. *DIN 52186:1978-06*; Testing of Wood; Bending Test. Deutsches Institut für Normung: Berlin, Germany, 1978.
51. *DIN 52185:1976-09*; Testing of Wood; Compression Test Parallel to Grain. Deutsches Institut für Normung: Berlin, Germany, 1976.
52. Sell, J.; für das Holz Lignum, S.A. *Eigenschaften und Kenngrössen von Holzarten*; 4., überarb. und erw. Aufl.; Sell, J., Ed.; Baufachverl: Dietikon, Switzerland, 1997; ISBN 3855652236.
53. *DIN 52188:1979-05*; Testing of Wood; Determination of Ultimate Tensile Stress Parallel to Grain. Deutsches Institut für Normung: Berlin, Germany, 1979.
54. *EN 320:2011*; Particleboards and Fibreboards-Determination of Resistance to Axial Withdrawal of Screws. European Committee for Standardization: Brussels, Belgium, 2011.
55. Akyildiz, M.H. Screw-nail withdrawal and bonding strength of paulownia (*Paulownia tomentosa* Steud.) wood. *J. Wood Sci.* **2014**, *60*, 201–206. [CrossRef]

Communication

In Situ Detection of the Flexural Fracture Behaviors of Inner and Outer Bamboo-Based Composites

Xiu Hao [1], Yanglun Yu [2], Chunmei Yang [1,*] and Wenji Yu [1,2,*]

1   College of Mechanical and Electrical Engineering, Northeast Forestry University, Harbin 150006, China
2   Research Institute of Wood Industry, Chinese Academy of Forestry, Beijing 100091, China
*   Correspondence: ycmnefu@126.com (C.Y.); yuwenji@caf.ac.cn (W.Y.); Tel.: +86-188-4515-4869 (C.Y.); +86-010-8288-9427 (W.Y.)

Abstract: This paper investigated the fracture toughness and enhancement mechanism for each component in bamboo-based composites at the cellular level. In situ characterization techniques identified the fracture behaviors of bamboo-based composites in three-point bending tests, and scanning electron microscope (SEM) further visualized the crack propagation of the fracture surface. In addition, the improvement mechanism of bamboo-based composites was illustrated by mechanical properties at the cellular level assisted with nanoindentation tests. Our in situ test results showed that the bamboo-based composites exhibited a longer deformation and higher bending load compared with bamboo. The fracture was non-catastrophic, and crack propagated in a tortuous manner in bamboo-based composites. Microstructural analysis revealed that phenol-formaldehyde (PF) resin pulled out and middle lamella (ML) breaking rather than transverse transwall fracturing occurred in parenchymal cells. The higher density of fibers in the bamboo-based composites triggered massive interfacial delamination in the middle lamella (ML), which was a weak mechanical interface. Furthermore, indented modulus and hardness illustrated that phenol-formaldehyde (PF) resin improved the mechanical strength of cell walls, especially parenchymal cells. The crosslinks of PF resin with the cell walls and massive fibers were the primary mechanisms responsible for the fracture toughness of bamboo-based composites, which could be helpful for advanced composites.

Keywords: fracture behaviors; fibers; in situ bending tests; mechanical properties; parenchymal cells; bamboo-based composites

Citation: Hao, X.; Yu, Y.; Yang, C.; Yu, W. In Situ Detection of the Flexural Fracture Behaviors of Inner and Outer Bamboo-Based Composites. *Forests* **2023**, *14*, 515. https://doi.org/10.3390/f14030515

Academic Editors: Roman Reh, Lubos Kristak, Muhammad Adly Rahandi Lubis, Seng Hua Lee and Petar Antov

Received: 31 January 2023
Revised: 25 February 2023
Accepted: 28 February 2023
Published: 6 March 2023

1. Introduction

Bamboo is a biodegradable, rapidly renewable, and high-yielding natural material on Earth [1,2]. Bamboo is composed of fiber cells, parenchymal cells, and vessels. The stiff fibers embed into the soft parenchyma ground exhibiting remarkable mechanical strength and fracture toughness, which has traditionally been used in the building industry, such as flooring, roofing, walls, and scaffolding [3,4]. However, bamboo is highly variable in mechanical properties owing to its hollow structure [5,6]. These unique characteristics require great attention to bamboo-based panels or engineered products (bamboo plywood, laminated bamboo lumber, bamboo particleboard, bamboo scrimber), in which component preparation and adhesives bonding are critical to the manufacturing process to reduce the environmental load [7,8]. Bamboo is a natural functional hierarchical structure material composed of orientated stiff fibers and a soft parenchyma matrix [9–11]. At a macroscopic scale, vascular bundles display a decisive role in the mechanical properties of bamboo; however, the parenchyma tissues contribute to transferring the stress. At the microscopic level, fiber cell walls have a multilayered wall structure with a tiny lumen, and the parenchymal cells have a thin wall and larger lumen [12–17]. Getting a deep study on the improvement mechanism of different components in the bamboo-based panel could be helpful in designing biomimetic polymeric composites.

More research efforts have been made recently in studying the mechanical properties and toughness mechanism by observing the fracture behavior of the hierarchical structure in bamboo with the assistance of in situ mechanical testing, particularly at the cellular level [18,19]. Among these efforts, most studies indicated that fiber gradient distribution and parenchymal tissue displayed different fracture behavior under bending tests [9,10]. Parenchymal cells were weak points [18]; however, the oriented stiff fibers were attributed to inhibiting crack growth as the bamboo fractured [19].

Bamboo scrimber is composed of bamboo fibers arranged in parallel with adhesive, showing excellent strength and quality by condensation of fiber and bonding resin [20–35]. The study focused on investigating the structure–property relationships in the bamboo scrimber. To preserve the morphology of fibers and parenchymal cells, the bamboo-based composites in this study were prepared by the inside and outside thin bamboo strips. Specially, we focus on the bending performance and fracture toughness of bamboo-based composites. In situ bending tests were conducted to prevent the fracture behaviors of bamboo-based composites. In addition, the fracture characteristics were observed by Field-emission scanning electron microscope (SEM). Finally, nanoindentation tests were used to illustrate the mechanical properties at the cellular level.

2. Materials and Methods

2.1. Samples Preparation

The 4 years old moso bamboo (*Phyllostachys pubescens* Mazel ex H.de Lehaie) was collected from Zhejiang province in China for this study. The bamboo wall was 8–10 mm thick, and the diameter at breast height was 120 mm. Samples were taken from the internodes located at the culm height of 2 m. Then, the specimens were kept in an environment of 65% relative humidity and a conditioning chamber (25 °C) until an EMC of 10%–12%.

The distribution of vascular bundles in bamboo followed a gradually increased order from the inner to the outer part along the radial direction showing a hierarchical characteristic, as presented in Figure 1A. Bamboo was sliced into two pieces of ~1 mm thickness strands from the inner and outer parts (Figure 1B) to prepare the bamboo-based composites made from the different zone of the bamboo portion. The densities of inner and outer bamboo were 0.65 g/cm^3 and 0.9 g/cm^3, respectively. The making process of bamboo-based composites was as follow: firstly, the bamboo strips from the inner and outer part with a dimension of 100 (l) × 10 (t) × 1 (r) mm^3 were immersed in a 15% aqueous solution PF resin for 12 h. Then, samples were kept under ambient conditions for 12 h and were then oven dried at 50 °C for 2 h. After oven drying, the specimens were conditioned at 60% relative humidity (20 °C) to regulate moisture content. Four oven-dried resin-impregnated specimens of each part were parallel glued and pressed at 150 °C under 3 MPa to obtain bamboo-based composites with a dimension of 100 (l) × 10 (t) × 2 (r) mm^3 (Figure 1C), and the density of bamboo-based composites is 1.10–1.15 g/cm^3. The specimens were kept in an environment of 60% relative humidity and a conditioning chamber to EMC of 10%–12%.

2.2. In Situ Mechanical Tests

The in situ bending tests were conducted in an SEM chamber (Quanta 2000, FEI, Hillsboro, OR, USA) with a miniature mechanical testing device (Microtest 2000, Deben Ltd., East Grinstead, UK) equipped with a maximum load capacity of 660 N (Figure 1D). For the three-point bending tests, five samples were cut from the middle of the scrimber (Figure 1E). The inner and outer parts of bamboo were also obtained as a control group (Figure 1E). Then, the samples were polished by a sliding microtome. The dimension of inner bamboo, outer bamboo, and bamboo-based composites was 30 (l) × 2 (t) × 4 (r) mm^3, 30 (l) × 2 (t) × 4 (r) mm^3, and 30 (l) × 2 (t) × 2 (r) mm^3 to obtain the load–displacement curves. To obtain the microscopic images with equal precision, the dimension of inner bamboo, outer bamboo, and bamboo-based composites was 30 (l) × 2 (t) × 2 (r) mm^3. The three-point bending tests were conducted according to the Chinese standard of GB/T 15780-1995 and the fracture process of the samples during bending tests was examined in real-time. The crack initiation and propagation mechanism in bamboo-based composites and bamboo could be characterized simultaneously by the SEM images. The in situ tests were taken at room temperature with an acceleration voltage of 5 kV.

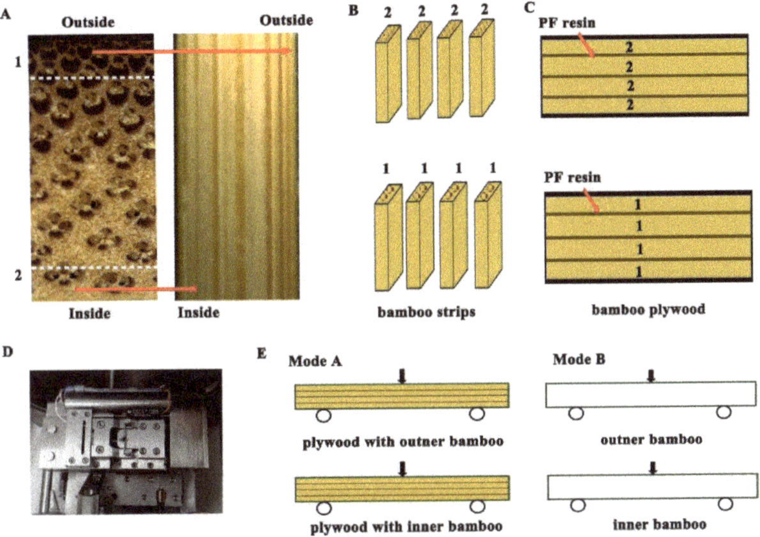

Figure 1. Schematic of test samples preparation. (**A**) images of bamboo culm; (**B**) bamboo strips from the inside and outside; (**C**) bamboo-based composites; (**D**) bending tests; (**E**) flexural bending configures of bamboo-based composites (mode A) and bamboo strips (mode B).

2.3. Morphology Characterization

Bending-induced fracture appearances of specimens were produced by a field-emission scanning electron microscope (SEM, Hitachi, S-4800,Tokyo, Japan). Five specimens with a dimension of 5 (l) $\times$ 2 (t) $\times$ 2 (r) mm^3 were plated with gold, and the acceleration voltage was 10 kV. The SEM experiments were operated at ambient conditions.

2.4. Nanoindentation and Dynamic Modulus Analysis

The nanoindentation and dynamic modulus maps tests (Hysitron TI 980 TriboIndenter, Bruker, Germany) measured the mechanical properties of the cell walls in the bamboo-based composites. Five samples with a size of 5 (l) $\times$ 1 (r) $\times$ 1 (t) mm^3 were prepared by grinding and polishing to obtain a smooth surface. The Berkovich tip was loaded to a 10 mN peak at a 20 uN·s^{-1} rate and then held at a constant load for 5 s. The hardness and reduced elastic modulus were calculated from the positions on each wall layer and middle lamellae (ML) in fiber cells and parenchymal cells in bamboo-based composites. After indentation, the in situ scanning probe microscopy images observed the indents' position and morphological characteristics. The nanoindentation tests were taken at room temperature.

3. Results

3.1. The Fracture Behavior of Inner Bamboo-Based Composites

The load–displacement curves for in situ bending tests of bamboo-based composites from inner bamboo and the inner part of bamboo were plotted in Figure 2. The results revealed that the fracture of bamboo-based composites during bending was non-catastrophic and more complicated than the failure of bamboo. The bending force in bamboo-based composites with 1.15 g/cm^3 was ~50 N, which was higher than that of bamboo with 0.65 g/cm^3.

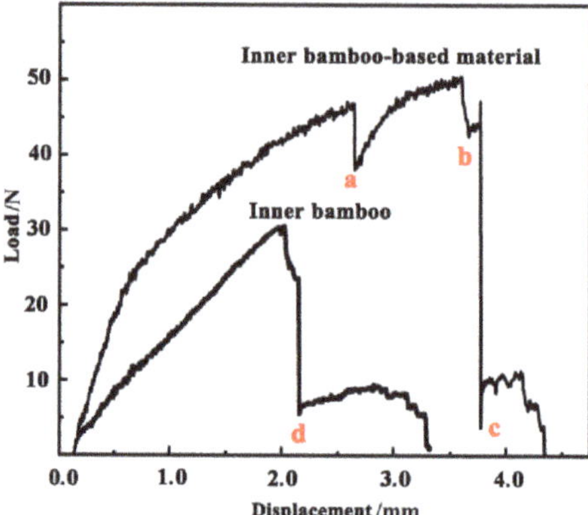

Figure 2. The represented load–displacement curves and loading stage for in situ bending tests on the specimens from the inner bamboo-based composites (a–c) and the inner bamboo (d).

A serious of SEM images corresponding to the bending strain detected by in situ tests were also shown in Figure 3. As shown in Figure 3B, the initial cracks occurred in the outer of the bamboo-based composites, and the cracks propagated across the parenchyma tissue in a tortuous manner, which was accompanied by the first load drop (a) in Figure 2. In the following process, the cracks reached fiber bundles and deviated from the initial direction to the interface of fiber bundles. Once the cracks induce dramatic interface delamination of fibers, the cracks developed in the parenchymal tissues on the side of the fiber, which also showed a tortuous route, where the loading was indicated by points b–c. In accordance with the in situ snapshots in Figure 3C, parenchymal tissue fracture and interfacial failure were the dominant fracture mode in the bamboo-based composites. Meanwhile the inner part of the bamboo failure displayed almost a brittle fracture, as indicated by points d in Figure 2. The visible cracks initiated quickly within the bottom area in the parenchyma cells (Figure 3E), which sharply propagated and resulted in a transverse wall failure of the parenchymal tissues (Figure 3F).

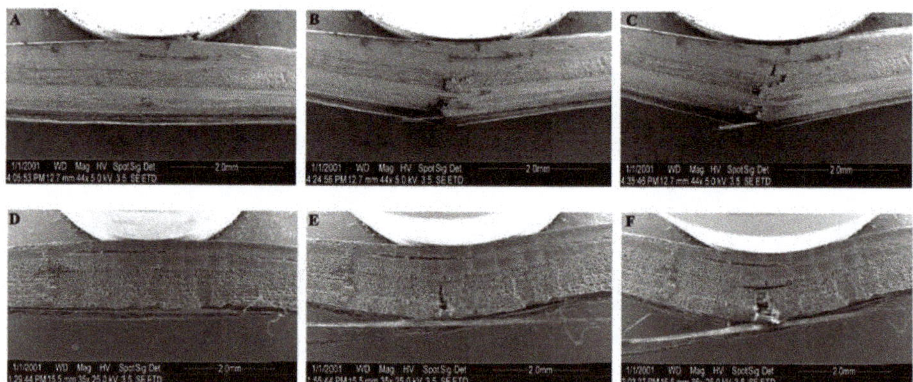

Figure 3. In situ three-point bending tests of the inner bamboo-based composites (**A–C**) and the inner bamboo (**D–F**).

Figure 4 further presents SEM images of the fracture surfaces and crack growth of failure samples. According to the microstructure in Figure 4A, the bamboo-based composites displayed almost ML breaking and pullout of the PF resin, exhibiting a rough fracture surface. In addition, the crack propagated along the ML interface of parenchymal cells in a tortuous manner. Meanwhile, SEM images in Figure 4B illustrated that repeated ML breaking and transverse wall of parenchyma cells were the most prominent feature in the inner bamboo fracture surface with a straight fracture path. This could be PF resin filled in the parenchymal cells and crosslinked with the cell wall and improved the mechanical properties of cell walls. Therefore, the PF resin could help dissipate the stress and prevented the parenchymal cells from failing.

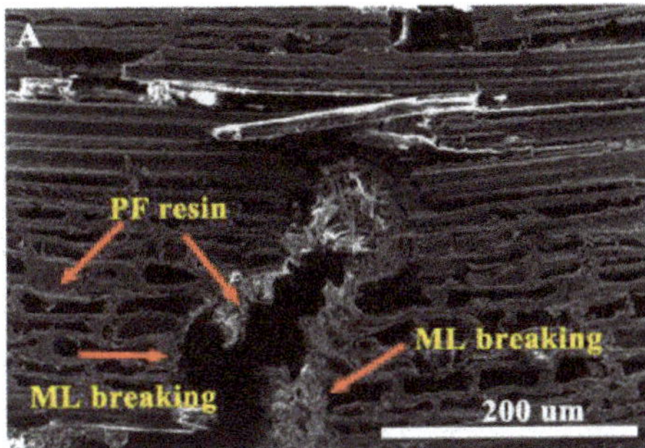

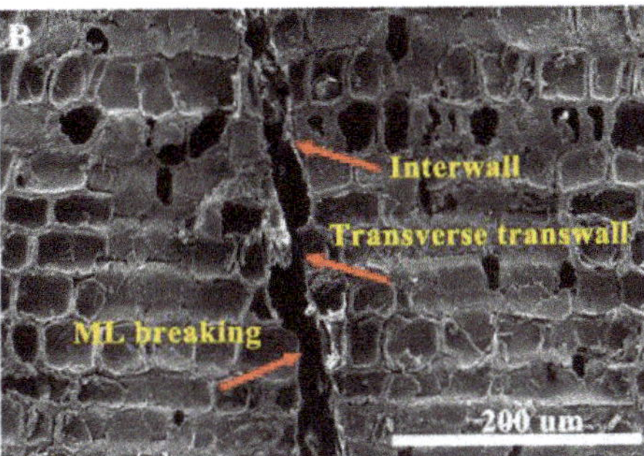

Figure 4. SEM images of three-point bending tests of the inner bamboo-based composites (**A**) and inner bamboo (**B**).

3.2. Fracture Behavior of Outer Bamboo-Based Composites

In this study, the simultaneous detection of load–displacement behaviors of the bamboo-based composites from outer bamboo and the outer bamboo is shown in Figure 5. According to the results, the bamboo-based composites exhibited a long deformation with a maximum load of ~80 N, while the samples of bamboo indicated a short elongation with a maximum load of ~40 N. During the bending process, the bending behaviors of bamboo-based composites were more complicated, displaying a stepwise fracture process from the higher force field (point a) to the lower force field (point d).

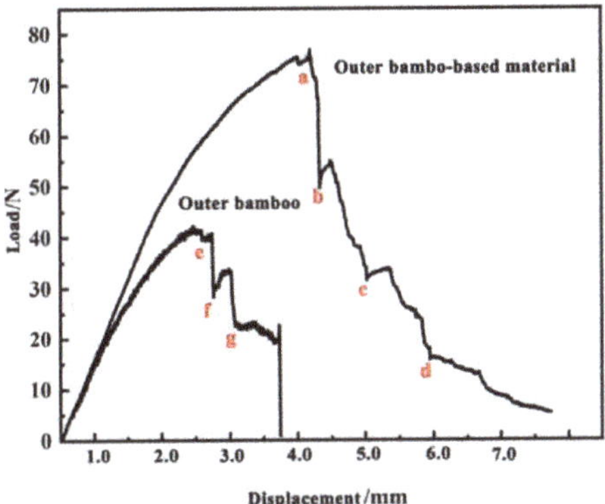

Figure 5. The represented load-displacement curves and loading stage for in situ bending tests on the specimens from the outer bamboo-based composites (a–d) and the outer bamboo (e–g).

Figure 6 presents a series of SEM images of the instant crack initiation and propagation process of the specimens in bending tests, corresponding to the various stages in Figure 5. At first, when the bending load reached the maximized value, visible cracks were initiated and occurred in the bottom area of the bamboo-based composites, which sharply propagated to the closer fiber layer (Figure 6B), and an obvious yield point (a) took place in the load–displacement curve as shown in Figure 5. Afterward, the closest fiber obstructed the crack extension, and the crack tip propagated along the interface of the fiber bundles, which induced debonding of the fibers (Figure 6C). In the following process, the stiff fiber bundles bore a high load, which resulted in the repeated process of crack propagating and interface delamination with further increasing loading (Figure 6D). The cracks propagated in a layer-by-layer path resulting in successive drop load from point a to b (Figure 5). Furthermore, the bamboo-based composites with a higher density of fiber bundles were able to sustain the higher bending loads.

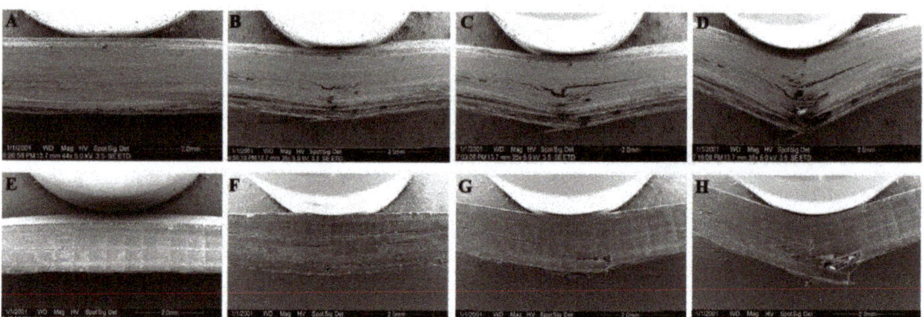

Figure 6. In situ three-point bending tests of the outer bamboo-based composites (**A–D**) and the outer bamboo (**E–H**).

In the case of bamboo (Figure 6E), the initial cracks occurred in the bottom (Figure 6F) and propagated perpendicular to the closest fiber layer (Figure 6G), which triggered a drop from point e to g in Figure 5. In the flowing process, the cracks also resulted in interface delamination in the fiber bundles (Figure 6H). The results showed that the degree of deformation in bamboo-based composites was obviously larger than that in outer bamboo, which can be attributed to more fiber density.

To further visualized the crack growth mode, Figure 7 showed the SEM images of the bamboo-based composites and the outer part of bamboo specimens in bending. Microstructural images presented in Figure 7A revealed that the bamboo-based composites with higher fibers exhibited longitudinal crack propagation, and the stiff fiber bundles obstructed crack tip growing by fiber bridging and ML breaking, resulting in interface delamination. In addition, the fiber bundles displayed a rough fracture appearance attributed to PF resin coating. By contrast, the bending tests of bamboo with lower fiber triggered more fiber failure displaying fiber pullout and bridging (Figure 7B). Compared with crack propagation in the outer of the bamboo, the influence of higher fiber and PF resin in bamboo-based composites blunted the transverse crack growth.

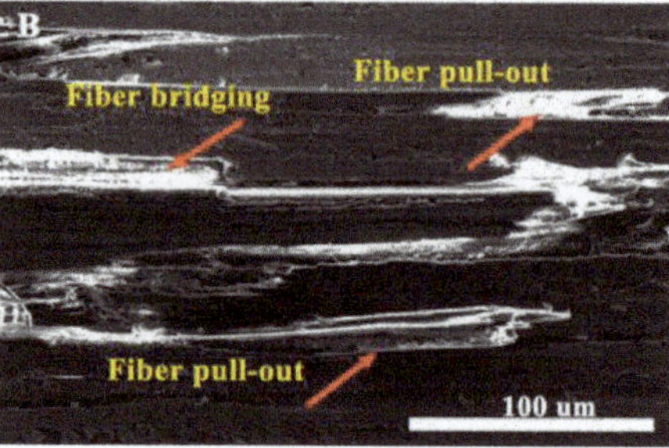

Figure 7. Micro-images of the outer bamboo-based composites (**A**) and the outer bamboo (**B**).

3.3. Micro-Mechanical Properties of Cell Walls in Bamboo-Based Composites

Reduced modulus and hardness were obtained to investigate the mechanical properties of fiber and parenchymal cells in bamboo-based composites, which were showed in Figure 8. In the fiber and parenchymal cells of bamboo-based composites, the structural difference in cell wall layers resulted in the reduced modulus and hardness values varying from the secondary cell wall to ML corresponding to bamboo. The storage modulus of the s2, s1, and ML layers in the fiber cell was 27 GPa, 24 GPa, and 18 GPa, respectively, which were higher than that in bamboo in our previous study [15]. Furthermore, the storage modulus of the parenchymal cell walls and ML layer was 15 GPa and 13 GPa in the bamboo-based composites, which were more than those in bamboo. The hardness in the fibers (s2: 0.78 GPa, s1: 0.66 GPa, ML: 0.67 GPa) and parenchymal cells (s: 0.66 GPa, ML: 0.64 GPa) also exhibited increment compared to that in bamboo. These results may be due to the cross-linking reaction between PF resin and cell walls in fiber and parenchymal cells, which improved the mechanical strength of cell walls. Comparing with the mechanical strength values of fiber and parenchyma cells, it was found that the increment rate of indentation modulus and hardness values of parenchyma cells was greater than that of fiber cells, which indicated more PF resin permeating in the cell walls in parenchymal cells. Hence, cell wall failure of parenchymal cells in the bamboo-based composites occurred an ML and PF resin breaking rather than transverse transwall fracture. Meanwhile, as shown in Figure 8, the storage modulus of the ML in fiber and parenchyma cells was lower than those of the cell walls, which were the weak interface in the bamboo-based composites. Hence, interface delamination of fibers and ML-PF resin pullout in parenchymal cells were the prominent fracture characteristic in the bamboo-based composites. While compared with the fiber cells, the reduced modulus value of the s layer in parenchymal cells was much lower than that in fiber cell walls. This may be accounting for the fiber cells being the strongest interface in graded bamboo structure, which prevented fiber cells from bucking under the external force.

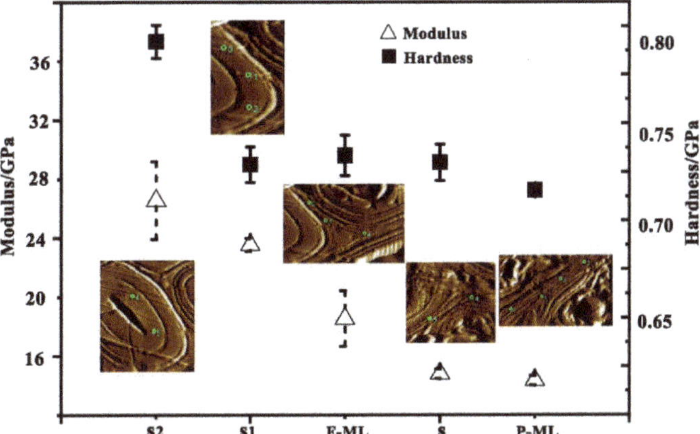

Figure 8. The modulus and hardness of the fiber and parenchymal cell walls and different layers in bamboo-based composites (s2, s1: Fiber cell wall; F-ML: ML in fiber cell; s: Parenchymal cell wall; P-ML:ML in parenchymal cell; 0–5: nanoindentation test number in fiber cells and parenchymal cells).

4. Discussion

Bamboo scrimber is prepared by bonding bamboo trips with PF resin, which show well quality and strength compared with natural bamboo. A lot of attempts have been made to illustrate the preparation and physical and mechanical properties of bamboo scrimber. Meanwhile, very few studies focused on the improvement mechanism within bamboo scrimber, especially at the cellular level. In situ detection of the fracture behaviors of moso bamboo in bending tests illustrated that the synergistic effects of the fiber bundles and parenchyma ground played an important role in restraining crack propagation, which was mainly responsible for the fracture toughness of bamboo [17,21]. In this study, we performed in situ bending tests to directly observe the flexural response of the structural components in bamboo-based composites. Since the fiber density increases from the inside to the outside of the bamboo stem, the bamboo-based composites were composed of thin inner and outer bamboo strips arranged in parallel with adhesive in this work to get an understanding of the fracture behavior of the fiber and parenchymal cells. The load–displacement curves revealed that inner and outer bamboo-based composites owned higher fracture loads than that of natural bamboo. The failure load of inner bamboo-based composites was two times that of inner bamboo. Compared with inner bamboo, the failure process of inner bamboo-based composites was complicated, and the load–displacement curve showed stepwise. The cured PF resin restrained the vertical crack propagation as it acted in the inner bamboo, and it triggered the peeling and delamination of parenchymal cell walls in the inner bamboo-based composites, which extended the crack growth path. Compared with the outer bamboo, the fracture load of outer bamboo-based composites increased with the fiber bundle density increasing. As the load increased, the fiber bundles were peeled off from the outer layer to the inner layer until the composite was completely destroyed, and the crack propagation path showed a zigzag crack growth mode, which was similar to the fracture behaviors of bamboo [15,17]. Meanwhile, in the in situ bending tests of outer bamboo-based composites, the fiber bundles with cured PF resin restrained crack propagation, and the fiber bundles mainly acted as interfacial debonding and fiber bridging, which showed few fiber bundle pull-out compared with the outer bamboo. As shown in a series of in situ bending tests, the bamboo-based composites exhibited distinct asymmetric crack propagation manners induced by the fiber and parenchymal cells as they acted in the bending-induced fracture behavior of bamboo in the previous study [14,15,17,19]. In this work, the cured PF resin exhibited a significant effect on crack propagation in fibers and parenchyma cells. The crack propagation behaviors in bamboo-based composites from inner bamboo showed that cracks generally occurred in the ML layer and PF resin, which caused the peeling of the parenchymal cells and the tearing of the PF resin. Crack propagation behaviors within bamboo-based composites made of outer bamboo showed that the fiber bundles played an important role in restraining crack propagation, acting as crack stoppers. In situ bending tests and the SEM micro-images illustrated the fracture toughness attributed to the higher fiber density and PF resin crosslink with the cell wall in fiber cells and parenchymal cells.

The nanoindentation characterization was conducted on the cell wall layer and ML layer in the fiber cells and parenchymal cells. The elastic modulus of bamboo presented a peak in the s2 (23.52 GPa) and declines in s1 (18.02 GPa), ML (15.91 GPa), and lumen (14.63 GPa), in that order [22]. The elastic modulus in the parenchymal cell layers exhibited a downward trend of s (10 GPa) and ML (5.57 GPa) [22]. The storage in the fiber cells (s2: 27 GPa, s1: 24 GPa, ML: 18 GPa) and (s: 15 GPa, ML: 13 GPa) parenchymal cells within bamboo-based composites exhibited higher values compared to natural bamboo. We confirmed that fibers were indeed the strongest phase and ML was comparatively weaker, while PF resin improved the mechanical properties of cell wall layers in bamboo. By learning from the cellular structure of bamboo-based composites during real-time bending deformation, it is possible to reveal the structure–property relationships of bamboo-based composites and could be helpful in understanding and designing the advanced biological composites.

5. Conclusions

In summary, the bamboo-based composites exhibited greater fracture toughness than bamboo, showing a longer deformation and less damage to fiber and parenchymal cell walls. The improvement mechanism was highly attributed to the reaction of PF resin with cell walls and fiber density. In situ bending tests and SEM observing of bamboo-based composites from inner bamboo illustrated that cracks generally occurred in the ML layer and PF resin, which caused the peeling of the parenchymal cells and the tearing of the PF resin. The cracks propagated in a tortuous manner, leading to massive interfacial debonding rather than transverse transwall fracture behavior for parenchymal cell walls in bamboo-based composites.

Crack propagation behavior within bamboo-based composites made of outer bamboo showed that the fiber bundles played an important role in restraining crack propagation, acting as crack stoppers. More fiber bundles in the bamboo-based composites trigged massive interfacial delamination, exhibiting ML breaking and fiber bridging. The cracks propagated in a zigzag manner within the fiber bundles under bending, leading to interfacial delamination. The occurrence of parenchymal cell pealing, PF resin tearing, and interfacial delamination was regarded as dissipating the crack energy.

Nanoindentation tests were conducted to reveal the improvement mechanism of fiber and parenchymal cell walls. The fiber cell and parenchymal cells within bamboo-based composites exhibited higher storage modulus compared to natural bamboo, which indicated that the PF resin improved the mechanical properties of fiber and parenchymal cells in bamboo-based composites. The effects of PF resin were regarded to be mainly responsible for preventing cell walls from destruction, especially parenchymal cells.

Author Contributions: X.H.: Data curation, Investigation, Writing—original draft. Y.Y.: Formal analysis. C.Y.: supervision, Writing—review and editing. W.Y.: Conceptualization, Funding acquisition, Methodology, Project administration. All authors have read and agreed to the published version of the manuscript.

Funding: The authors gratefully acknowledge the Nature Science Foundation of China (No. 31870550) and the financial support of Fundamental Research Funds for the Central Universities (No. 2572020AW40).

Data Availability Statement: All relevant data are within the paper.

Conflicts of Interest: The authors declare that they have no known competing financial interests or personal relationships that could have appeared to influence the work reported in this paper.

References

1. Lugt, V.; Dobbelsteen, V.D. An environmental, economic and practical assessment of bamboo as a building material for supporting structures. *Constr. Build. Mater.* **2006**, *20*, 648–656. [CrossRef]
2. Li, Y.; Xu, b.; Zhang, Q. Present situation and the countermeasure analysis of bamboo timber processing industry in China. *J. For. Eng.* **2016**, *1*, 2–7. (In Chinese)
3. Sharma, B.; Gatoo, A.; Bock, M. Engineered bamboo for structural applications. *Constr. Build. Mater.* **2015**, *81*, 66–73. [CrossRef]
4. Wu, B.; Gatoo, A.; Bock, M. Engineered bamboo for structural applications. *J. Nanjing For. Univ. (Nat. Sci. Ed.).* **2014**, *38*, 115–120. (In Chinese)
5. Qi, J.; Xie, J.; Huang, X.; Yu, W.; Chen, S. Influence of characteristic inhomogeneity of bamboo culm on mechanical properties of bamboo plywood: Effect of culm height. *J. Wood. Sci.* **2014**, *60*, 396–402. [CrossRef]
6. Zhan, T.; Sun, F.; Chao, L. Moisture diffusion properties of graded hierarchical structure of bamboo: Longitudinal and radial variations. *Constr. Build. Mater.* **2020**, *259*, 119641. [CrossRef]
7. Nogata, F.; Takahashi, H. Intelligent functionally graded material: Bamboo. *Composites.* **1995**, *5*, 743–751. [CrossRef]
8. Nkeuwa, W.; Zhang, J.; Semple, K.; Chen, M.; Xia, Y.; Dai, C. Bamboo-based composites: A review on fundamentals and processes of bamboo bonding. *Composites* **2022**, *235*, 109776. [CrossRef]
9. Dixon, P.G.; Gibson, L.J.; Bock, M. The structure and mechanics of moso bamboo material. *Sci. Rep.* **2014**, *11*, 20140321. [CrossRef]
10. Wang, X.; Ren, H.; Zhang, B.; Fei, B.; Burgert, I. Cell wall structure and formation of maturing fibres of moso bamboo (Phyllostachys pubescens) increase buckling resistance. *J. R. Soc. Interface* **2012**, *9*, 988–996. [CrossRef]
11. Lian, C.; Liu, R.; Zhang, S.; Yuan, J.; Luo, J.; Yang, F.; Fei, B. Ultrastructure of parenchyma cell wall in bamboo (Phyllostachys edulis) culms. *Cellulose* **2020**, *27*, 7321–7329. [CrossRef]
12. Parameswaran, N.; Liese, W. On the fine structure of bamboo fibres. *Wood Sci. Technol.* **1976**, *10*, 231–246. [CrossRef]
13. Shao, Z.; Fang, C.; Huang, S. Tensile properties of moso bamboo (Phyllostachys pubescens) and its components with respect to its fiber-reinforced composite structure. *Wood Sci. Technol.* **2010**, *44*, 655–666. [CrossRef]
14. Habibi, M.; Lu, Y. Cracks propagation in bamboo's hierarchical cellular structure. *Sci. Rep.* **2014**, *4*, 5598. [CrossRef] [PubMed]
15. Liu, H.; Wang, X.; Zhang, X.; Sun, Z.; Jiang, Z. In situ detection of the fracture behaviour of moso bamboo (phyllostachys pubescens) by scanning electron microscopy. *Holzforschung* **2016**, *70*, 1183–1190. [CrossRef]
16. Wang, F.; Shao, Z.; Bock, M. Study on the variation law of bamboo fibers' tensile properties and the organization structure on the radial direction of bamboo stem. *Ind. Crops Prod.* **2020**, *152*, 112521. [CrossRef]

17.   Wang, D.; Lin, L.; Fu, F. Fracture mechanisms of moso bamboo (*Phyllostachys pubescens*) under longitudinal tensile loading. *Ind. Crops Prod.* **2020**, *153*, 112574. [CrossRef]
18.   Obataya, E.; Kitin, P.; Yamauchi, H. Bending characteristics of bamboo (Phyllostachys pubescens) with respect to its fiber–foam composite structure. *Wood Sci. Technol.* **2007**, *41*, 385–400. [CrossRef]
19.   Chen, M.; Ye, L.; Wang, G.; Fang, C.; Dai, C.; Fei, B. Fracture modes of bamboo fiber bundles in three-point bending. *Cellulose* **2019**, *26*, 8101–8108. [CrossRef]
20.   Hao, X.; Tian, X.; Li, S.; Yang, C.; Yu, Y.; Yu, W. The separation mechanism of bamboo bundles at the cellular level. *Forests* **2022**, *13*, 1897. [CrossRef]
21.   Yu, W.; Yu, Y. Development and prospect of wood and bamboo scrimber industry in China. *China Wood Ind.* **2013**, *27*, 5–8. (In Chinese)
22.   Zhang, Y.; Huang, Y.; Zhang, Y.; Yu, Y.; Yu, W. Scrimber board (sb) manufacturing by a new method and characterization of sb's mechanical properties and dimensional stability. *Holzforschung* **2018**, *72*, 283–289. [CrossRef]
23.   Wei, J.; Rao, F.; Zhang, Y.; Li, C.; Yu, W. Effect of veneer crushing process on physical and mechanical properties of scrimbers made of pinus radiata and populus tomentosa. *China Wood Ind.* **2019**, *33*, 55–58. (In Chinese)
24.   Yu, Y. Manufacturing technology and mechanism of high performance bamboo-based fiber composites. Ph.D. Thesis, Chinese Academy of Forestry, Beijing, China, 2014. (In Chinese).
25.   Bao, M.; Huang, X.; Zhang, Y.; Yu, W.; Yu, Y. Effect of density on the hygroscopicity and surface characteristics of hybrid poplar compreg. *Wood Sci. Technol.* **2016**, *62*, 441–451. [CrossRef]
26.   Li, J.; Wu, Z.; He, S.; Huang, C.; Chen, Y. Effect of bundle diameter on water absorption properties of bamboo scrimber. *J. Zhejiang For. Sci. Technol.* **2016**, *36*, 67–70. (In Chinese)
27.   Li, Q.; Wang, K.; Yang, W.; Hua, X.; Weng, F.; He, Q. Research on technology of remaking bamboo glued board. *J. Bamboo. Re.* **2003**, *22*, 56–60. (In Chinese)
28.   Fu, Y.; Fang, H.; Dai, F. Study on the properties of the recombinant bamboo by finite element method. *Composites* **2017**, *115*, 151–159. [CrossRef]
29.   Kumar, A.; Vlach, T.; Laiblova, L. Engineered bamboo scrimber: Influence of density on the mechanical and water absorption properties. *Constr. Build. Mater.* **2016**, *127*, 815–827. [CrossRef]
30.   Li, Z.; Chen, C.; Mi, R.; Gan, W.; Dai, J.; Jiao, M.; Xie, H.; Yao, Y.; Xiao, S.; Hu, L. A Strong, Tough, and Scalable Structural Material from Fast growing Bamboo. *Adv. Mater.* **2020**, *32*, 1906308. [CrossRef]
31.   Yu, Y.; Liu, R.; Huang, Y.; Meng, F.; Yu, W. Preparation, physical, mechanical, and interfacial morphological properties of engineered bamboo scrimber. *Constr. Build. Mater.* **2017**, *157*, 1032–1039. [CrossRef]
32.   Yu, Y.; Huang, X.; Yu, W. A novel process to improve yield and mechanical performance of bamboo fiber reinforced composite via mechanical treatments. *Composites* **2014**, *56*, 48–53. [CrossRef]
33.   Meng, F. Study on the bonding interface and mechanism of bamboo-based fiber composites. Ph.D. Thesis, Beijing Forestry University, Beijing, China, 2017. (In Chinese).
34.   Shams, M.; Kagemori, N.; Yano, H. Compressive deformation of wood impregnated with low molecular weight phenol formaldehyde (PF) resin IV: Species dependency. *J. Wood Sci.* **2006**, *52*, 179–183. [CrossRef]
35.   Wang, X.; Luo, X.; Ren, H. Bending failure mechanism of bamboo scrimber. *Constr. Build. Mater.* **2022**, *4*, 326. [CrossRef]

Article

Effects of Adhesive Types and Structural Configurations on Shear Performance of Laminated Board from Two Gigantochloa Bamboos

Norwahyuni Mohd Yusof [1], Paridah Md Tahir [1,2,*], Seng Hua Lee [3,*], Mohd Khairun Anwar Uyup [4], Redzuan Mohammad Suffian James [1], Syeed Saifulazry Osman Al-Edrus [1], Lubos Kristak [5], Roman Reh [5] and Muhammad Adly Rahandi Lubis [6,7]

1   Institute of Tropical Forestry and Forest Products, Universiti Putra Malaysia, UPM, Serdang 43400, Selangor, Malaysia
2   Faculty of Forestry and Environment, Universiti Putra Malaysia, UPM, Serdang 43400, Selangor, Malaysia
3   Department of Wood Industry, Faculty of Applied Sciences, Universiti Teknologi MARA (UiTM) Cawangan Pahang Kampus Jengka, Bandar Tun Razak 26400, Pahang, Malaysia
4   Forest Product Division, Forest Research Institute Malaysia, Kuala Lumpur 52109, Kuala Lumpur, Malaysia
5   Faculty of Wood Sciences and Technology, Technical University in Zvolen, 96001 Zvolen, Slovakia
6   Research Center for Biomass and Bioproducts, National Research and Innovation Agency (BRIN), Bogor 16911, Indonesia
7   Research Collaboration Center for Biomass and Biorefinery between BRIN and Universitas Padjadjaran, National Research and Innovation Agency, Jatinangor 45363, Indonesia
*   Correspondence: parida.introp@gmail.com (P.M.T.); leesenghua@hotmail.com (S.H.L.)

Citation: Mohd Yusof, N.; Md Tahir, P.; Lee, S.H.; Anwar Uyup, M.K.; James, R.M.S.; Osman Al-Edrus, S.S.; Kristak, L.; Reh, R.; Lubis, M.A.R. Effects of Adhesive Types and Structural Configurations on Shear Performance of Laminated Board from Two Gigantochloa Bamboos. *Forests* **2023**, *14*, 460. https://doi.org/10.3390/f14030460

Academic Editor: Christian Brischke

Received: 11 January 2023
Revised: 7 February 2023
Accepted: 21 February 2023
Published: 23 February 2023

Abstract: Semantan (*Gigantochloa scortechinii*) and beting (*Gigantochloa levis*) bamboo are the two Malaysian bamboo that are suitable to be converted into laminated bamboo boards. One of the main criteria for laminated board is its good bondability, which is determined by shear performance. The shear performance of laminated board is influenced by several factors such as the species used, adhesive types and lamination configurations. Therefore, in this study, laminated bamboo boards were produced using Semantan and Beting bamboo bonded with phenol–resorcinol–formaldehyde (PRF) and polyurethane (PUR) adhesives. Different configurations (lay-up patterns and strip arrangements) were used during the consolidation of the laminated boards. The bamboo strips were arranged in three different arrangements, namely vertical, horizontal and mixed, and then assembled into a three-layered structure with two lay-up patterns, which are perpendicular and parallel. Shear performances, such as shear strength, bamboo failure and delamination of the boards, were evaluated. The results revealed that the adhesive type and lay-up pattern were the most influential factors on the shear performance. PRF-bonded laminated bamboo boards outperformed PUR-bonded laminated bamboo boards in terms of shear strength and bamboo failure but PUR bonding had better bond durability as indicated by its low delamination. Boards laminated parallelly significantly outperformed those bonded perpendicularly. As for strip arrangement, PRF-bonded laminated boards were less influenced by it compared to PUR-bonded laminated boards. The results suggested that PRF is a better adhesive for bamboo lamination due to its higher shear performance and more consistent performance across structural configurations (lay-up patterns and strip arrangements).

Keywords: semantan bamboo; beting bamboo; shear strength; strip arrangement; lay-up pattern

1. Introduction

Laminated bamboo board is an excellent substitute for wood, with performance comparable to or exceeding that of wood [1,2]. While maintaining the benefits of round bamboo, studies have shown that laminated bamboo board can overcome round bamboo's disadvantages such as size limitations and dimensional inconsistencies [3,4]. In Malaysia, there are approximately 70 identified bamboo species [5], with 13 of them being commercially used [6]. Beting bamboo (*Gigantochloa levis*) and Semantan bamboo (*Gigantochloa scortechinii*) are two bamboo species used commercially in Malaysia [6]. In Malaysia, *G. scortechinii* has been widely used by researchers to produce laminated bamboo board [4,7], plybamboo [8] and polymeric bamboo composites [9,10]. Meanwhile, *G. levis* has also been used to produce glued laminated bamboo lumber [11], plybamboo [12] and bamboo/epoxy composite [13]. Given the abundance of these bamboos and their exceptional strength properties, both *G. scortechinii* and *G. levis* could be viable commercial candidates in the laminated bamboo board industry.

The performance of laminated bamboo boards is influenced by several factors such as the layer structure [14], adhesive type, adhesive spreading rate and clamping pressure and time [15]. Adhesive is critical in the formation of high-quality, long-lasting bonds as well as in achieving the proper interface bond and penetration between the fibre and laminas [16]. Adhesives are generally classified based on their chemistry, according to Stoeckel et al. [17]. Adhesives are classified into two types based on this criterion: in situ polymerised adhesives and pre-polymerised adhesives. Aminoplastic adhesives, phenol–resorcinol–formaldehyde (PRF), and polymeric methylene–diphenyl–diisocyanate (pMDI) are examples of in situ polymerised adhesives containing relatively rigid, highly crosslinked polymers. Meanwhile, polyurethane (PUR) and polyvinyl acetate (PVAc) are examples of pre-polymerised adhesives with flexible polymers. The ability of these two groups to distribute moisture-induced stress in an adhesive bond varies significantly, resulting in different failure mechanisms. However, the chemistry of the adhesives is not the only factor to consider when categorising adhesives. One of the primary concerns is the mechanical response of the adhesives [18]. Previous research by Guan et al. [19] discovered that the type of adhesive has a significant impact on the shear bond strength of bamboo materials. The penetration of the adhesive into the bamboo cell walls alters the bonding mechanism and has a significant impact on the mechanical properties of laminated bamboo materials from various bamboo species and densities. Dong et al. [20] studied the bonding performance of cross-laminated timber–bamboo composites and concluded that the adhesive type is the most important factor that affects its performance.

In comparison with other mechanical properties, such as compressive, tensile and bending resistance, the bonding shear strength of the laminated materials, particularly bamboo, has not been fully addressed [16]. This study focuses on investigating the effects of adhesive types and assembly configurations on the bonding shear performance of laminated bamboo made from two local bamboo species, *G. scortechinii* and *G. levis*. Phenol–resorcinol–formaldehyde (PRF) and polyurethane (PUR) were used as adhesive for the laminated bamboo boards' fabrication. The laminated bamboos were consolidated using different configurations, that is, different lay-up patterns (parallel and perpendicular) and strip arrangements. Three strip arrangements, namely horizontal, vertical and mixed, were used during the consolidation of the laminated bamboo boards. The bamboo strips were consolidated into a three-layer structure horizontally, vertically and a combination of horizontal and vertical, called a mixed pattern, in this study. Shear strength, bamboo failure and delamination were evaluated as functions of the above-mentioned parameters.

2. Materials and Method

2.1. Preparation of Bamboo Strips

Gigantochloa levis, locally called Beting bamboo, and *Gigantochloa scortechinii*, locally called Semantan bamboo were used to make 3-layer laminated bamboo in this study. Three-year old bamboo culms were selected from a bamboo plantation near Nami, Kedah. *G. scortechinii* and *G. levis* have average densities of 700 kg/m^3 and 751 kg/m^3, respectively. The modulus of rupture and modulus of elasticity of the former are 125 N/mm^2 and 10,039 N/mm^2, while those of the latter are 163 N/mm^2 and 13,185 N/mm^2, respectively. *G. levis* is a large species of bamboo with an average culm size of 11–13 cm and a wall thickness of 11–15 mm, with estimated height and length of 18–23 m and 35 cm, respectively. The height of *G. scortechinii* is between 17 and 20 m. This bamboo's internode is 42 cm long and has a culm length of 9–11 cm. The wall thickness ranges from 6 to 10 mm. The culms were transported to Saudagar Bamboo Enterprise located in Kuala Nerang, Kedah, for further processing. The culms were cut to 2 m long then ripped into splits of 22 mm wide (Figure 1). The bamboo splits were then flattened and shaped using thicknesser machine before being planed to a final thickness of 5 mm using a double-sided planer. The final dimensions of the bamboo strips was 2000 mm long × 20 mm wide × 5 mm thick. The bamboo strips were then soaked in 5% boric acid solution for 24 h and kiln dried to a $12 \pm 2\%$ moisture content. The densities of the bamboo strips after conditioning are 685.51 kg/m^3 and 689.91 kg/m^3 for *G. scortechinii* and *G. levis*, respectively.

Figure 1. Splits of: (**a**) *Gigantochloa scortechinii* and (**b**) *Gigantochloa levis*.

2.2. Fabrication of 3-Layer Laminated Bamboo Boards

Prior to fabrication of laminated bamboo boards, the bamboo strips were sorted into three categories, which are: (i) straight and square strips for edge bonding, (ii) slightly curved and square strips for face bonding and (iii) reject strips due to being highly curved, bent or not meet the desired size. The bamboo strips were consolidated by using different configurations (lay-up patterns and strip arrangements) as shown in Figure 2. Three strip arrangements, namely horizontal, vertical and mixed, were used during the consolidation of the laminated bamboo boards. The bamboo strips were consolidated into 3-layer structures horizontally, vertically and a combination of horizontal and vertical, called mixed pattern, in this study. In the mixed arrangement pattern, the bamboo strips were assembled horizontally in the 2 outer layers of the boards and vertically in middle layer. All arrangement patterns were laid parallelly (oriented at 0° to the adjacent layer) and perpendicularly (oriented at right angles to the adjacent layer) (Figure 2). Therefore, laminated bamboo boards with a total of 6 configurations were fabricated.

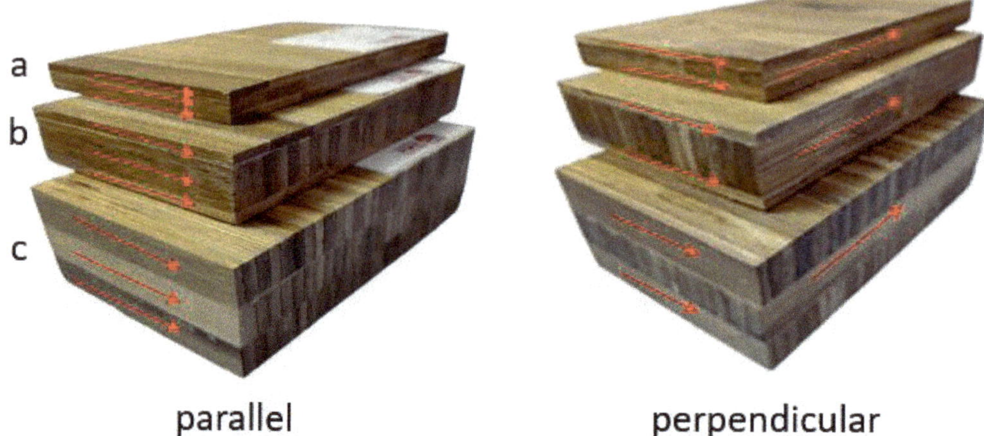

Figure 2. Parallel and perpendicular lay-up patterns of bamboo strips with different arrangements: (a) horizontal; (b) mixed; and (c) vertical.

Two cold setting adhesives, phenol–resorcinol–formaldehyde (PRF) and polyurethane (PUR) (supplied by AkzoNobel Sdn. Bhd., Petaling Jaya, Selangor), were used as bonding agents for the laminated bamboo boards. The glue spread rate used was 250 g/m² and 200 g/m² for PRF and PUR, respectively, as recommended by the supplier. Each layer of bamboo strips was edge glued firstly using a pressing pressure of 75 kg/cm². The formed layers were then face glued into a 3-layer structure using a pressing pressure of 125 kg/cm². Edge trimming and sanding were performed on the 3-layer laminated bamboo boards to remove squeezed out adhesive for smooth and flat surfaces. A total of 144 boards (2 species × 2 adhesives × 6 configurations × 6 replications) with dimensions of 300 wide × 1220 mm long were fabricated. The thicknesses of the final boards were different according to different configurations, which were 54 mm vertically, 13 mm horizontally and 27 mm mixed.

2.3. Evaluation of Shear Strength of Laminated Bamboo Boards

The shear strengths of the glue lines of the laminated bamboo boards produced in this study were evaluated based on British Standard (BS) EN 392: 1995—Glued laminated timber—Shear test of glue lines. Samples with dimensions of 40 mm wide × 40 mm long with various thicknesses were prepared. Prior to testing, the samples were conditioned at a temperature of 20 ± 2 °C and relative humidity 65 ± 5% until contact mass was achieved. The conditioned samples were placed on a shear machine and load was applied at the glue line between the laminations of the laminated bamboo until failure occurred (Figure 3). The load was applied under a displacement control rate of 3 mm/min, ensuring failure after no less than 20 s. The shear strength fv was determined for every tested glue line and was calculated in accordance with the following formula:

$$fv = k\frac{Fu}{A} \tag{1}$$

where:

Fu is the ultimate load (in N);
A is the sheared area (in mm²);
k is factor: $kv = 0.78 + 0.0044\ t$;
t is thickness (in mm).

Figure 3. Shearing machine set up for shear strength test.

Twelve replications were tested for each species, adhesive and configuration. A total of 288 samples (2 × 2 × 6 × 12) were tested. The shear strengths of the glue lines obtained were then compared with EN14080:2013—Timber structures—Glued laminated timber and glued solid timber—Requirements for parallel arrangement and EN16351:2021—Timber structures—Cross laminated timber—Requirements for perpendicular arrangement. It is stipulated in the standard that the minimum requirements for parallelly arranged laminated boards shall be at least 4 N/mm² (EN14080) and at least 2 N/mm² for laminated boards arranged perpendicularly.

2.4. Estimation of Bamboo Failure

The estimation of bamboo failure was performed on the sheared area specimens as shown in Figure 4. Each failure surface was measured and the results were averaged. The bamboo failure percentage was estimated to the nearest 5% by examining the total area covered by the bamboo fibre (signifying bamboo failures) on the sheared area in comparison to the area covered with glue failure.

Figure 4. Sheared samples for failure estimation after shear test.

2.5. Evaluation of Delamination of Laminated Bamboo Boards

Three-layer laminated bamboo samples with dimensions of 75 mm wide × 130 mm long and various thicknesses were prepared for delamination evaluation. The delamination test was conducted according to the procedures specified in EN 391:2002—Glued laminated timber—Delamination test of glue lines. The test cycle for method B was chosen. The samples were placed in a pressure vessel as shown in Figure 5 and submerged in water at ambient temperature. A vacuum of 60 kPa was applied and held for 30 min. Subsequently, the vacuum was released and a pressure of 550 kPa was applied and retained for 2 h. Once the vacuuming was completed, the test pieces were dried for a period of approximately 10–15 h in a circulating oven at 65 ± 10 °C. Delamination was observed and recorded when the mass of the test pieces had returned to within 100% to 110% of the original mass. After removal from the oven, the samples were examined for the occurrence of delamination or open glue lines. The length of the open glue lines was determined by first inserting a thin metal probe between the two delaminated surfaces. Measurements count if the depth of the delamination is less than 2.5 mm and more than 5 mm. Two attributes were determined, (i) total delamination and (ii) maximum delamination of a test piece, according to Equations (2) and (3):

$$Delam\ tot = 100\ \frac{ltot,\ delam}{ltot,\ glueline}\ (\%) \tag{2}$$

$$Delam\ max = 100\ \frac{lmax,\ delam}{2l,\ glueline}\ (\%) \tag{3}$$

where:

$ltot,\ delam$ = the total delamination length (in mm);
$ltot,\ glueline$ = the sum of the perimeter of all glue lines in a delamination specimen (in mm);
$lmax,\ delam$ = the maximum delamination length (in mm);
$l,\ glueline$ = the perimeter of one glue line in a delamination specimen (in mm).

Figure 5. Samples placed inside a pressure vessel for delamination tests.

2.6. Statistical Analysis

Analysis of the obtained data were performed using statistical software IBM-SPSS version 25.0 by employing one-way analysis of variance (ANOVA). Meanwhile, mean separation was carried out using the Least Significance Difference (LSD) method.

3. Results and Discussion

3.1. Shear Strength and Bamboo Failure

Table 1 summarises the analysis of variance (ANOVA) results for the effects of species, adhesive and lay-up on the shear strength and failure of laminated bamboo arranged vertically, horizontally and in a mixed pattern. The shear strength of the laminated bamboo in the horizontal and mixed arrangements is affected by the species ($p \leq 0.05$). The bamboo species had no effect on shear strength in the vertical arrangement. Meanwhile, the shear strength of the laminated bamboo in all arrangements is significantly affected by adhesive type and lay-up ($p \leq 0.01$). On the other hand, adhesive type was discovered to have a significant effect on bamboo failure, whereas species had no influence on bamboo failure. Only the bamboo failure of laminated boards manufactured in a mixed arrangement was significantly influenced by the lay-up pattern.

Table 1. Analysis of variance (ANOVA) for the effects of species, adhesive and lay-up on the shear strength and failure of laminated bamboo arranged vertically, horizontally and in a mixed pattern.

Source	*p*-Value					
	Vertical		**Horizontal**		**Mixed**	
	Shear Strength	Bamboo Failure	Shear Strength	Bamboo Failure	Shear Strength	Bamboo Failure
species	0.4477 ns	0.7913 ns	0.0135 *	0.0041 **	0.0347 *	0.470 1ns
adhesive	<0.0001 **	<0.0001 **	<0.0001 **	<0.0001 **	<0.0001 **	<0.0001 **
lay-up	<0.0001 **	0.3030 ns	0.0068 **	0.0674 ns	<0.0001 **	0.0095 **

Note: ns not significant. * Significantly different at $p \leq 0.05$. ** Significantly different at $p \leq 0.01$.

Shear strength and bamboo failure of laminated bamboo of different configurations bonded with PRF and PUR adhesives are tabulated in Table 2.

Table 2. Shear strength of laminated bamboo boards fabricated with different species, adhesives and configurations.

Label	Variable			Shear Strength (N/mm^2)		
	Species	**Adhesive**	**Lay-Up**	**Vertical**	**Horizontal**	**Mixed**
BPA	Beting	PRF	Parallel	5.52 [B] (1.58)	4.21 [B] (1.97)	7.99 [A] (2.42)
BPB	Beting	PRF	Perpendicular	2.34 [D] (0.82)	3.07 [C] (1.68)	2.36 [C] (0.80)
BUA	Beting	PUR	Parallel	4.05 [C] (1.78)	7.55 [A] (3.33)	5.64 [B] (2.35)
BUB	Beting	PUR	Perpendicular	3.31 [C] (0.39)	6.87 [A] (1.77)	2.63 [C] (1.07)
SPA	Semantan	PRF	Parallel	7.91 [A] (1.08)	6.69 [A] (3.58)	6.41 [B] (2.57)
SPB	Semantan	PRF	Perpendicular	2.44 [D] (0.50)	1.54 [D] (0.63)	3.32 [C] (0.59)
SUA	Semantan	PUR	Parallel	3.71 [C] (1.51)	3.47 [C] (0.97)	3.16 [C] (1.91)
SUB	Semantan	PUR	Perpendicular	1.86 [D] (0.31)	5.6 [B] (1.20)	2.66 [C] (0.99)

Note: Mean followed by the different letters [A, B, C, D] in the same column are significantly different at $p \leq 0.05$ according to LSD; Values in parentheses ( ) are standard deviation.

Laminated bamboo boards of vertical, horizontal and mixed arrangements recorded shear strength values ranging from 1.86 to 7.91 N/mm², 1.54 to 7.55 N/mm² and 2.36 to 7.99 N/mm², respectively (Figure 6). The highest shear strength was observed in the samples fabricated parallelly with a mixed arrangement using *G. levis* (Beting bamboo) bonded with PRF resin. Meanwhile, the lowest shear strength was recorded in the laminated bamboo boards fabricated perpendicularly with horizontal arrangement using *G. scortechinii* (Semantan bamboo) bonded with PRF resin.

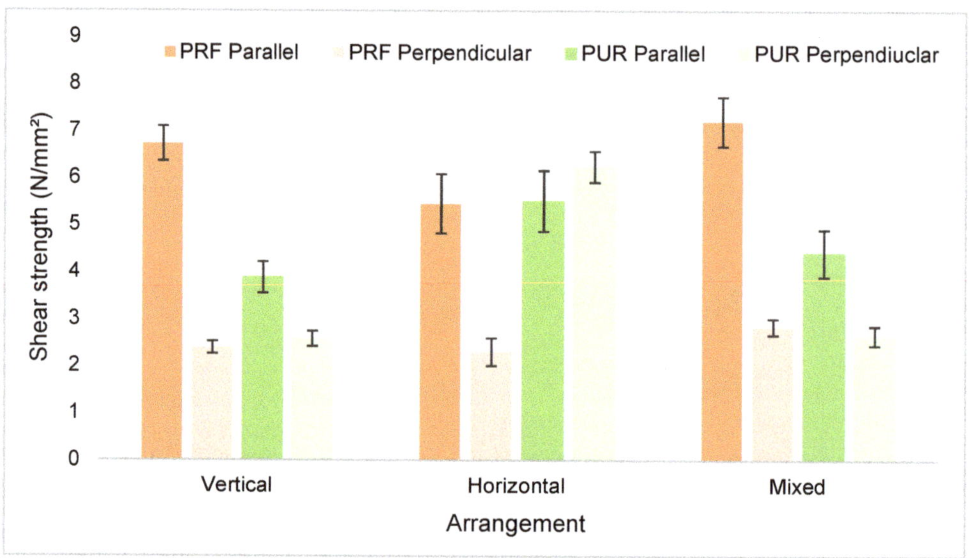

Figure 6. Effects of adhesive and lay-up pattern on the glue line shear for three different strip arrangements. Note: Vertical line in every bar represents the standard deviation.

It was observed that the shear strength of the laminated bamboo boards with parallel arrangements were greater than the minimum requirements (≥ 4 N/mm²) as specified in EN14080, with the exception of SUA (Semantan bamboo bonded parallelly with PUR resin) for all arrangements (vertical, horizontal and mixed). Meanwhile, the shear strengths of the laminated bamboo with perpendicular arrangements were relatively lower compared with those using parallel arrangements, but still fulfilled the minimum requirements of shear strength specified in EN16351, which is ≥ 2 N/mm². The only exception was found in SUB (Semantan bamboo bonded perpendicularly with PUR resin) in the vertical arrangement and SPB (Semantan bamboo bonded perpendicularly with PRF resin) in the horizontal arrangement.

The ANOVA results indicated that the interaction between adhesive and lay-up pattern are significant on the shear strength of laminated bamboo boards. Figure 6 shows the effects of adhesive and lay-up pattern on the shear strength of the boards.

The bamboo failure of laminated bamboo boards in a vertical arrangement ranged between 3.75 and 80.67% (Table 3). For the horizontal arrangement, the bamboo failure ranged between 5.08 and 92.0%, while for the mixed arrangement, the bamboo failure ranged between 0.83 and 39.67%. The bamboo failure was higher in the laminated bamboo boards bonded with PRF compared to that of PUR in all configurations. The average bamboo failure was 22%–92% for laminated bamboo bonded with PRF and 3%–22% for laminated bamboo bonded with PUR. Sikora et al. [21] discovered that the PRF has a very good durability performance compared to that of PUR resin. The visual appearance of the sheared samples is shown in Figure 7.

Table 3. Bamboo failure of laminated bamboo boards fabricated with different species, adhesives and configurations.

Label	Variable			Bamboo Failure (%)		
	Species	Adhesive	Lay-Up	Vertical	Horizontal	Mixed
BPA	Beting	PRF	Parallel	76.17 [A] (17.21)	77.67 [B] (13.54)	28.25 [A] (14.85)
BPB	Beting	PRF	Perpendicular	75.75 [A] (13.01)	87.42 [A] (5.16)	39.67 [A] (17.79)
BUA	Beting	PUR	Parallel	3.33 [B] (6.16)	5.08 [D] (6.19)	4.17 [C] (11.45)
BUB	Beting	PUR	Perpendicular	7.5 [B] (9.31)	16.33 [C] (7.11)	18.08 [B] (23.48)
SPA	Semantan	PRF	Parallel	75.75 [A] (18.9)	86.33 [A] (10.46)	35 [A] (15)
SPB	Semantan	PRF	Perpendicular	80.67 [A] (10.55)	92 [A] (4.97)	22.25 [B] (13.14)
SUA	Semantan	PUR	Parallel	3.75 [B] (5.28)	21.92 [C] (14.81)	0.83 [C] (1.95)
SUB	Semantan	PUR	Perpendicular	5.17 [B] (7.1)	10.5 [D] (12.71)	22.67 [B] (20.02)

Note: Mean followed by the different letters [A, B, C, D] in the same column are significantly different at $p \leq 0.05$ according to LSD. Values in parenthesis ( ) are standard deviation.

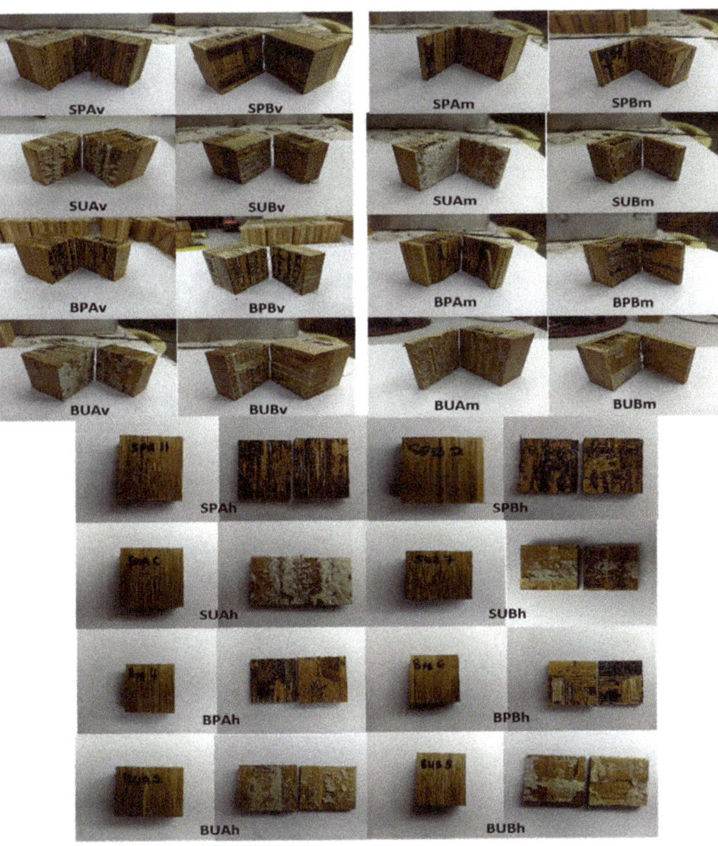

Figure 7. Sheared samples of laminated bamboo boards fabricated with different species, adhesives and configurations. Note: SPA—Semantan PRF parallel; SPB—Semantan PRF perpendicular/cross. SUA—Semantan PUR parallel; SUB—Semantan PUR perpendicular/cross; BPA—Beting PRF parallel; BPB—Beting PRF perpendicular/cross; BUA—Beting PUR parallel; BUB—Beting PUR perpendicular/cross; v—vertical arrangement; h—horizontal arrangement; m—mixed arrangement.

3.2. Delamination

Table 4 tabulates the analysis of variance (ANOVA) of the effect of species, adhesive and lay-up pattern on the delamination of laminated bamboo boards. Species exert slightly significant ($p \leq 0.05$) effects on the shear strength of vertically and horizontally arranged boards, and the effects on mixed arrangement boards were highly significant ($p \leq 0.01$). Meanwhile, adhesive types had significant effects on the delamination of horizontally and mixed arranged boards. The lay-up pattern was found to be significantly affected vertically ($p \leq 0.01$) and horizontally ($p \leq 0.05$) arranged boards.

Table 4. The analysis of variance (ANOVA) for the effects of species and adhesive for the delamination of parallel laminated bamboo in different configurations.

Source	df	*p*-Value		
		Vertical	Horizontal	Mixed
species	1	0.0126 *	0.0343 *	0.0025 **
adhesive	1	0.0657 ns	<0.0001 **	0.0050 **
lay-up	1	<0.0001 **	0.0106 *	0.1720 ns

Note: ns not significant. * Significantly different at $p \leq 0.05$. ** Significantly different at $p \leq 0.01$.

Table 5 shows the effect of species, adhesive and lay-up on the delamination of laminated bamboo in three different arrangements. The delamination for vertically arranged boards ranged from 0.78% to 26.66% The highest delamination was found in SUB (Semantan bamboo bonded perpendicularly with PUR resin) samples while the lowest was observed in SUA (Semantan bamboo bonded parallelly with PUR resin) samples. For horizontally arranged samples, the delamination varied between 1.16% and 38.01%. The highest delamination was observed in BPB (Beting bamboo bonded perpendicularly with PRF resin), and the lowest was observed in BUB (Beting bamboo bonded perpendicularly with PUR resin) samples. Meanwhile, for the mixed arrangement, the delamination ranged from 6.34% to 24.27%. The highest delamination was recorded in BUA (Beting bamboo bonded parallelly with PUR resin) samples while the lowest delamination was recorded in SPA (semantan bamboo bonded parallelly with PRF resin). Generally, the vertical arrangement has better glue line durability compared to those of the mixed and horizontal arrangements, as shown by lower delamination values.

Table 5. Delamination of parallel laminated bamboo boards fabricated with different configurations.

Label	Variable			Shear Strength (N/mm²)		
	Species	Adhesive	Lay-Up	Vertical	Horizontal	Mixed
BPA	Beting	PRF	Parallel	6.31 [B] (5.49)	28.32 [B] (6.54)	16.83 [B] (9.62)
BPB	Beting	PRF	Perpendicular	7.64 [B] (3.94)	38.01 [A] (18.11)	17.28 [B] (4.8)
BUA	Beting	PUR	Parallel	2.73 [C] (3.6)	4.27 [D] (3.62)	6.34 [C] (3.23)
BUB	Beting	PUR	Perpendicular	10.42 [B] (5.34)	1.16 [D] (2.71)	16.9 [B] (9.75)
SPA	Semantan	PRF	Parallel	5.42 [B] (6.4)	16.85 [C] (6)	24.27 [A] (10.17)
SPB	Semantan	PRF	Perpendicular	9.82 [B] (6.11)	31.3 [B] (4.5)	19.12 [A] (10.38)
SUA	Semantan	PUR	Parallel	0.78 [C] (1.62)	6.98 [D] (6.52)	15.75 [B] (5.92)
SUB	Semantan	PUR	Perpendicular	26.66 [A] (16.75)	2.78 [D] (2.9)	19.14 [A] (8.68)

Note: Mean followed by the different letters [A, B, C, D] in the same variable category are significantly different at $p \leq 0.05$ according to LSD.

Based on ANOVA, the interaction between adhesive types and lay-up were found significantly affected the delamination percentage of boards in all arrangements. Figure 8 illustrates the effects of adhesive and lay-up pattern on the delamination of laminated bamboo of all arrangements.

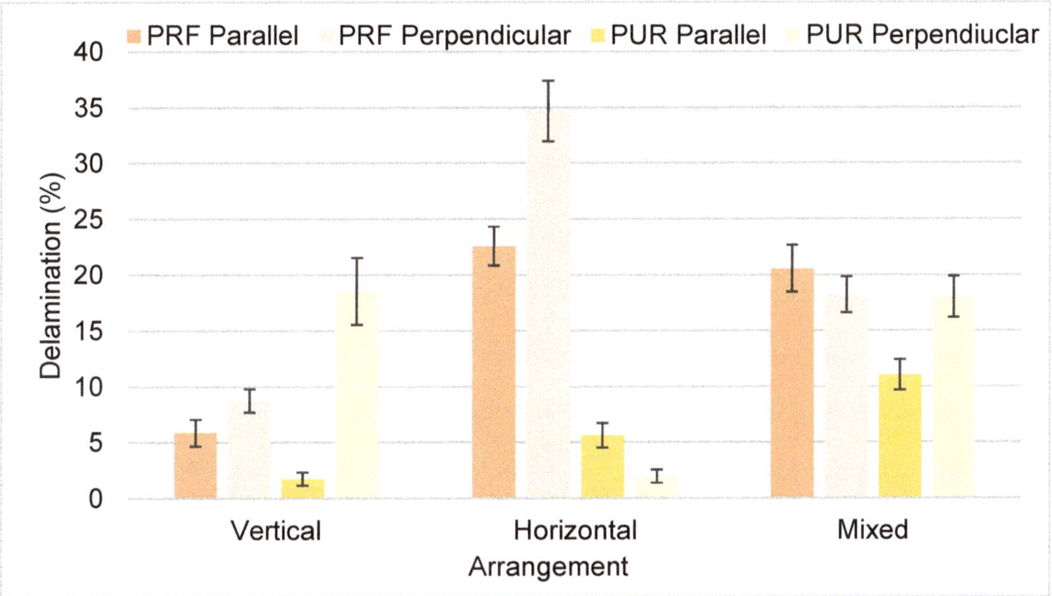

Figure 8. Effects of adhesive and lay-up on the delamination of laminated bamboo boards of all arrangements. Note: Vertical line in every bar represents the standard deviation.

Adhesive types had the most influence on the delamination of samples. PRF-bonded laminated bamboo has a lower delamination percentage as compared with PUR-bonded laminated bamboo in vertical arrangements. However, in the horizontal and mixed arrangements, this relationship was reversed. The parallel lay-up pattern had a generally lower percentage of delamination as compared with the perpendicular lay-up in all arrangements. Although the delamination results varied significantly between samples, it was observed that the mechanism causing delamination in the glue line was the same for all specimens, i.e., delamination occurred in a single glue line on one side for vertical arrangements and both sides for horizontal and mixed arrangements. Overall, PUR adhesive has a more durable bond than PRF adhesive, despite better shear strength and bamboo failure shown by PRF resin. A study by Lu et al. [22] also made a similar observation that PUR-bonded samples displayed the lowest delamination. PUR is a flexible polymer; therefore, its capability to withstand moisture-induced stress might be better than PRF resin and thus lead to better bond durability.

3.3. Effects of Variables on the Shear Strength of Laminated Bamboo Board

Table 6 compares the effects of single variables on the shear strength of laminated bamboo boards. It can be observed that laminated bamboo boards made from Beting bamboo have significantly higher shear strengths than those made of Semantan. Both Semantan and Beting bamboo are members of the Gigantochloa family and thus have similar anatomical structures. Both Semantan and Beting bamboo have type III vascular bundles, according to Abdullah Siam et al. [6]. The study by Abdullah Siam et al. [6] stated that Beting bamboo had higher density and mechanical properties compared to Semantan bamboo. It was discovered that wood density positively affected the apparent shear strength of the wood [23]. However, the difference in density between the bamboo strips used in this study is very small (685.51 kg/m^3 for Semantan and 689.91 kg/m^3 for Beting). Therefore, the density did not play a prominent role in this study and the difference in shear strength could be caused by other factors that require further investigation.

Table 6. Comparison between effects of variables on the properties of laminated bamboo board.

Variable	Glue Line Shear (MPa)
Species	
Beting (*Gigantochloa levis*)	4.63 [A]
Semantan (*Gigantochloa scortechinii*)	4.06 [B]
Adhesive	
PRF	4.48 [A]
PUR	4.21 [A]
Lay-up	
Parallel	5.53 [A]
Perpendicular	3.17 [B]
Arrangement	
Vertical	5.19 [B]
Horizontal	6.50 [A]
Mixed	5.70 [B]

Note: Mean followed by the different letters [A,B] in the same variable category are significantly different at $p \leq 0.05$ according to LSD.

Similarly, the effects of adhesive types also did not exert significant effects on the shear strength of laminated bamboo boards. However, laminated bamboo boards bonded with PRF resin was found to perform slightly better than those bonded with PUR resin. Studies have shown that PRF performed better than PUR in bonding materials owing to superior gap-filling properties of PRF resin [24]. Studies by several researchers also reported the same observation that the PRF-bonded samples exhibited superior shear performance than those of PUR-bonded samples [25–27]. Konnerth et al. [28] stated that PRF resin is able to penetrate into the wood cell wall, therefore resulting in better bonding performance compared to PUR resin, which is unable to penetrate into the wood cell wall.

On the other hand, the lay-up pattern of the bamboo strips was found to have significantly affected the shear strength of the laminated board. Boards arranged parallelly performed significantly better than those arranged perpendicularly. Because wood is anisotropic, the fibre direction of the bonding interface has a significant impact on shear strength. According to Qin [29], the shear strength of a perpendicular fibre direction is two-thirds to three-quarters that of a parallel fibre direction. Several studies, for instance, Ashaari et al. [30] and Rabi'atol Adawiah et al. [31], also found that the laminates assembled parallelly displayed significantly superior mechanical performance than those assembled perpendicularly. In terms of arrangement, laminated bamboo boards that were assembled horizontally outperformed those assembled vertically and in a mixed pattern. The shear strength for laminated boards constructed by arranging the strips horizontally is generally high despite having lower thicknesses (13 mm) than the vertical (54 mm) and mixed (27 mm) arrangements. This could be due to the size effect, where laminated bamboo boards in vertical and mixed arrangements with higher thicknesses had larger bamboo volumes loaded under shear stresses and thus experienced more critical strength-reducing defects than laminated bamboo boards arranged horizontally [32]. Therefore, the shear strength for parallel lay-up boards is notably higher than perpendicular ones. The same observation was also made by several researchers. For instance, Sikora et al. [33] stated that the thickness of cross-laminated timber (CLT) has adverse effects on the rolling shear strength of the CLT panels. A study by Li (2017) also confirmed that the CLT panels with higher thicknesses had lower rolling shear strengths compared to those with lower thicknesses.

Figure 9 depicts the shear strength ratios of each arrangement in order to examine the effects of strips arrangements in greater detail. The shear strength ratios of V/H, V/M and H/M ranged from 0.41 to 2.36. Shear ratios revealed that strip arrangements had less effect on PRF-bonded laminated bamboo boards, with values ranging from 0.76 to 1.23, all distributed around 1.0. Meanwhile, the arrangement had a significant impact on PUR-bonded laminated bamboo boards, with strength ratios ranging from 0.41 to 2.36. Surprisingly, strip arrangements have very different effects on PRF- and PUR-bonded laminated bamboo boards. The mixed arrangement has the highest shear resistance capacity for PRF-bonded laminated bamboo boards, followed by the vertical arrangement and the lowest being horizontal. The strip arrangement had little effect on the lay-up pattern for PRF-bonded laminated bamboo boards, as both parallelly and perpendicularly assembled boards showed a fairly consistent shear strength ratio. The high variation in the shear strength ratio in PUR-bonded boards, on the other hand, revealed that the effect of strip arrangement on the board's shear strength is significant. The horizontal arrangement of PUR-bonded laminated bamboo demonstrated the highest shear resistance capacity, followed by mixed and then vertical. In comparison to parallel boards, perpendicularly assembled laminated boards were heavily influenced by strip arrangement, as shown by their extremely wide range of shear strength ratio. Xing et al. [16] speculated that the low compatibility of PUR adhesive with bamboo materials and the inferior strength of the resin itself may have contributed to the abnormal trend in PUR-bonded laminated bamboo boards.

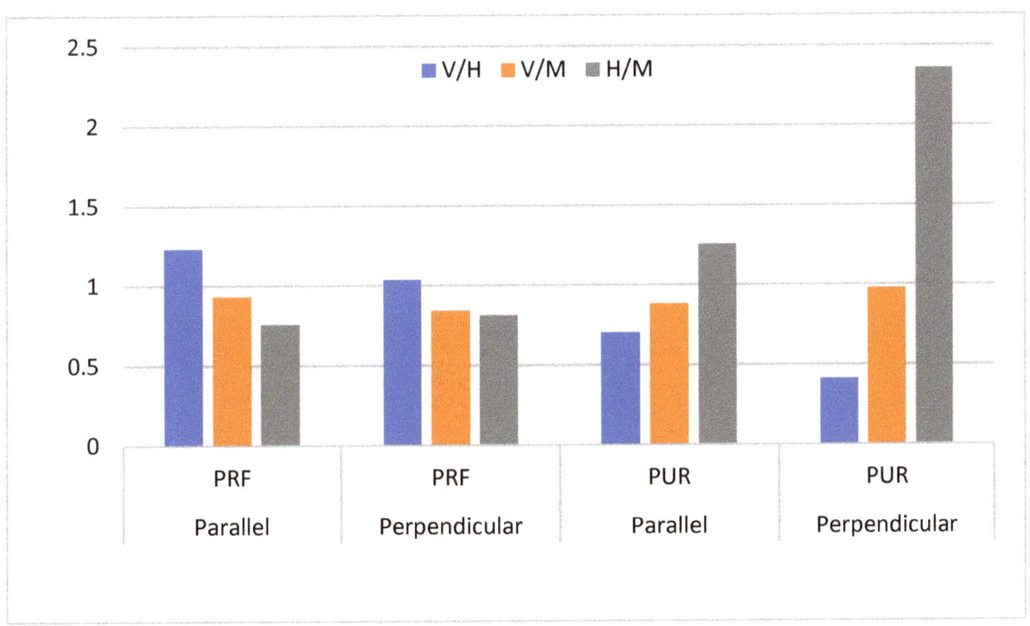

Figure 9. Shear strength ratios of each arrangement of vertical (V), horizontal (H) and mixed (M).

4. Conclusions

In this study, laminated bamboo boards with various structural configurations were fabricated from two local bamboo species, *G. scortechinii* and *G. levis*, and bonded with PRF and PUR adhesives. The shear performance of laminated bamboo boards is heavily influenced by the adhesive types and lay-up pattern (parallel and perpendicular). PRF-bonded laminated bamboo boards outperformed PUR-bonded laminated bamboo boards in terms of shear strength and bamboo failure. However, PUR-bonded laminated bamboo boards showed less delamination than PRF-bonded boards, indicating that PUR-bonded boards have better bonding durability. Furthermore, boards laminated parallelly outperformed those bonded perpendicularly. It should be noted that the arrangement of bamboo strips (vertical, horizontal and mixed) had no significant effect on the shear performance of PRF-bonded laminated boards. The strip arrangements, on the other hand, had a significant influence on PUR-bonded boards, with those assembled horizontally showing the highest shear strength compared to the other two arrangements (vertical and mixed). Furthermore, the effect is stronger in those laminated perpendicularly. According to the findings, PRF is a better adhesive for bamboo lamination due to its higher shear performance and more consistent performance across structural configurations (lay-up patterns and strip arrangements).

Author Contributions: Conceptualization, S.H.L. and P.M.T.; methodology, P.M.T.; software, S.S.O.A.-E.; validation, M.K.A.U., L.K. and R.R.; formal analysis, M.K.A.U.; investigation, N.M.Y., R.M.S.J.; resources, P.M.T.; data curation, N.M.Y., R.M.S.J. and S.H.L.; writing—original draft preparation, N.M.Y.; writing—review and editing, S.H.L., L.K., R.R. and M.A.R.L.; visualization, S.S.O.A.-E.; supervision, S.H.L., M.K.A.U. and P.M.T.; project administration, S.H.L., M.K.A.U. and P.M.T.; funding acquisition, S.H.L. and P.M.T. All authors have read and agreed to the published version of the manuscript.

Funding: The presented work is supported by the Higher Institution Centre of Excellence (HICoE) grant Phase 2 (Project title: Bamboo for industrial application) by the Ministry of Higher Education Malaysia. The work is also supported by the RMK11 collaborative with Malaysian Timber Industry Board (MTIB) (grant number: UPMH/UPMCS/C/SB/Lantikan/20191001-1) and Slovak Research and Development Agency under contracts No. APVV-19-0269 and No. SK-CZ-RD-21-0100.

Data Availability Statement: Not applicable.

Acknowledgments: The authors also wish to express gratitude to Faculty of Forestry and Environment, Universiti Putra Malaysia, for providing the mechanical testing facilities.

Conflicts of Interest: The authors declare no conflict of interest.

References

1. Sulastiningsih, I.M.; Nurwati. Physical and mechanical properties of laminated bamboo board. *J. Trop. For. Sci.* **2009**, *21*, 246–251.
2. Sharma, B.; Gatóo, A.; Bock, M.; Ramage, M. Engineered bamboo for structural applications. *Constr. Build. Mater.* **2015**, *81*, 66–73. [CrossRef]
3. Li, H.T.; Zhang, Q.S.; Huang, D.S.; Deeks, A.J. Compressive performance of laminated bamboo. *Compos. Part B Eng.* **2013**, *54*, 319–328. [CrossRef]
4. Bakar, E.S.; Nazip, M.N.M.; Anokye, R.; Lee, S.H. Comparison of three processing methods for laminated bamboo timber production. *J. For. Res.* **2019**, *30*, 363–369. [CrossRef]
5. Nordahlia, A.S.; Anwar, U.M.K.; Hamdan, H.; Zaidon, A.; Paridah, M.T.; Razak, O.A. Effects of age and height on selected properties of Malaysian bamboo (*Gigantochloa levis*). *J. Trop. For. Sci.* **2012**, *24*, 102–109.
6. Nordahlia, A.S.; Anwar, U.M.K.; Hamdan, H.; Mohmod, A.L.; Awalludin, M.F. Anatomical, physical, and mechanical properties of thirteen Malaysian bamboo species. *BioResources* **2019**, *14*, 3925–3943.
7. Ashaari, Z.; Lee, S.H.; Zahali, M.R. Performance of compreg laminated bamboo/wood hybrid using phenolic-resin-treated strips as core layer. *Eur. J. Wood Wood Prod.* **2016**, *74*, 621–624. [CrossRef]
8. Fadhlia, F.A.R.N.; Paridah, M.T.; Anwar, U.M.K.; Juliana, A.H.; Zaidon, A. Enhancing mechanical properties and dimensional stability of phenolic resin-treated plybamboo. *J. Trop. For. Sci.* **2017**, *29*, 19–29.
9. Rassiah, K.; Ahmad, M.M.; Ali, A.; Nagapan, S. Mechanical properties of layered laminated woven bamboo Gigantochloa scortechinii/epoxy composites. *J. Polym. Environ.* **2018**, *26*, 1328–1342. [CrossRef]
10. Ali, A.; Rassiah, K.; Othman, F.; Lee, H.P.; Tay, T.E.; Hazin, M.S.; Ahmad, M.M.H.M. Fatigue and fracture properties of laminated bamboo strips from *Gigantochloa scortechinii* polyester composites. *BioResources* **2016**, *11*, 9142–9153. [CrossRef]
11. Sulastiningsih, I.M.; Trisatya, D.R.; Indrawan, D.A.; Malik, J.; Pari, R. Physical and mechanical properties of glued laminated bamboo lumber. *J. Trop. For. Sci.* **2021**, *33*, 290–297. [CrossRef]
12. Asniza, M.; Alia, S.Y.; Lee, S.H.; Anwar, U.M.K. Properties of plybamboo manufactured from two Malaysian bamboo species—. *Phys. Sci. Rev.* **2022**; *published online.* [CrossRef]
13. Kamarudin, M.H.; Saringat, M.S.; Sulaiman, N.H. The study of mechanical properties of laminated bamboo strip (LBS) from Gigantochloa levis type mixed with epoxy. *J. Teknol.* **2016**, *78*, 59–64. [CrossRef]

14. Lee, C.H.; Chung, M.J.; Lin, C.H.; Yang, T.H. Effects of layered structure on the physical and mechanical properties of laminated moso bamboo (*Phyllosachys edulis*) flooring. *Constr. Build. Mater.* **2012**, *28*, 31–35. [CrossRef]
15. Ayanleye, S.; Udele, K.; Nasir, V.; Zhang, X.; Militz, H. Durability and protection of mass timber structures: A review. *J. Build. Eng.* **2022**, *46*, 103731. [CrossRef]
16. Xing, W.; Hao, J.; Sikora, K.S. Shear performance of adhesive bonding of cross-laminated bamboo. *J. Mater. Civ. Eng.* **2019**, *31*, 04019201. [CrossRef]
17. Stoeckel, F.; Konnerth, J.; Gindl-Altmutter, W. Mechanical properties of adhesives for bonding wood—A review. *Int. J. Adhes. Adhes.* **2013**, *45*, 32–41. [CrossRef]
18. Frihart, C.R. Adhesive groups and how they relate to the durability of bonded wood. *J. Adhes. Sci. Technol.* **2009**, *23*, 601–617. [CrossRef]
19. Guan, M.; Huang, Z.; Zeng, D. Shear strength and microscopic characterization of a bamboo bonding interface with phenol formaldehyde resins modified with Larch thanaka and urea. *BioResources* **2015**, *11*, 492–502. [CrossRef]
20. Dong, W.; Wang, Z.; Chen, G.; Wang, Y.; Huang, Q.; Gong, M. Bonding performance of cross-laminated timber-bamboo composites. *J. Build. Eng.* **2023**, *63*, 105526. [CrossRef]
21. Sikora, K.S.; McPolin, D.O.; Harte, A.M. Shear strength and durability testing of adhesive bonds in cross-laminated timber. *J. Adhes.* **2016**, *92*, 758–777. [CrossRef]
22. Lu, Z.; Zhou, H.; Liao, Y.; Hu, C. Effects of surface treatment and adhesives on bond performance and mechanical properties of cross-laminated timber (CLT) made from small diameter Eucalyptus timber. *Constr. Build. Mater.* **2018**, *161*, 9–15. [CrossRef]
23. Hernández, R.E.; Almeida, G. Effects of wood density and interlocked grain on the shear strength of three Amazonian tropical hardwoods. *Wood Fiber Sci.* **2003**, *35*, 154–166.
24. Kurt, R. The strength of press-glued and screw-glued wood-plywood joints. *Holz als Roh-und Werkstoff* **2003**, *61*, 269–272. [CrossRef]
25. Ammann, S.; Niemz, P. Mixed-mode fracture toughness of bond lines of PRF and PUR adhesives in European beech wood. *Holzforschung* **2015**, *69*, 415–420. [CrossRef]
26. Yusof, N.M.; Tahir, P.M.; Roseley, A.S.M.; Lee, S.H.; Halip, J.A.; James, R.M.S.; Ashaari, Z. Bond integrity of cross laminated timber from Acacia mangium wood as affected by adhesive types, pressing pressures and loading direction. *Int. J. Adhes. Adhes.* **2019**, *94*, 24–28. [CrossRef]
27. Yusof, N.M.; Tahir, P.M.; Lee, S.H.; Khan, M.A.; James, R.M.S. Mechanical and physical properties of Cross-Laminated Timber made from *Acacia mangium* wood as function of adhesive types. *J. Wood Sci.* **2019**, *65*, 20. [CrossRef]
28. Konnerth, J.; Harper, D.; Lee, S.; Rials, T.G.; Gindl, W. Adhesive penetration of wood cell walls investigated by scanning thermal microscopy. *Holzforschung* **2008**, *62*, 91–98. [CrossRef]
29. Qin, Z.Y. Study on Wood and Adhesive Surface/Interface Wettability Characterization and Influencing Factors. Ph.D. Thesis, Beijing Forestry University, Beijing, China, 2014.
30. Ashaari, Z.; Lee, S.H.; Nabil, F.L.; Bakar, E.S.; Ghani, A.; Rais, M.R. Physico-mechanical properties of laminates made from Semantan bamboo and Sesenduk wood derived from Malaysia's secondary forest. *Int. For. Rev.* **2017**, *19*, 1–8.
31. Adawiah, M.R.A.; Zaidon, A.; Izreen, F.N.; Bakar, E.S.; Hamami, S.M.; Paridah, M.T. Addition of urea as formaldehyde scavenger for low molecular weight phenol formaldehyde-treated compreg wood. *J. Trop. For. Sci.* **2012**, *24*, 348–357.
32. Li, M. Evaluating rolling shear strength properties of cross-laminated timber by short-span bending tests and modified planar shear tests. *J. Wood Sci.* **2017**, *63*, 331–337. [CrossRef]
33. Sikora, K.S.; McPolin, D.O.; Harte, A.M. Effects of the thickness of cross laminated timber (CLT) panels made from Irish Sitka spruce on mechanical performance in bending and shear. *Constr. Build. Mater.* **2016**, *116*, 141–150. [CrossRef]

Article

Influence of Isocyanate Content and Hot-Pressing Temperatures on the Physical–Mechanical Properties of Particleboard Bonded with a Hybrid Urea–Formaldehyde/Isocyanate Adhesive

Apri Heri Iswanto [1,*], Jajang Sutiawan [1], Atmawi Darwis [2], Muhammad Adly Rahandi Lubis [3,4], Marta Pędzik [5,6], Tomasz Rogoziński [6,*] and Widya Fatriasari [3]

1 Department of Forest Products, Faculty of Forestry, Universitas Sumatera Utara, Medan 20155, Indonesia
2 School of Life Sciences and Technology, Institut Teknologi Bandung, Bandung 40132, Indonesia
3 Research Center for Biomass and Bioproduct, National Research and Innovation of Indonesia, Cibinong 16911, Indonesia
4 Research Collaboration Center for Biomass and Biorefinery between BRIN and Universitas Padjadjaran, Jatinangor 45363, Indonesia
5 Center of Wood Technology, Łukasiewicz Research Network, Poznan Institute of Technology, 60-654 Poznan, Poland
6 Department of Furniture Design, Faculty of Forestry and Wood Technology, Poznan University of Life Sciences, 60-627 Poznan, Poland
* Correspondence: apri@usu.ac.id (A.H.I.); tomasz.rogozinski@up.poznan.pl (T.R.)

Citation: Iswanto, A.H.; Sutiawan, J.; Darwis, A.; Lubis, M.A.R.; Pędzik, M.; Rogoziński, T.; Fatriasari, W. Influence of Isocyanate Content and Hot-Pressing Temperatures on the Physical–Mechanical Properties of Particleboard Bonded with a Hybrid Urea–Formaldehyde/Isocyanate Adhesive. *Forests* **2023**, *14*, 320. https://doi.org/10.3390/f14020320

Academic Editor: Zeki Candan

Received: 4 January 2023
Revised: 31 January 2023
Accepted: 2 February 2023
Published: 6 February 2023

Abstract: Particleboard (PB) is mainly produced using urea–formaldehyde (UF) adhesive. However, the low hydrolytic stability of UF leads to poor water resistance by the PB. This research aimed to analyze the effect of hot-pressing temperatures and the addition of methylene diphenyl diisocyanate (MDI) in UF adhesive on the physical and mechanical properties of PB. The first experiment focused on pressing temperature treatments including 130, 140, 150, and 160 °C. The particles were bonded using a combination of UF and MDI resin at a ratio of 70/30 (%w/w). Furthermore, the second experiment focused on UF/MDI ratio treatment, including 100/0, 85/15, 70/30, and 55/45 (%w/w), and the particles were pressed at 140°C. All of the single-layer particleboard in this research were produced in 250 × 250 mm, with a target thickness and density of 10 mm and 750 kg/m^3, respectively. This research used 12% resin content based on oven-dry weight wood shaving. The pressing time and pressing pressure were determined to be 10 min and 2.5 N/mm^2, respectively. Before the tests, the board was conditioned for 7 days. When studying the effect of treatment temperature, good physical properties (thickness swelling and water absorption) and mechanical properties (MOR and MOE) were obtained at 140 °C. However, no significant difference was observed in the UF/MDI ratio between 85/15 and 55/45 using the same temperature. The increase in the MDI adhesive ratio improves the MOE and MOR values. However, the internal bond was the contrary. This study suggests that a combination of UF/MDI at a ratio of 85/15 and hot-pressing temperature at 140 °C could produce a PB panel that meets a type 8 particleboard according to the JIS A5908-2003 standard and type P2 according to the EN 312-2010 standard.

Keywords: basic properties; wood shaving; composite; composites materials; adhesive combination

1. Introduction

Urea–formaldehyde (UF) is an adhesive widely used to manufacture particleboard. An amount greater than 70% of the total UF resin produced is used in the particleboard and medium-density fiberboard industry [1], and its low price is the main reason this adhesive is used. Furthermore, it has disadvantages in terms of low dimensional stability, it is not moisture-resistant, and it has poor durability (biological and weather). Therefore, it is only suitable for interior use. Mansouri et al. [2] and Guru et al. [3] stated that UF adhesive particleboard has low dimensional stability, which has also been supported by other studies [4–10].

The advantages of 4-4 diphenylmethane diisocyanate (MDI)-based resins over UF adhesive include good bonding performance, higher water resistance, aging resistance, and no formaldehyde emission concerns [11]. In addition, the isocyanate group of MDI resins may react with the hydroxyl groups in wood to generate a polyurethane bond, providing direct covalent connections between the adhesive and wood [2]. A number of studies have examined the wood–MDI cure using various techniques, including differential scanning calorimetry (DSC) [12], infrared (IR) spectroscopy [13], and nuclear magnetic resonance (NMR) spectroscopy [14]. In their research, they discovered that wood–MDI cure systems frequently produce biuret, polyuret, and polyurea formations. Only when extremely high doses of MDI were administered, however, were urethane linkages found to form.

Several studies have been conducted on improving composite board quality using UF adhesives. Hybrid resin is one of the studies to improve the performance of UF adhesives. Furthermore, melamine–formaldehyde (MF) has been fortified with melamine urea formaldehyde (MUF) to reduce the weakness of UF adhesives. The UF adhesive was modified by adding isocyanates to improve the thickness swelling properties of the board [11–17]. Mansouri et al. [2] reported that adding a small amount of MDI into the UF improved the performance of the adhesive. This kind of adhesive combination increases the bonding quality of beech (Fagus sylvatica) plywood after immersion in hot (boiling) water. Particleboards made with wheat straw and 4% MDI addition had better mechanical properties and less thickness swelling than resin panels with UF, SPI (soybean protein isolate), and SF (soybean flour). According to [18], three-layer UF-bonded particleboards with 30% waste paper content and pMDI-bonded panels with up to 50% waste paper content in the core layer in terms of mechanical properties meet the requirements of European Standard EN312 for P2-type panels for furniture applications.

Meanwhile, the difference between this study and several others is the application of hybrid adhesive with different ratios during the single-layer board manufacturing process. This is accomplished by spraying the particles individually, with the UF adhesive sprayed first, followed by the MDI adhesive. Iswanto et al. [5] reported that this application technique improved the physical and mechanical properties of particleboards. Therefore, this study aims to analyze the influence of pressing temperature and the ratio of UF and MDI adhesive mixtures on the physical and mechanical properties of particleboards.

2. Material and Methods

2.1. Materials

Sengon (*Paraserianthes falcataria*) wood shavings were obtained from the wood industry in Medan, Indonesia. Furthermore, the commercial UF adhesive (solids content 65.7% and viscosity 210.5 mPa · s) was obtained from PT. Pamolite Adhesive Industry, Probolinggo, Indonesia. The commercial MDI adhesive (solids content 99.5% and viscosity 212.4 mPa · s) was obtained from PT. Polychemie Asia Pacific Permai, Jakarta, Indonesia.

2.2. Characterization of the Hybrid Adhesive Properties

The properties of the hybrid UF/MDI adhesives were determined according to the published methods [19,20]. The viscosity of hybrid UF/MDI adhesives was analyzed using a rotational rheometer (RheolabQC, AntonPaar, Graz, Austria) with a No. 27 spindle at 100 RPM and 25 °C. The gel time of neat UF resins and UF/MDI adhesives was measured at 100 °C using a gel time meter (Techne GT6, Colepalmer, Vernon Hills, IL, USA). The non-volatile solids content of the hybrid UF/MDI adhesives was determined by drying 2 g of the sample in an oven at 105 °C for three hours and dividing the oven-dried weight by the initial weight. Each experiment was repeated in triplicate.

The curing temperature (Tp) of the hybrid UF/MDI adhesives was scanned using differential scanning calorimetry (DSC4000, Perkin Elmer, Hopkinton, MA, USA) from 30 °C to 200 °C with a heating rate of 10 °C/min under 40 mL/min of nitrogen gas. The spectra of hybrid UF/MDI adhesives were also recorded using Fourier transform infrared (FTIR) spectroscopy (SpectrumTwo, PerkinElmer Inc., Hopkinton, MA, USA) with the universal attenuated total reflectance (UATR) method in the range of 400–4000 cm^{-1} at room temperature to detect any changes in the adhesives' functional groups.

2.3. Determination of Slenderness Ratio and Aspect Ratio

One hundred samples of wood shavings were obtained, and the length, width, and thickness were randomly measured. The slenderness ratio (SR) value was determined based on the length and thickness of the particles, while the aspect ratio (AR) value was determined based on the ratio of the width and thickness.

2.4. Particleboard Manufacturing and Testing

The particles in the form of wood shavings were oven-dried at a temperature of $103 \pm 2\,^{\circ}C$ to reach a moisture content of 5%. Furthermore, the adhesive content determined was 12% based on the dry weight of the particles. Single-layer particleboards were produced with a nominal density of 750 kg/m^3 and dimensions of $250 \times 250 \times 10$ mm. The boards were made using a specific pressure of 2.5 N/mm^2 and a pressing time of 10 min. The treatment in the manufacture of the boards was divided into two stages, namely: the influence of pressing temperature on board properties and the influence of the UF/MDI ratio on board properties. After the pressing process, the boards were conditioned for 7 days at ambient temperature. The test sample was cut before testing after the board conditioning process, and it was consistent with the JIS A5908-2003 and EN 312-2011 standards. The samples were air-conditioned under standard defined humidity and temperature conditions until a constant weight was achieved. The test parameters included the physical and mechanical properties of the board. The physical properties include the density, moisture content, thickness swelling, and water absorption. Meanwhile, the mechanical properties include the modulus of rupture and the modulus of elasticity in bending regarding EN 310 and internal bond regarding EN 319.

2.4.1. Influence of Pressing Temperature on Board Properties

In the first stage of the study, the PB was manufactured at different hot-pressing temperatures including 130, 140, 150, and 160 °C. The adhesive used was a mixture of UF and MDI with a ratio of 70/30 UF/MDI (%*w/w*) based on the determined content of 12%. Furthermore, the use of adhesive content of 12% refers to Iswanto et al. [5]. The application was conducted by spraying UF and MDI adhesive separately.

2.4.2. Influence of UF/MDI Ratio on Board Properties

At this stage, the board manufacturing process was treated as a mixture ratio of UF and MDI adhesives consisting of 100/0, 85/15, 70/30, and 55/45 (%*w/w*) based on the determined adhesive content of 12%. The pressing temperature used was the best in the first stage of the study, which was 140 °C with the same pressing pressure and time as the previous stage. The application was conducted by spraying UF and MDI adhesive separately.

2.5. Scanning Electron Microscopy (SEM) Analysis

The morphological observation of the particleboard surface area was conducted by scanning electron microscopy (SEM) JSM-6360 (JEOL Ltd., Tokyo, Japan). Previous imaging demonstrated that the sample surface was coated with an 80 nm gold layer using a sputter coater and then it operated at an accelerating voltage of 15 kV with a magnification of $500\times$.

2.6. Data Analysis

A non-factorial, completely randomized design was used, and the first stage of treatment was pressing temperatures consisting of 130, 140, 150, and 160 °C. Furthermore, the second stage of treatment was in the form of a comparison of UF/MDI adhesives consisting of 100/0, 85/15, 70/30, and 55/45 (%*w*/*w*), and the number of board repetitions for each treatment was three replications.

3. Result and Discussion

3.1. Properties of the Hybrid UF/MDI Adhesives

The characteristics of UF/MDI adhesives at various ratios are shown in Table 1. The basic characteristics including non-volatile solids content, gel duration, and viscosity influence the performance of the adhesives in wood-based panels. In general, the solids content and viscosity of UF/MDI adhesives increased as a function of MDI content while the gel time decreased. The increase in solids content is related to the addition of the MDI, which has a solids content of 99.5%. The high solids content of MDI indicates a high content of active materials for bonding, resulting in the more excellent adhesion and cohesion strength of the MDI adhesive [11–14]. Furthermore, the presence of MDI in UF resins markedly increased the viscosity of hybrid adhesives. The increase in viscosity is probably because of the reaction between –NCO groups and –CH_2OH groups of the UF [21]. As a result, the gel time of the UF/MDI hybrid adhesive decreased with a higher MDI content, which in practice means that a higher MDI content makes the life of the UF/MDI hybrid adhesive shorter.

Table 1. Basic properties of hybrid UF/MDI adhesives.

Properties	UF/MDI Ratio (%*w*/*w*)			
	100/0	85/15	70/30	55/45
Non-volatile solids content (%)	65.7 ± 0.32	67.4 ± 0.23	69.3 ± 0.24	70.6 ± 0.34
Viscosity (mPa · s)	210.5 ± 6.48	230.4 ± 3.91	275.7 ± 8.68	332.4 ± 6.98
Gel time (s)	202.0 ± 2.88	195.6 ± 2.89	180.2 ± 4.04	165.8 ± 3.21

The value after ± indicates the standard deviation.

The DSC analysis showed the curing temperature (Tp) of the hybrid UF/MDI adhesives as an exothermic reaction (arrow) (Figure 1a). The neat UF adhesive with 0% MDI had a Tp around 136.2 °C. Regardless of the MDI content, incorporating MDI into the UF adhesive decreased the Tp to 96.8 to 123.5 °C (asterisk). The decrease in Tp was probably because of the reaction between the –CH_2OH groups of the UF and the –NCO groups of the MDI to form urethane bonds [21]. The –NCO groups of MDI are known to have greater reactivity than other adhesive functional groups [12,14]. The ATR-FTIR spectra of hybrid UF/MDI adhesives are displayed in Figure 1b. The spectra reveal several specific functional groups, such as free –NCO groups and the C–C aromatic of MDI that were observed at 2250 cm^{-1} and 1525 cm^{-1}, respectively, while the neat UF adhesive had C=O groups at 1650 cm^{-1}. Regardless of the MDI content, adding MDI into the UF adhesive increased the intensity of N–H at 3300 cm^{-1} and C–H at 2948 cm^{-1} and 2880 cm^{-1}. The addition of MDI into the UF adhesive formed urethane linkages at 1730 cm^{-1} due to the reaction between the –CH_2OH groups of the UF and the –NCO groups of the MDI. The hybrid adhesives containing higher MDI content had –NCO groups at wavenumber 2270 cm^{-1} from the excess MDI. The free –NCO groups could further react with the –OH of wood [12–14].

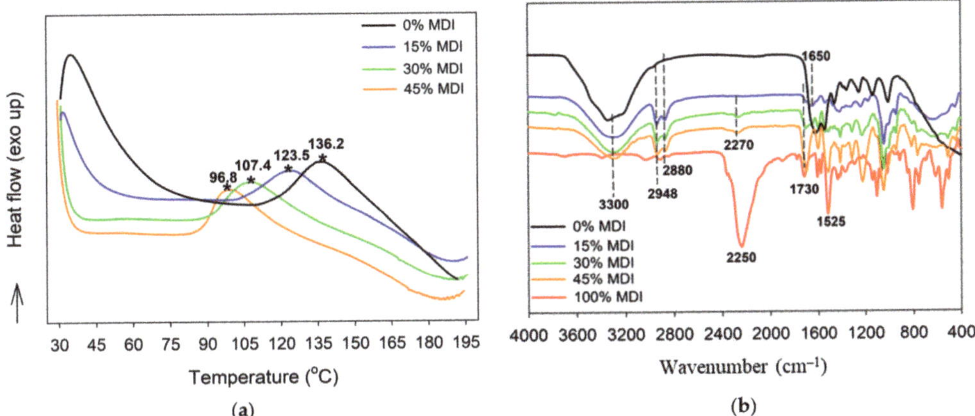

Figure 1. Characteristics of the UF resin and the UF/MDI hybrid adhesives (**a**) DSC thermogram of the hybrid adhesives and (**b**) ATR-FTIR spectra of the hybrid adhesives.

3.2. Slenderness Ratio and Aspect Ratio of Particles

The distribution of SR and AR particles and the calculation results of these two values are displayed in Figure 2. The results indicate that the SR value in wood shavings is dominated in 20–30, while the AR is still below 2. The SR value is included in the low category, and Maloney [21] stated that the ideal SR and AR values are 150 and 3, respectively. The low SR value is one of the causes of the low bending value of the resulting board. According to Maloney [21], particle geometry is one factor that influences the panel modulus of rupture value. This particle geometry deals with size, slenderness, and aspect ratio. The particles with a high slenderness ratio will be more accessible to orient since the resulting board strength will increase and need less adhesive per surface area to bond the particles.

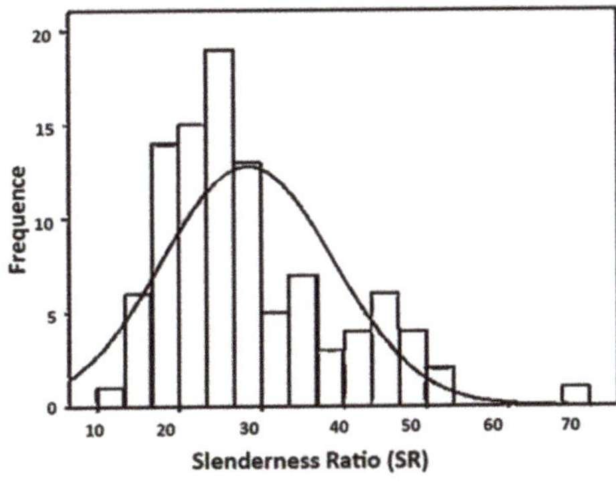
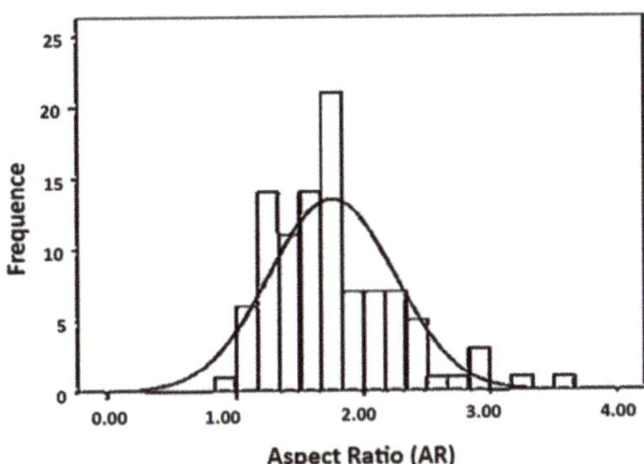

Figure 2. The distribution of the particles' slenderness ratio and aspect ratio.

3.3. Density

Figure 3 shows that the board density value for the pressing temperature treatment ranges from 545–689 kg/m^3. The board produced the lowest and highest density value with a pressing temperature treatment of 140 °C and 150 °C. In addition, the density values ranged from 545–640 kg/m^3 at different UF/MDI ratios. The lowest and highest were obtained at UF/MDI ratios of 70/30 and 100/0. The resulting board density values for the treatments, pressing temperatures, and the overall UF/MDI ratio were still below the target. This is because the particles were removed during the board manufacturing process. Bufalino et al. [22] stated that the low-density value was caused by the loss of particles during the manufacturing process. Similarly, Kelly [23] stated that the factors that influence the board density value include the type of wood (wood density), the amount of pressing pressure, the number of wood particles in the plinth, the adhesive content, and other additives.

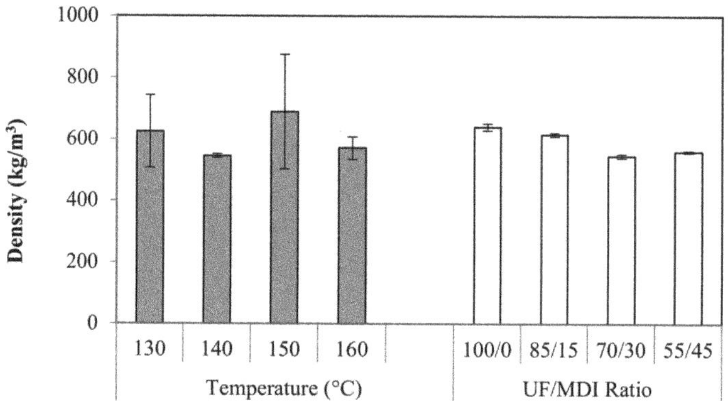

Figure 3. The density of the particleboard. The bars indicate the standard deviation.

The statistical analysis showed that at the 95% confidence interval, the pressing temperature treatment did not significantly affect the density parameter. Meanwhile, the UF/MDI ratio treatment gave a significantly different effect. The final board's density value complies with JIS A 5908-2003 standard requirements, which range from 400 to 900 kg/m^3 [24] and EN 312-2010, whose requirements indicate only ±10% tolerance on the mean density within a board [25].

3.4. Moisture Content

Figure 4 shows that the value of the board's moisture content for the pressing temperature treatment ranges from 5.5 to 6.6%. The particleboard produces the lowest and highest moisture content values with a pressing temperature treatment of 160 °C and 130 °C, respectively. The pressing process at a higher temperature causes a greater decrease in the moisture content of the resulting board. Meanwhile, the temperature of 160 °C showed a drastic decrease in moisture content. This was presumably due to the influence of lignin melting at that temperature. This can lead to a partial closure of the cell cavities in the wood particles and may result in limited water and water vapor accessibility. Ferra et al. [26] stated that the pressing temperature would influence changes in chemical components, such as the liquefaction of lignin. Furthermore, Iswanto et al. [27] reported that pressing at higher temperatures on sorghum bagasse particleboard and jatropha rind produced a lower response to moisture content.

The moisture content value for the treatment of the UF/MDI ratio ranged from 5.8 to 6.5%, where the lowest and highest values were obtained on the boards with a ratio of 100/0 and 55/45, respectively. Figure 3 shows that slightly increasing the MDI adhesive ratio causes an increase in the moisture content value. The MDI proportion and the board density value are inversely related when viewed from the board density value. Furthermore, the low-density value contributed to the increase in the board's moisture content. The statistical analysis showed that at the 95% confidence interval, the pressing temperature treatment and the UF/MDI ratio significantly differed in moisture content parameters. Generally, the moisture content of the resulting boards meets the requirements of JIS A 5908-2003 and EN 312-2010 standards, which specify a moisture content in the range of 5–13% [21,24].

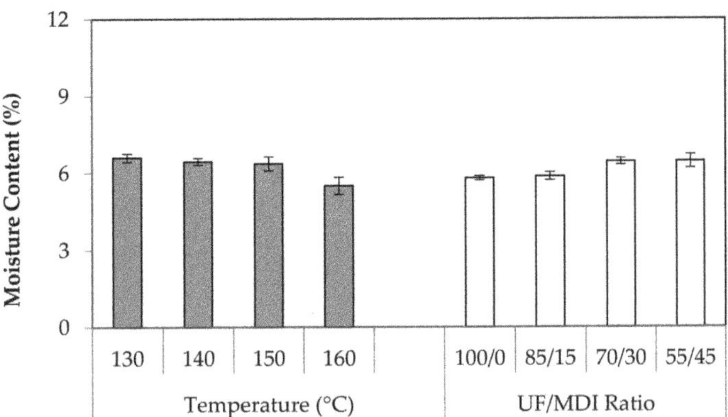

Figure 4. Moisture content of the particleboard. The bars indicate the standard deviation.

3.5. Water Absorption

Figure 5 shows that the water absorption value of the board for the pressing temperature treatment ranged from 46.1 to 61.6%. The board produced the lowest and highest water absorption values with a pressing temperature treatment of 140 °C and 160 °C, respectively. The temperature and water absorption capacity of the board is directly proportional. In the water absorption test, the board was immersed in water at a temperature 20 ± 1 °C for 24 h. Therefore, the water absorption ability is largely determined by the adhesives' performance. In the pressing temperature treatment, the particleboard uses a mixture of UF and MDI adhesives with a ratio of 70/30, where the dominance is on the UF adhesive. As previously reported, the ideal UF adhesive works at low temperatures. Therefore, over-curing is impacted when the temperature increases during the same pressing period, which influences the decreased adhesive ability. This is evidenced in the parameters of the internal bonding of the boards produced for the treatment at a temperature of 160 °C, which has the lowest value. Winandy and Krzysik [28] reported that increased pressing time and temperature did not hinder the ability to absorb water from the panels.

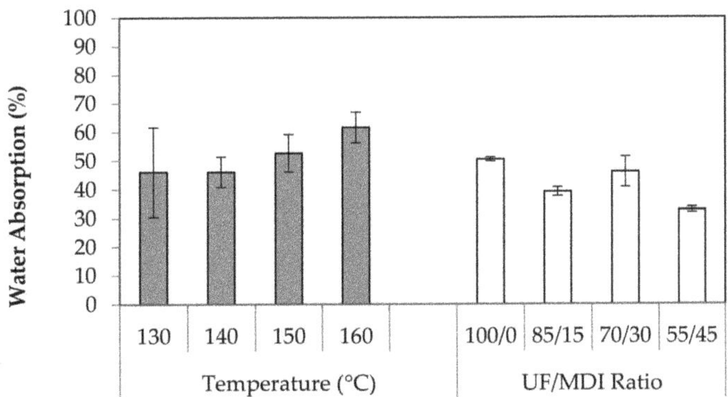

Figure 5. Water absorption of the particleboard. The bars indicate the standard deviation.

Meanwhile, for the treatment of the UF/MDI ratio, the water absorption value ranged from 32.9–50.5%, where the lowest and highest values were obtained on the boards with a ratio of 55/45 and 100/0, respectively. The MDI adhesive ratio and water absorption value of the board are inversely proportional. This is consistent with the trend in the development of board thickness. The statistical analysis showed that at the 95% confidence interval, the pressing temperature treatment did not have a significantly different influence on the water absorption parameters. Meanwhile, the UF/MDI ratio treatment produced a significantly different influence.

3.6. Thickness Swelling

Figure 6 shows that the value of the board thickness swelling for the pressing temperature treatment ranged from 12.1–19.9%. The board produced the lowest and the highest thickness swelling values with a pressing temperature treatment of 140 °C and 160 °C, respectively. An increase in temperature causes a higher thickness swelling value. The UF adhesive underwent over-curing at a higher temperature for the same pressing period. In the hot pressing process of particleboard manufacturing with the UF adhesive, several studies used temperatures varying from 120 °C to 160 °C with time variations between 4–10 min [29–33].

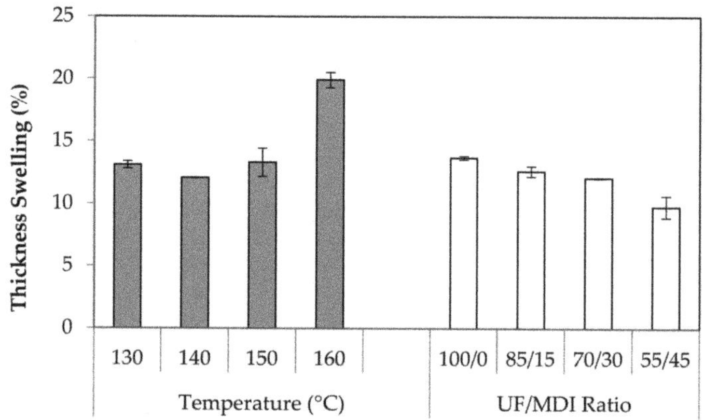

Figure 6. Thickness swelling of the particleboard. The bars indicate the standard deviation.

For the treatment of the UF/MDI ratio, the thickness swelling value ranged from 9.8 to 13.7%, where the lowest and highest values were obtained on boards with a ratio of 55/45 and 100/0, respectively. Figure 6 showed that the MDI adhesive ratio and the thickness swelling value were inversely related. The MDI adhesive can bond chemically. It is stronger than other exterior adhesives, such as PF, which only have mechanical bonding capabilities. Furthermore, Veigel [34] and Mara [35] stated that there is a chemical bond between the MDI adhesive and lignocellulosic material. It produces higher strength and is more stable than mechanical bonds such as PF and UF adhesives. Isocyanates react chemically with hydroxyl groups to form urethane linkages between wood particles [36]. The combination of nonpolar and aromatic compounds from MDI produces resistance to hydrolysis reactions. In addition, isocyanates react physically with the water contained in the wood to form polyurethane. The MDI adhesive usually penetrates the wood surface to a depth of 1 mm [37]. The MDI should penetrate at least 0.3 mm for good adhesion to wood, and the penetration capability results in suitable thickness swelling properties.

The statistical analysis showed that at the 95% confidence interval, the pressing temperature treatment and the UF/MDI ratio produced significantly different effects on the parameters of the thickness of the board. Generally, the resulting thickness swelling value does not meet the JIS A 5908-2003 standard, except for boards with a UF/MDI ratio of 55/45. In contrast, according to EN 312-2010 standard, each board except that obtained by treatment at a pressing temperature of 160 °C meets the requirements (max. 17%) and, due to the value of swelling thickness, is qualified as type P3, i.e., non-load-bearing boards for use in humid conditions [25].

3.7. Modulus of Rupture and Modulus of Elasticity

Figure 7 shows the board's modulus of rupture (MOR) and modulus of elasticity. The MOR value for the pressing temperature treatment ranged from 4.9–8.3 N/mm^2, where the board produced the lowest and highest values with a pressing temperature treatment of 130 °C and 140 °C, respectively. The board's strength decreased at temperatures above 140 °C because the UF adhesive was over-cured at high temperatures. Paridah et al. [38] stated that the adhesive's polymerization rate would increase or decrease depending on the raw materials (wood and adhesive) used. This will directly influence the temperature and pressing time in particleboard manufacturing. The conduction of hot pressing for a long time can affect the over-curing of adhesives, and the strength may be negatively influenced [39].

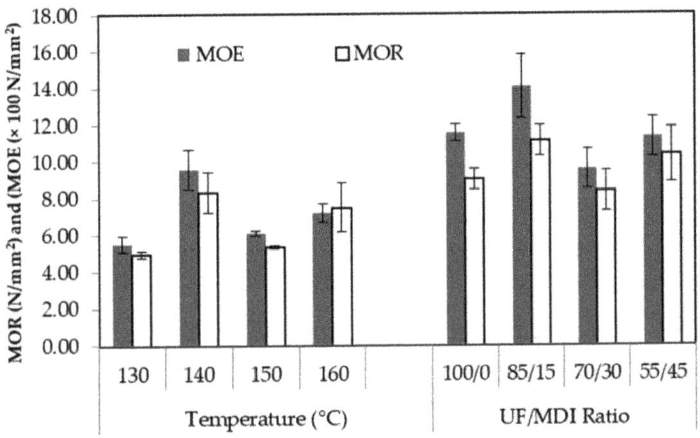

Figure 7. Modulus of rupture of the particleboard. The bars indicate the standard deviation.

For the treatment of the UF/MDI ratio, the MOR values ranged from 8.3–11.1 N/mm^2, where the lowest and highest values were obtained on boards with a ratio of 70/30 and 85/15, respectively. The MDI adhesive ratio and the MOR value of the board are inversely related. The statistical analysis showed that at the 95% confidence interval, the pressing temperature treatment and the UF/MDI ratio produced significantly different influences on the MOR parameters of the board. The pressing temperature treatment of 140 °C resulted in a MOR value consistent with the JIS A 5908-2003 standard. For the UF/MDI ratio, all of the boards produced were consistent with the standard whereby JIS A 5908-2003 requires a minimum MOR value of 8 N/mm^2 [24]. In terms of the EN 312-2010 standard, the board produced with a UF/MDI ratio of 85/15, whose MOR is 12.6 N/mm^2, meets the minimum requirements (11 N/mm^2) for board type P2, i.e., boards for interior fitments (including furniture) for use in dry conditions, and the board with a ratio of 55/45 for board type 1 (10.5 N/mm^2) [25].

The board's modulus of elasticity (MOE) for pressing temperature treatment ranges from 550–959 N/mm^2. The board produced the lowest and highest MOE values with a pressing temperature treatment of 130 °C and 140 °C, respectively. Meanwhile, for the treatment of the UF/MDI ratio, the MOE values ranged from 959–1406 N/mm^2. The lowest and highest values were obtained on boards with a ratio of 70/30 and 85/15, respectively. According to Maloney [21], various factors influence the MOE value, including the resin type, resin content, adhesive bond, and fiber length. The statistical analysis showed that at the 95% confidence interval, the pressing temperature treatment and the UF/MDI ratio significantly differed from the board's MOE parameters. Furthermore, the treatment did not produce boards that meet both standards as the JIS A 5908-2003 standard requires a minimum board MOE value of 2000 N/mm^2 [24] and the EN 312-2010 standard requires a minimum of 1800 N/mm^2 (type P2) [25].

3.8. Internal Bond

Figure 8 shows that the internal bond value of the board for the pressing temperature treatment ranged from 0.07 to 0.6 N/mm^2. The board produced the lowest and highest internal bond values with a pressing temperature treatment of 160 °C and 140 °C, respectively. Forging at a temperature of 140 °C was the optimal condition in this study and therefore, the bond between the UF/MDI mixed adhesive at a ratio of 70/30 with wood shavings particles. Temperatures below and above 140 °C are suspected of causing pre-curing and over-curing, respectively. Both of these conditions reduce the value of the board's internal ties. Nemli [40] states that increasing pressure, pressing temperature, time, and adhesive content are directly proportional to IB. This is closely related to resin maturation, decreased particle wettability, and increased surface area [40]. Furthermore, Ferra et al. [26] reported that the differences in the characteristics of IB using 10% UF adhesive at five different temperatures were explained by two approaches: (1) the temperature influence the UF bond on wood. It facilitates the movement of fluids in the wood and is accompanied by an accelerated diffusion of resin molecules. The low temperature decreases the resin dispersion in the wood, resulting in a decrease in mechanical interlocking. (2) The pressing temperature influences chemical substrate modifications such as lignin fusion and hydrogen bonds associated with increasing the strength values. Low temperatures can inhibit the mobility of the hydroxyl groups of the polymer. This will prevent the conversion of the methyl ether bridge into a methylene bridge and decrease the bond strength's value.

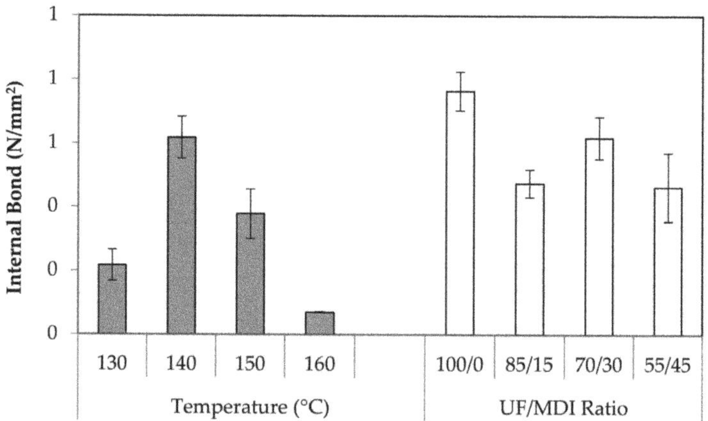

Figure 8. Internal bond of the particleboard. The bars indicate the standard deviation.

For the treatment of the UF/MDI ratio, the internal bond values ranged from 0.5–0.8 N/mm^2. The lowest and highest values were obtained on boards with a ratio of 55/45 and 100/0, respectively. The presence of the MDI adhesive resulted in a lower internal bond value when compared to that without the MDI mixture. This was due to the low moisture content of the particles used to accommodate the UF adhesive. This study was conducted using a moisture content of 5% particles.

Furthermore, the MDI performance will be better in sufficient water. This is because there will be a reaction with molecules containing active hydrogen to produce basic polyurethane and polyurea molecules. Active hydrogen sources can bind hydroxyl groups in wood, wood extractives, or wood resins and their moisture content. Wood has a chemical functional group known as a hydroxyl group. Meanwhile, the MDI on the isocyanate group (–N=C=O) reacts with the hydroxyl group to form a urethane chain. A combination of nonpolar factors and MDI aromatic components are resistant to hydrolysis [35]. The statistical analysis showed that at the 95% confidence interval, the two treatments produced significantly different influences on the internal bond parameters. Most of the IBs were consistent with the JIS A 5908-2003 standard which requires a minimum value of 0.15 N/mm^2 [24] except for the pressing temperature treatment of 160 °C. In terms of the EN 312-2010 standard, all of the boards manufactured in the UF/MDI ratio and the boards with the pressing temperature treatment of 140 °C meet the minimum IB requirements of 0.45 N/mm^2 for the board type P5, i.e., load-bearing for use in humid conditions, and the board in the 100/0 ratio even type P7 (heavy duty load-bearing boards for use in humid conditions)—0.75 N/mm^2 [25].

3.9. Scanning Electron Microscopy Analysis

The morphology of the particleboard surface using UF adhesive and UF/MDI adhesive combination was observed by scanning electron microscopy (SEM). The UF/MDI ratios observed in this research were 100/0, 85/15, and 70/30 (Figure 9). The hot-pressing process resulted in densification on the cell wall of the wood particles as shown in the triangle area in Figure 9a. This was described as the cells being flattened. The particleboard became more compact, and the density tended to increase. The particleboard bonded by UF only showed that UF was not distributed evenly as shown in the rectangle area and narrow in Figure 9b. A more even distribution of MDI in particles can be seen in Figure 9c,d. This analysis proved that adding MDI in a UF/MDI adhesive combination was required to create a more even distribution of adhesive by filling the cell cavities among the particles. It was concluded that the addition of MDI adhesive in the manufacture of a particleboard UF/MDI adhesive combination was able to improve its dimensional stability and bending properties.

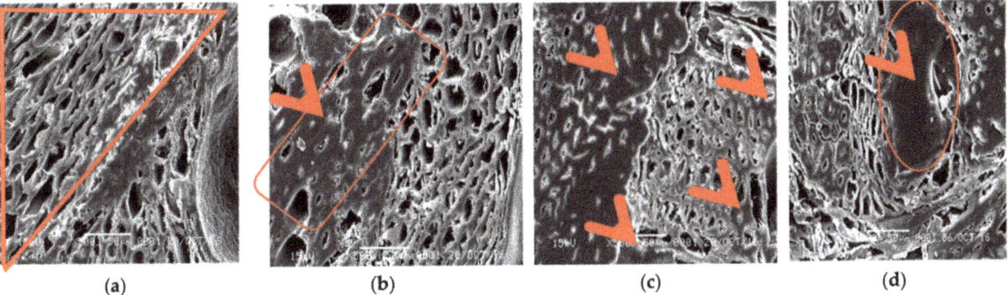

(a)   (b)   (c)   (d)

Figure 9. SEM analysis of the particleboards: (**a**) densification on the cell wall of wood particles and (**b**) UF/MDI adhesive combination ratios 100/0, (**c**) 85/15, and (**d**) 70/30 in 500× magnification.

4. Conclusions

In summary, the increased pressing temperature resulted in increased thickness swelling and water absorption. The resulting swelling thickness value ranges from 12.1–19.9% On the contrary, the MOE, MOR, and internal bond values decreased. The increase in the MDI ratio on the UF/MDI adhesive combination successfully improved the water absorption, thickness swelling, and modulus of rupture (MOR) values of the particleboards. Compared to several related studies, it was shown that the presence of the MDI adhesive in the UF/MDI adhesive combination showed an improvement in thickness swelling, MOE, MOR, and IB values. However, in this study, the IB value produced decreased. The optimum temperature to obtain the particleboard's physical and mechanical properties was 140 °C. The UF/MDI ratio of 85/15 was determined for an optimum combination at this temperature. The increasing ratio of MDI in the UF/MDI adhesive combination resulted in a more even distribution of adhesives. Consequently, the particleboard's water absorption, thickness swelling, and MOR properties were improved.

Author Contributions: Data curation, A.H.I.; methodology, A.H.I.; supervision, A.D., W.F. and T.R.; writing—original draft, A.H.I., J.S., M.A.R.L. and M.P.; writing—review and editing, W.F., A.H.I. and M.P. All authors have read and agreed to the published version of the manuscript.

Funding: This research was funded by the Directorate General of Higher Education Ministry of Research, Technology, and Higher Education, Republic Indonesia, for supporting the research fund through the grant of Penelitian Dasar Unggulan Perguruan Tinggi (PDUPT) for the year of 2020 (Number 11/AMD/E1/KP.PTNBH/2020).

Institutional Review Board Statement: Not applicable.

Informed Consent Statement: Not applicable.

Data Availability Statement: The data presented in this study are available on request from the corresponding author.

Acknowledgments: We would like to express our sincere gratitude to the Directorate General of Higher Education Ministry of Research, Technology, and Higher Education, Republic Indonesia, for supporting the research fund through to the grant of Penelitian Dasar Unggulan Perguruan Tinggi (PDUPT) for the year of 2020 (Number 11/AMD/E1/KP.PTNBH/2020).

Conflicts of Interest: The authors declare no conflict of interest.

References

1. Liu, M.; Wang, Y.; Wu, Y.; Wan, H. Hydrolysis and recycling of urea formaldehyde resin residues. *J. Hazard. Mater.* **2018**, *355*, 96–103. [CrossRef]
2. Mansouri, H.R.; Pizzi, A.; Leban, J.M. Improved water resistance of UF adhesives for plywood by small pMDI additions. *Holz Roh Werkst.* **2006**, *64*, 218–220. [CrossRef]
3. Gürü, M.; Tekeli, S.; Bilici, I. Manufacturing of urea-formaldehyde-based composite particleboard from almond shell. *Mater. Des.* **2006**, *27*, 1148–1151. [CrossRef]
4. Iswanto, A.H.; Febrianto, F.; Hadi, Y.S.; Ruhendi, S.; Hermawan, D.; Fatriasari, W. Effect of particle pre-treatment on properties of jatropha fruit hulls particleboard. *J. Korean Wood Sci. Technol.* **2018**, *46*, 155–165. [CrossRef]
5. Iswanto, A.H.; Simarmata, J.; Fatriasari, W.; Azhar, I.; Sucipto, T.; Hartono, R. Physical and mechanical properties of three-layer particleboards bonded with UF and UMF adhesives. *J. Korean Wood Sci. Technol.* **2017**, *45*, 787–796. [CrossRef]
6. Iswanto, A.H.; Sucipto, T.; Nadeak, S.S.D.; Fatriasari, W. Post-treatment effect of particleboard on dimensional stability and durability properties of particleboard made from sorghum bagasse. *IOP Conf. Ser. Mater. Sci. Eng.* **2017**, *180*, 012015. [CrossRef]
7. Ashori, A.; Nourbakhsh, A. Effect of press cycle time and resin content on physical and mechanical properties of particleboard panels made from the underutilized low-quality raw materials. *Ind. Crops Prod.* **2008**, *28*, 225–230. [CrossRef]
8. Lias, H.; Kasim, J.; Johari, N.A.N.; Mokhtar, I.L.M. Influence of board density and particle sizes on the homogenous particleboard properties from kelempayan (*Neolamarckia cadamba*). *Int. J. Latest Res. Sci. Technol.* **2014**, *3*, 173–176.
9. Mamza, P.A.P.; Ezeh, E.C.; Gimba, E.C.; Arthur, D.E. Comparative study of phenol formaldehyde and urea formaldehyde particleboards from wood waste for sustainable environment. *Int. J. Sci. Technol. Res.* **2014**, *3*, 53–61.
10. Elbadawi, M.; Osman, Z.; Paridah, T.; Nasroun, T.; Kantiner, W. Mechanical and physical properties of particleboards made from ailanthus wood and UF resin fortified by acacias tannins blend. *J. Mater. Environ. Sci.* **2015**, *6*, 1016–1021.
11. Lubis, M.A.R.; Park, B.D.; Lee, S.M. Modification of urea-formaldehyde resin adhesives with blocked isocyanates using sodium bisulfite. *Int. J. Adhes. Adhes.* **2017**, *73*, 118–124. [CrossRef]
12. Dziurka, D.; Mirski, R. UF-pMDI hybrid resin for waterproof particleboards manufactured at a shortened pressing time. *Drv. Ind.* **2010**, *61*, 245–249.
13. Dziurka, D.; Mirski, R. Properties of Liquid and Polycondensed UF Resin Modified with pMDI. *Drv. Ind.* **2014**, *65*, 115–119. [CrossRef]
14. Simon, C.; George, B.; Pizzi, A. UF/pMDI wood adhesives: Networks blend versus copolymerization. *Holzforschung* **2002**, *56*, 327–334. [CrossRef]
15. Yanhua, Z.; Jiyou, G.; Haiyan, T.; Xiangkai, J.; Junyou, S.; Yingfeng, Z.; Xiangli, W. Fabrication, performances, and reaction mechanism of urea-formaldehyde resin adhesive with isocyanate. *J. Adhes. Sci. Technol.* **2013**, *27*, 191–203. [CrossRef]
16. Lubis, M.A.R.; Labib, A.; Sudarmanto; Akbar, F.; Nuryawan, A.; Antov, P.; Kristak, L.; Papadopoulos, A.N.; Pizzi, A. Influence of Lignin Content and Pressing Time on Plywood Properties Bonded with Cold-Setting Adhesive Based on Poly (Vinyl Alcohol), Lignin, and Hexamine. *Polymers* **2022**, *14*, 2111. [CrossRef]
17. Mo, X.; Cheng, E.; Wang, D.; Sun, X.S. Physical properties of medium-density wheat straw particleboard using different adhesives. *Ind. Crops Prod.* **2003**, *18*, 47–53. [CrossRef]
18. Lykidis, C.; Parnavela, C.; Goulounis, N.; Grigoriou, A. Potential for utilizing waste corrugated paper containers into wood composites using UF and PMDI resin systems. *Eur. J. Wood Wood Prod.* **2012**, *70*, 811–818. [CrossRef]

19. Hidayat, W.; Aprilliana, N.; Asmara, S.; Bakri, S.; Hidayati, S.; Banuwa, I.S.; Lubis, M.A.R.; Iswanto, A.H. Performance of eco-friendly particleboard from agro-industrial residues bonded with formaldehyde-free natural rubber latex adhesive for interior applications. *Polym. Compos.* **2022**, *43*, 2222–2233. [CrossRef]
20. Lubis, M.A.R.; Park, B.D.; Lee, S.M. Microencapsulation of polymeric isocyanate for the modification of urea-formaldehyde resins. *Int. J. Adhes. Adhes.* **2020**, *100*, 102599. [CrossRef]
21. Maloney, T.M. *Modern Particleboard and Dry Process Fiberboard Manufacturing*; Miller Freeman Inc.: San Francisco, CA, USA, 1993.
22. Bufalino, L.; Albino, V.C.S.; De Sá, V.A.; Corrêa, A.A.R.; Mendes, L.M.; Almeida, N.A. Particleboards made from Australian red cedar: Processing variables and evaluation of mixed-species. *J. Trop. For. Sci.* **2012**, *24*, 162.
23. Kelly, M.W. *Critical Literature Review of Relationship between Processing Parameter and Physical Properties of Particleboard*; General Technical Report FPL-10 Department of Agriculture Forest; Wisconsin University: Madison, WI, USA, 1997.
24. Japanese Standards Association (JSA). *Japanese Industrial Standard (JIS) A 5908. Particleboard*; Japanese Standards Association: Tokyo, Japan, 2003.
25. *EN 312:2010*; Particleboards—Specifications. European Committee for Standardization: Brussels, Belgium, 2010.
26. Ferra, J.M.M.; Ohlmeyer, M.; Mendes, A.M.; Costa, M.R.N.; Carvalho, L.H.; Magalhes, F.D. Evaluation of urea-formaldehyde adhesives performance by recently developed mechanical tests. *Int. J. Adhes. Adhes.* **2011**, *31*, 127–134. [CrossRef]
27. Iswanto, A.H.; Febrianto, F.; Hadi, Y.S.; Ruhendi, S.; Hermawan, D. The effect of pressing temperature and time on the quality of particle board made from jatropha fruit hulls treated in acidic condition. *Makara Seri Teknol.* **2013**, *17*, 145–151. [CrossRef]
28. Winandy, J.E.; Krzysik, A.M. Thermal degradation of wood fibers during hot-pressing of MDF composites: Part I. Relative effects and benefits of thermal exposure. *Wood Fiber Sci.* **2007**, *39*, 450–461.
29. Clausen, C.A.; Cartal, S.N.; Muehl, J. Properties of particleboard made from recycled cca-treated wood. In Proceedings of the 31st Annual Meeting of IRG, Kona Surf, HI, USA, 14–19 May 2000.
30. Lee, Y.K.; Kim, S.; Yang, H.S.; Kim, H.J. Mechanical properties of rice husk flour-wood particleboard by urea-formaldehyde resin. *J. Mokchae Konghak* **2003**, *31*, 42–49.
31. Yang, H.S.; Kim, D.J.; Kim, H.J. Rice straw-wood particle composite for sound absorbing wooden construction materials. *J. Bioresour. Technol.* **2003**, *86*, 117–121. [CrossRef] [PubMed]
32. Sulastiningsih, I.M.; Novitasari; Turoso, A. Effect of resin portion on bamboo particleboard properties. *J. Penelit. Has. Hutan* **2006**, *24*, 1–8. [CrossRef]
33. García-Ortuño, T.; Andréu-Rodríguez, J.; Ferrández-García, M.T.; Ferrández-Villena, M.; Ferrández-García, C.E. Evaluation of the physical and mechanical properties of particleboard made from giant reed (*Arundo donax* L.). *BioResources* **2011**, *6*, 477–486. [CrossRef]
34. Veigel, S.; Rathke, J.; Weigl, M.; Gindl-Altmutter, W. Particle board and oriented strand board prepared with nanocellulose-reinforced adhesive. *J. Nanomater.* **2012**, *2012*, 1–8. [CrossRef]
35. Marra, A.A. *Technology of Wood Bonding Principles in Practise*; Van Nostrand Reinhold: New York, NY, USA, 1992.
36. Blomquist, R.F.; Christiansen, A.W.; Gillespie, R.H.; Myers, G.E. *Adhesive Bonding of Wood and Other Structural Materials*; Forest Product Technology USDA Forest Service; The University of Wisconsin: Madison, WI, USA, 1983.
37. Syamani, F.A.; Subiyanto, B.; Massijaya, M.Y. Termite resistant properties of sisal fiberboards. *Insects* **2011**, *2*, 462–468. [CrossRef]
38. Paridah, M.T.; Chin, A.M.E.; Zaidon, A. Bonding properties of *Azadirachta excelsa*. *J. Trop. For. Prod.* **2001**, *7*, 161–171.
39. Xing, C.; Zhang, S.Y.; Deng, J. Medium density fiberboard performance as affected by wood fiber acidity, bulk density and size distribution. *J. Wood Sci Technol.* **2006**, *40*, 637–646. [CrossRef]
40. Nemli, G. Factors affecting the production of E1 type particleboard. *Turkish J. Agric. For.* **2002**, *26*, 31–36. [CrossRef]

Article

Effects of Boric Acid Pretreatment on the Properties of Four Selected Malaysian Bamboo Strips

Norwahyuni Mohd Yusof [1], Lee Seng Hua [2,*], Paridah Md Tahir [1,3,*], Redzuan Mohammad Suffian James [1], Syeed Saifulazry Osman Al-Edrus [1], Rasdianah Dahali [1], Adlin Sabrina Muhammad Roseley [3], Widya Fatriasari [4], Lubos Kristak [5,*], Muhammad Adly Rahandi Lubis [4] and Roman Reh [5]

1 Institute of Tropical Forestry and Forest Products, Universiti Putra Malaysia, Serdang 43400, Selangor, Malaysia
2 Depatment of Wood Industry, Faculty of Applied Sciences, Universiti Teknologi MARA (UiTM) Cawangan Pahang Kampus Jengka, Lintasan Semarak, Bandar Jengka, Bandar Tun Razak 26400, Pahang, Malaysia
3 Faculty of Forestry and Environment, Universiti Putra Malaysia, Serdang 43400, Selangor, Malaysia
4 Research Center for Biomass and Bioproducts, National Research and Innovation Agency (BRIN), Jl Raya Bogor KM 46, Cibinong Bogor 16911, Indonesia
5 Faculty of Wood Sciences and Technology, Technical University in Zvolen, 96001 Zvolen, Slovakia
* Correspondence: leesenghua@hotmail.com (L.S.H.); parida.introp@gmail.com (P.M.T.); kristak@tuzvo.sk (L.K.)

Abstract: Bamboo requires treatment to extend its service life. However, as bamboo strips could serve as a suitable candidate for lamination, the treatment may affect its bendability. The current study investigated the effects of boric acid treatment on the physical, mechanical, adhesion, and morphological properties of bamboo strips. Owing to their availability and popularity in local industries, four Malaysian bamboo species were used in this study, namely *Gigantochloa scortechinii*, *Gigantochloa levis*, *Dendrocalamus asper*, and *Bambusa vulgaris*. These four species' bamboo strips were treated with 5% boric acid and their properties were evaluated. The findings revealed that the boric acid treatment had varying degrees of effect on the properties of the bamboo. Despite having lower treatability and stability, both *G. scortechinii* and *G. levis* have greatly superior mechanical properties that justify their use in the production of laminated products. The boric acid treatment was found to provide several benefits to bamboo strips intended for lamination, including increased wettability, dimensional stability, and mechanical strength.

Keywords: bamboo; boric acid; physical properties; mechanical properties; adhesion properties; morphological properties

Citation: Yusof, N.M.; Hua, L.S.; Tahir, P.M.; James, R.M.S.; Al-Edrus, S.S.O.; Dahali, R.; Roseley, A.S.M.; Fatriasari, W.; Kristak, L.; Lubis, M.A.R.; et al. Effects of Boric Acid Pretreatment on the Properties of Four Selected Malaysian Bamboo Strips. *Forests* **2023**, *14*, 196. https://doi.org/10.3390/f14020196

Academic Editor: Diego Elustondo

Received: 12 December 2022
Revised: 14 January 2023
Accepted: 17 January 2023
Published: 19 January 2023

1. Introduction

Bamboo is regarded as an eco-friendly plant that grows and matures quickly, has a versatile use, a unique appearance, is efficient at photosynthesis, and has great potential as a substitute material for wood with the various fibre products that can be produced [1–3]. Because of its higher physical and mechanical properties, in comparison to some commercial wood species, it has also been used for engineered products, offering higher value addition and market potential [4–6]. Due to its high strength and rapid growth, bamboo is the best alternative for replacing timber [7]. Bamboo is no longer restricted to round forms, but has been extended to splits or strips, which have been encouraged to form in engineered products such as composites, laminated bamboo boards, and plybamboo, which can be used in structural applications [8–12].

In Malaysia, at least 70 bamboo species from 10 genera have been identified, with 59 of them found in Peninsular Malaysia. The top three most studied bamboo species in Malaysia are *Gigantochloa scortechinii* (buluh semantan), *Bambusa vulgaris* (buluh minyak), and *Dendrocalamus asper* (buluh betong), with *G. scortechinii* dominating the list. The basic properties of these bamboos, particularly *G. scortechinii*, have been extensively researched. Local researchers have extensively studied the anatomical, physical and mechanical properties, machining properties, and chemical content of *G. scortechinii* [13–15]. Several researchers [16,17] have also reported on the chemical composition, morphology, physical and mechanical strength of *B. vulgaris*. In the meantime, there are few reports on the basic properties of *D. asper* in comparison to the previous two bamboo species. *D. asper*, on the other hand, is a favorite bamboo species among local researchers for the production of wood-based products such as particleboard [18,19]. *G. levis* (buluh beting), on the other hand, has received very little attention in the country. However, Siam et al. [20] found *G. levis* to be a very promising bamboo species for laminated bamboo production. *G. levis* has a culm wall thickness of 12 mm, a high density of 750 kg/m^3, and superior mechanical strength, making it a suitable candidate for laminated products. These four bamboo species have been found in abundance in the surrounding forest. Furthermore, these are the bamboo species most commonly used in the local industry and community for the production of various bamboo-based household products.

Bamboo culms, on the other hand, are easily impacted by fungus and insects due to their high carbohydrate, including starch content. Improving bamboo's mechanical properties and fungi resistance is critical for increasing its outdoor use. Bamboo, like most lignocellulosic materials, is vulnerable to biological degradation agents [21]. As a result, treating the bamboo is required to extend the product's service life. Chemical treatments are the best option, because they ensure that the bamboo products have a longer service life while also maintaining their quality [22]. The preservatives used determine the success of the bamboo treatment. As many wood preservatives have already been banned due to arsenic and/or chromium content, boron-based preservatives with lower toxicity, such as boric acid, have become a popular alternative [23]. Boric acid is a water-based preservative that is much less toxic than arsenic or chromium compounds. Boric acid is well-known in wood preservation for its insecticide and fungicide properties [24].

However, while boric acid treatment improves biological durability, it may also change the physical and mechanical properties of the bamboo. According to some studies, wood treated with boric acid has lower mechanical strength than untreated wood [25,26]. Simsek and Baysal [27], on the other hand, discovered that boric acid treatment increased the surface roughness, density, and modulus of elasticity (MOE) of the treated wood. Because bamboo has the potential to be used in lamination, it is necessary to investigate the effects of boric acid treatment on the adhesion and bonding characteristics of bamboo, as well as its buffering capacity [28,29], which measures the resistance of wood to changing pH levels in acid or alkaline liquid. Density, moisture content, anatomical structure, buffering capacity of the bamboo surface, and species, are all factors that influence bonding quality. Therefore, the purpose of this study was to look into the effects of boric acid treatment on the physical, mechanical, and adhesion properties of bamboo strips that will be used to make laminated bamboo boards. This study used four Malaysian bamboo species based on their abundance and importance in the local industry: *G. scortechinii*, *G. levis*, *D. asper* and *B. vulgaris*.

2. Materials and Methods

2.1. Materials Preparation

Selected culms from a matured (3–5 years old) bamboo stand were harvested in a forest area near Sik, Kedah. Four bamboo species, namely *B. vulgaris* (buluh minyak), *D. asper* (buluh betong), *G. levis* (buluh beting), and *G. scortechinii* (buluh semantan) were used in this study for the production of bamboo strips (Figure 1). These bamboos were chosen due to their extensive use in the local bamboo industry and widespread distribution in Malaysia.

To obtain shorter and more workable pieces, the harvested culms were cut to lengths of 2 meters using a cross-cutting machine (CYM-001, Chang-hua Hsien, Taiwan). Before being ripped into splits, the culms were further cut to 1 meter lengths. The hollow bamboo culms were longitudinally split into 22 mm width segments with a thickness of 7–9 mm. The splitting of cylindrical bamboo stems resulted in slightly curved strips as shown in Figure 1. The bamboo strips were flattened and shaped using a thicknesser machine before being planed with a 2-side removal machine (PBM-TSP-001, Nagpur, Maharashtra, India). The process removed the bamboo green and bamboo yellow to ensure the middle part of the splits, called the bamboo timber, were extracted [30]. The strips had to be rectangular, flat on each side, and without an inner or outer layer to ensure proper bonding on all four sides. To ensure better bonding and no voids in the laminated bamboo sample, the bamboo S4S machine (MBXD-10B, Hangzhou, Zhejiang, China) was used to shape the strips into a fixed width and thickness (20 mm width × 5 mm thickness).

(a)	(b)	(c)	(d)

Figure 1. Splits of (**a**) *Bambusa vulgaris* (buluh minyak), (**b**) *Gigantochloa levis* (buluh beting), (**c**) *Gigantochloa asper* (buluh betong), and (**d**) *Gigantochloa scortechinii* (buluh semantan).

2.2. Treatment of Bamboo Strips

A 5% boric acid solution was prepared for the treatment of the bamboo strips. The treatment's goal was to provide bamboo strips with short-term protection against biodegrading agents, such as borers, termites, and fungi, before they were laminated. For 24 h, the bamboo strips were immersed in a 1 m × 1 m tank containing a boric acid solution. The strips were removed from the treatment tank after treatment, and the excess water was drained. The strips were then kiln dried for three days at 60 °C and 40% RH until the moisture content reached around 12%. After that, the strips were conditioned at 20 ± 2 °C and 65 ± 5% RH before testing

2.3. Evaluation of Physical Properties

The conditioned bamboo strips were then cut into samples with dimensions suitable for the different properties' evaluations. Physical properties, such as moisture content, specific gravity, moisture excluding efficiency, water absorption, thickness swelling, dry salt retention, and weight percent gain, were determined according to the formula listed in Table 1.

2.4. Surface Morphology Due to Boric Acid Treatment

The presence of boric acid in the bamboo strips was observed by using a Scanning Electron Microscope (SEM, Leo 1455VP, Thornwood, NY, USA). The samples were taken from both the treated and untreated bamboo strips. Cubes of 10 × 10 × 5 mm were prepared using a sharp knife. Morphological investigations were performed on the untreated and treated bamboo (boric acid) with a Scanning Electron Microscope (SEM, Leo 1455VP, Thornwood, NY, USA) machine. The SEM instrument was used at an emission current of

58 μA and an acceleration voltage of 5.0 kV; the working distance was set at 6.2 mm. Before the SEM analysis, the samples were coated with gold.

Table 1. Experimental test methods of bamboo strips for the determination of physical properties.

Test Item	Dimension (mm)	Formula
Moisture content (MC)	$20 \times 20 \times 5$	$MC\ (\%) = [(M_i - M_o)/M_o] \times 100$ where, M_i = initial weight of sample, g M_o = weight of sample after oven dry, g
Density (ρ)	$20 \times 20 \times 5$	$\rho\ (kg/m^3) = m/V \times 1{,}000{,}000$ where, m = weight of the sample, g V = volume of the sample, mm^3
Specific gravity (SG)	$20 \times 20 \times 5$	SG = density of sample / density of water
Moisture excluding efficiency (MEE)	$20 \times 20 \times 5$	$MEE\ (\%) = [(E_u - E_t)/E_t] \times 100$ where, E_u = equilibirum moisture content of untreated sample, % E_t = equilibirum moisture content of treated sample, %
Water absorption (WA) and thickness swellling (TS)	$20 \times 20 \times 5$	$WA\ (\%) = [(W_a - W_i)/W_i] \times 100$ $TS\ (\%) = [(T_a - T_i)/T_i] \times 100$ where, W_i = weight of sample before soaking, g W_a = weight of sample after soaking, g T_i = thickness of sample before soaking, mm T_o = thickness of sample after soaking, mm
Dry salt retention (DSR)	$1000 \times 20 \times 5$	$DSR\ (kg/m^3) = [(l/V) \times (\text{concentration of boric acid solution}/100)] \times 1{,}000{,}000$ where, l = boric acid solution uptake, g V = volume of bamboo culm, mm^3
Weight percent gain (WPG)	$1000 \times 20 \times 5$	$WPG\ (\%) = [(M_2 - M_1)/M_1] \times 100$ where, M_1 = weight of sample before treatment, g M_2 = weight of sample after treatment, g

2.5. Evaluation of Adhesion Properties

2.5.1. Evaluation of Surface Wettability by Contact Angle

The contact angle measurements were performed on bamboo green (previously removed during planing) samples measuring 60 mm in length, 20 mm in width, and 4 mm in thickness. A Theta Lite (TL100 and TL101, Västra Frölunda, Sweden) surface wettability tester was used to measure the contact angle of the specimens. A microscope with an attached camera was used to observe the contact angle. The specimens were conditioned at 20 °C and 65% RH for 24 h. Distilled water was dropped onto the surface at room temperature. To calculate the contact angle, the height and diameter of each droplet was recorded, and the contact angle was observed and measured. Thirty replicated surfaces for each species were tested.

2.5.2. Buffering Capacity

Samples of untreated bamboo strips were ground and dried. After that, fifteen grams of dry bamboo particles were boiled in a 200 mL flask for 30 min. The mixture was filtered using a glass crucible with a filter porosity (40–100 μm) (Carl Roth E562.1, Karlsruhe, Baden-Wuerttemberg, Germany) equipped with an aspirator vacuum (A-1000S, Jalan Bukit Merah, Singapore) and the filtrates were diluted with distilled water to a final volume of 200 mL each. After that, the dilute extract was cooled at 20 °C for 24 h. A digital Mettler Toledo Delta 320 pH Meter (MPN: 320, Columbus, Ohio, USA) was used to measure the actual pH of the diluted sample before proceeding to the aqueous extraction. The aqueous extract was manually titrated with 0.1 N sodium hydroxide (NaOH) and 0.1 N hydrochloride (HCl) solution until pH 10.0 and pH 3.0, respectively. The pH value was recorded after the addition of every 0.2 mL of titrant. After that, a graph of the pH vs. titrant volume (mL) was plotted to examine the change in pH [31].

2.6. Mechanical Properties

The samples were stacked in a conditioning chamber at $20 \pm 2\ ^\circ\text{C}$ and a relative humidity of $65 \pm 3\%$ (EMC 12%) for a week prior to testing. The samples for static bending and compression parallel to the grain were prepared in accordance with BS EN373:1957 [32]. A total of 240 specimens (30 samples $\times$ 4 species $\times$ 2 treatments) were tested. The test was performed on an Instron Universal Testing Machine (UTM, Instron-3366, Norwood, MA, USA) with a capacity of 100 kN. The loading rate and formula used to calculate the MOR, MOE, and compressive strength parallel to the grain, are given in Table 2. Bamboo strips were placed on the supporting span for the static bending test, with the bamboo green surface facing the load of the testing machine.

Table 2. Testing parameters of bamboo strips for the static bending and compression parallel to the grain tests.

Test Item	Dimension (mm)	Total Specimens	Loading Rate (in/min)	Formula
Bending	$300 \times 20 \times 5$	160 (Control) 160 (Treated)	0.26	Modulus of rupture (N/mm^2) = 3FL/2bd^2 where, F = load at a given point on the load deflection curve, N L = support span, mm b = width of test specimens, mm d = depth of test specimens, mm Modulus of elasticity (N/mm^2) = PL^3/4wbh^3 where, P = is an increment of load in N on the straight-line portion of the load deflection curve, mm w = is the increment of deflection at mid-length corresponding to P L = span length, mm b = width of test specimens, mm h = depth of the test specimens, mm
Compression parallel to the grain	$60 \times 20 \times 5$	160 (Control) 160 (Treated)	0.025	Compression (N/mm^2) = P/bt where, P = maximum crushing load, N b = width of test specimens, mm t = thickness of the test specimens, mm

2.7. Experimental Design and Statistical Analysis

The collected data were analyzed using an Analysis of Variance (ANOVA). Meanwhile, the mean separation was carried out using the Least Significant Difference (LSD) method.

3. Results and Discussion

3.1. Treatability of Bamboo

3.1.1. Weight Percent Gain (WPG)

Weight percent gain (WPG) represents the extent of chemical penetration into a material and indicates its treatability. The WPG of bamboo strips is tabulated in Table 3. The variation of WPG values is expected as it is highly dependent on the wood species. In this study, *G. levis* has a higher treatability with a WPG of 3.61% or 25.34 kg/m^3 compared to *D. asper*, which has a WPG of only 1.47% or 8.97 kg/m^3. Meanwhile, both *B. vulgaris* and *G. scortechinii* have moderate treatability, as evidenced by their WPG values, which were 2.89% or 19.06 kg/m^3 and 1.98% or 13.51 kg/m^3, respectively. The gain in weight is due to the precipitation of boric acid. Boric acid dissolves in water primarily in the chemical form of B(OH)$_3$, with only a trace of boric acid dissociating into H$^+$ and B(OH)$_4$ [33]. Yamauchi [34] discovered that B(OH)$_3$ is the dominant form of boric acid. According to their research, B(OH)$_3$ localization occurs during the air-drying process. The analysis of the

Fourier-transform (FT)-Raman spectra revealed that a portion of microcrystalline $B(OH)_3$ precipitated in the lumen, while another portion of the $B(OH)_3$ units penetrated the cell wall of the samples [34].

Table 3. Weight percent gain of untreated and treated bamboo strips of four selected species.

Bamboo Species	Weight Gain	
	kg/m^3 Mean $\pm$ SD	% Mean $\pm$ SD
Dendrocalamus asper	8.97 $\pm$ 4.27 ^{D}	1.47 $\pm$ 0.71 ^{D}
Bambusa vulgaris	19.06 $\pm$ 9.53 ^{C}	2.89 $\pm$ 1.41 ^{B}
Gigantochloa scortechinii	13.51 $\pm$ 10.51 ^{A}	1.98 $\pm$ 1.57 ^{C}
Gigantochloa levis	25.34 $\pm$ 10.16 ^{B}	3.61 $\pm$ 1.28 ^{A}

Means followed with the same letters ^{A,B,C,D} in the same column were not significantly different at $p \leq 0.05$.

3.1.2. Dry Salt Retention (DSR)

The mean dry salt retention (DSR) of bamboo with different species of bamboo are tabulated in Figure 2. Overall, the DSR of bamboo strips after boric acid treatment ranged from 13–14 kg/m^3. The highest DSR was recorded on the bamboo strips from *D. asper* followed by *B. vulgaris*, *G. levis* and *G. scortechinii*, respectively. Comparing the WPG and DSR, there are some discrepancies that occurred in terms of ranking treatability of the four bamboo species. In WPG, the rank (descending order) was *G. levis*, *B. vulgaris*, *G. scortechinii* and *D. asper* whereas in DSR, the rank was *D. asper*, *B. vulgaris*, *G. levis*, and *G. scortechinii*.

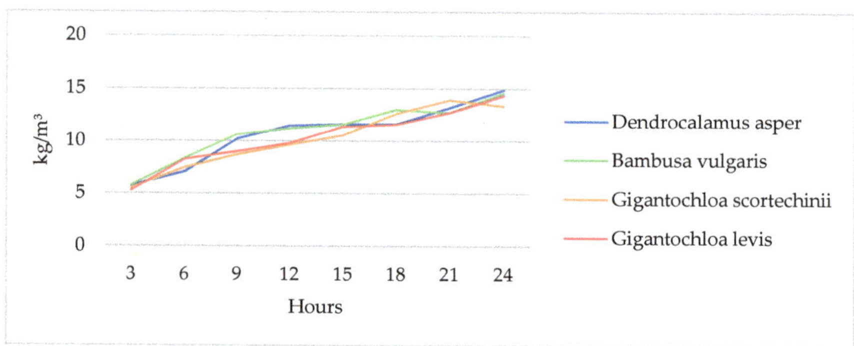

Figure 2. Mean dry salt retention of untreated and treated bamboo strips of four selected species.

3.2. Surface Morphological Due to Boric Acid Treatment

Figures 3 and 4 show the SEM images of untreated and treated bamboo strips. The bamboo strips' fibre lumen and parenchyma cells appear to have been filled with boric acid (Figure 4). Gigantochloa genus has a complete vascular bundle structure that comprises of three parts: a central, vascular strand, and two fibre strands, on each side of the central strand [35]. Dendrocalamus genus contained two parts, a central vascular strand and one fibre strand inside the central strand, while *Bambusa vulgaris* contained two parts, one fibre strand at the bottom of the main vascular bundle and two fibre strands located above and below the main vascular bundle [36]. This explained the reason why semantan and beting generally behave similarly (in almost all of the properties mentioned above) as compared to minyak and betong. Boric acid can remove the starch found in parenchyma, and improve the bonding properties of bamboo strips.

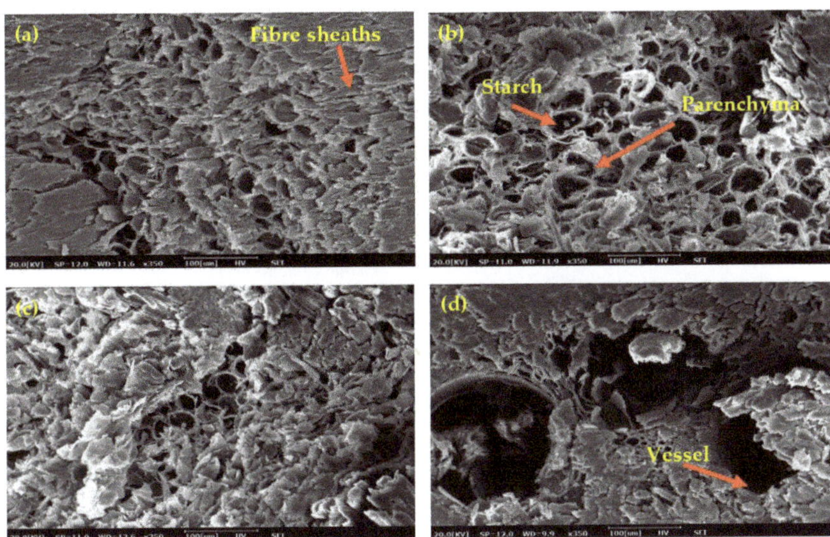

Figure 3. SEM for untreated bamboo strips of (**a**) *Dendrocalamus asper*, (**b**) *Bambusa vulgaris*, (**c**) *Gigantochloa scortechinii*, and (**d**) *Gigantochloa levis* (350×).

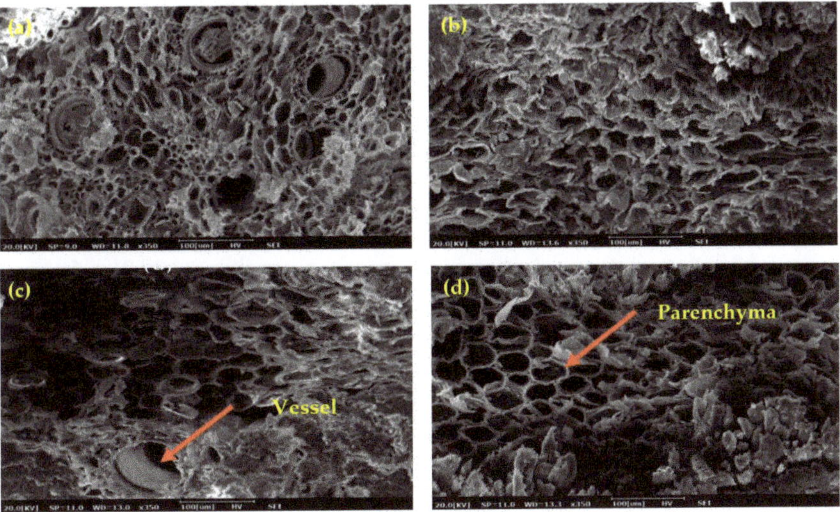

Figure 4. SEM for treated bamboo strips of (**a**) *Dendrocalamus asper*, (**b**) *Bambusa vulgaris*, (**c**) *Gigantochloa scortechinii*, and (**d**) *Gigantochloa levis* (350×).

3.3. Physical Properties of Untreated and Treated Bamboo Strips

After a week of conditioning, the moisture content of the bamboo strips was within the range of 7%–9.5%. The density and specific gravity (SG) for the treated and untreated bamboo strips are shown in Table 4. The density of the strips varied between the four species, ranging from 623.17 to 701.7 kg/m^3 for treated and 604.63 to 689.91 kg/m^3 for untreated. The culms' specific gravity ranged from 0.62 to 0.70 for treated and 0.60 to 0.69 for untreated. The moisture content of the bamboo strips ranged from 6.9 to 9.5% for treated and 7.2 to 8.0% for untreated bamboo strips. The density of *G. scortechinii* was higher than the others, with an average density of 713.35 kg/m^3, closely followed by *G. levis* (696.28 kg/m^3) and *B. vulgaris* (672.24 kg/m^3), while *D. asper* had the lowest density at 625.68 kg/m^3. The density result was close to previous research performed by Siam et al. [20]. In general, the boric acid treatment increased both the density and specific gravity, slightly. The increment in density and SG is mainly contributed to by the increment in WPG mentioned above. The treatment, however, had the greatest impact on *D. asper*.

Table 4. Density and specific gravity of treated and untreated bamboo strips of four selected species.

Treatment	Bamboo Species	Density (kg/m^3)	MC (%)	SG
Untreated	*Dendrocalamus asper*	604.63 [B] (49.37)	8 [B] (1.3)	0.6 [B] (0.05)
	Bambusa vulgaris	675.32 [A] (78.1)	7.3 [B] (2.2)	0.68 [A] (0.08)
	Gigantochloa scortechinii	685.51 [A] (48.81)	7.5 [B] (0.95)	0.69 [A] (0.04)
	Gigantochloa levis	689.91 [A] (103.33)	7.2 [B] (1.9)	0.69 [A] (0.1)
Treated	*Dendrocalamus asper*	623.17 [B] (54.35)	9.5 [A] (0.98)	0.62 [B] (0.05)
	Bambusa vulgaris	682.1 [A] (62.71)	7.2 [B] (0.87)	0.68 [A] (0.06)
	Gigantochloa scortechinii	691.86 [A] (38.34)	7.1 [B] (1.55)	0.69 [A] (0.05)
	Gigantochloa levis	701.7 [A] (101.42)	6.9 [C] (1.7)	0.7 [A] (0.1)

Note: Values in ( ) are the standard deviation. Means followed with the same letters [A,B,C] in the same column were not significantly different at $p \leq 0.05$.

The moisture excluding efficiency (MEE) was calculated from the EMC to evaluate the hydrophobic or hydrophilic properties of the material at high humidity [37]. Figure 5 depicts the effects of boric acid treatment on the MEE values. According to Figure 5, the MEE for treated bamboo strips was higher than for untreated strips. For *D. asper*, *B. vulgaris*, *G. scortechinii* and *G. levis*, the results were 7.98%, 4.34%, 5.62%, and 3.15% for untreated strips and 9.52%, 7.08%, 5.60%, and 3.64% for treated strips. The results show that there was no significant difference between the untreated and treated strips of *G. scortechinii* and *G. levis* when compared to *D. asper* and *B. vulgaris*. However, bamboo strips treated with boric acid become more hydrophobic when compared to untreated strips by an average of 12%–20%. *D. asper* had the highest MEE in this study, indicating that it is more resistant to moisture uptake than the other species. All of the MEE values of the bamboo strips increased after being treated with boric acid, regardless of species. Although *D. asper* maintained its superiority, *B. vulgaris* improved significantly in resisting moisture uptake. This characteristic may have a significant impact on bamboo bonding and coating.

The water absorption and thickness swelling of the bamboo strips for the four bamboo species are shown in Table 5. A different pattern was observed for both water absorption and thickness swelling, with the treated strips having the lowest values. After soaking in water for 24 h, the untreated strips absorbed 2%–5% more water and swelled 1%–2% more than the treated strips. Meanwhile, treated bamboo strips from *D. asper*, *G. scortechinii*, *B. vulgaris*, and *G. levis* outperformed the untreated bamboo strips by 5.52%, 4.13%, 3.31% and 2.04%, respectively. The bamboo strips with a higher density had a lower WA value and a higher TS value. Biswas et al. [38] found similar trends in both WA and TS for bamboo strips, which supported this finding. A study by Borthakur and Gogoi [39] stated that the bulking of the cell wall confers better dimension stabilization in bamboo. *D. asper* had the least swelling, which could be attributed to its lower density. As evidenced by the high WA of *D. asper*, low density bamboo has more void volume and, thus, can accommodate more water. The reduction in WA could be attributed to the precipitation of boric acid on the bamboo and the reduced availability of voids for water absorption [40]. The reduction of thickness swelling and water absorption of the treated bamboo strips could confer better dimensional stability to the laminated products.

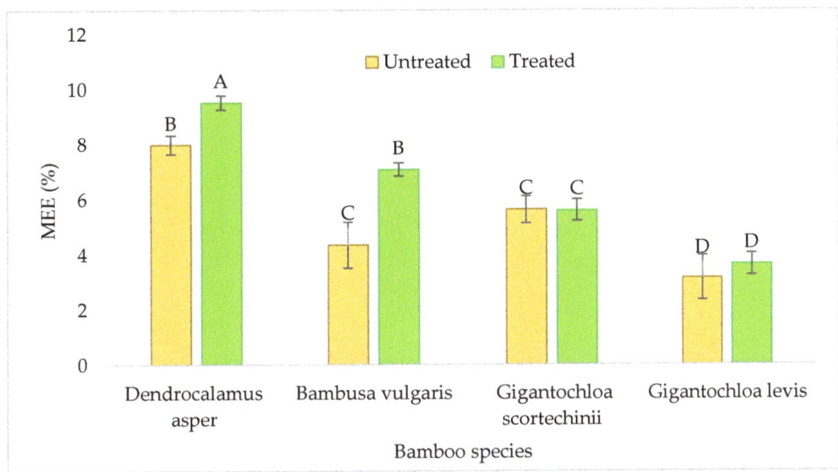

Figure 5. Moisture Excluding Efficiency (MEE) of untreated and treated bamboo strips for four selected species. Note: Bars of mean followed with the same letters A,B,C,D were not significantly different at $p \leq 0.05$.

Table 5. Water absorption and thickness swelling of untreated and treated bamboo strips for four selected species.

Treatment	Bamboo Species	Untreated	Treated with 5% Boric Acid	% Reduction
	Dendrocalamus asper	53.89 [A]	48.37 [A]	5.52
	Bambusa vulgaris	48.98 [A]	45.67 [B]	3.31
Water absorption (%)	*Gigantochloa scortechinii*	41.51 [B]	37.38 [C]	4.13
	Gigantochloa levis	39.55 [C]	37.51 [C]	2.04
	Dendrocalamus asper	5.01 [C]	3.95 [C]	1.06
	Bambusa vulgaris	7.89 [A]	6.56 [B]	1.33
Thickness swelling (%)	*Gigantochloa scortechinii*	6.44 [B]	4.16 [C]	2.28
	Gigantochloa levis	5.49 [B]	4.11 [C]	1.38

Note: Means followed with the same letters [A,B,C] in the same column were not significantly different at $p \leq 0.05$.

3.4. Adhesion Properties of Untreated and Treated Bamboo Strips

3.4.1. Wettability

The contact angle is formed between the surface and a liquid, and this provides useful information about how well an adhesive wets, spreads, and penetrates into wood samples, according to Paridah et al. [41]. In this study, all of the contact angles reached less than 30° after 60 s (Figure 6). Based on the smaller contact angle after 60 s, treated bamboo strips appear to have better wettability than untreated bamboo strips, implying that the surface of treated bamboo strips is more wettable and hydrophilic [42]. Boric acid treatment increased the surface roughness of the bamboo, most likely due to raised fibres on the bamboo surface. Higher roughness causes increased wettability [31], which explains the increased wettability in treated bamboo. Lesar et al. [43] discovered that wood treated with boric acid is more hygroscopic than untreated wood. When compared to other bamboo species, *G. levis* appears to be the least wettable, while *B. vulgaris* appears to be the most wettable. It took more than 40 s for *D. asper* to reach the contact angle obtained for other species. The increased wettability of the bamboo strips after boric acid treatment may benefit the bonding quality of laminated bamboo boards by allowing the adhesive to spread more easily.

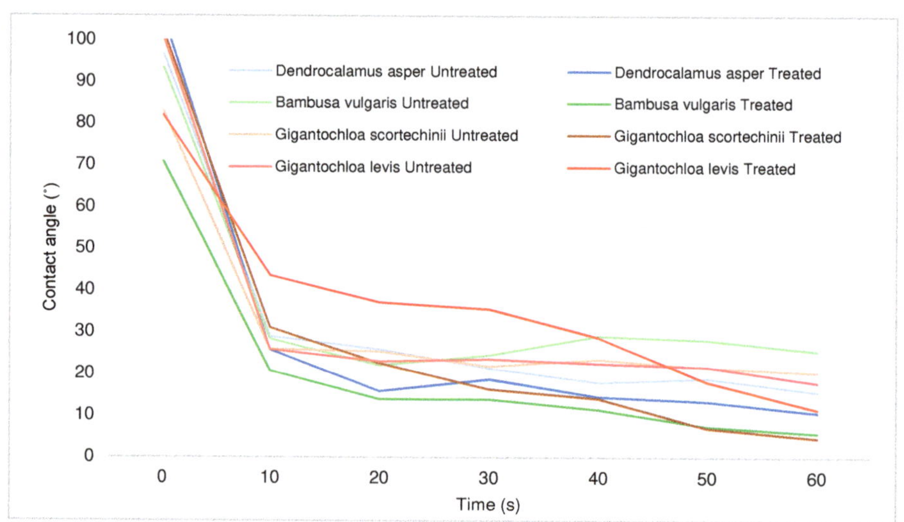

Figure 6. Changes in the contact angle of untreated and treated bamboo strips for four selected species.

3.4.2. Buffering Capacity of Bamboo Strips

Buffering capacity is the resistance of wood and non-wood materials to changes in pH. Understanding the pH value and buffering capacity is critical because they influence the curing behavior of resin [31,41,42]. The pH and buffering capacity values of untreated and treated strips from the four bamboo species are shown in Table 6. As shown in Table 6 and Figures 7 and 8, all of the untreated bamboo appeared to be more stable in acidic conditions than in alkaline conditions because this required more HCl (20–40 mL) to reach pH 3.0 rather than NaOH (12–27 mL) to reach pH 10.0. In contrast, treated bamboo is more stable in alkaline conditions than in acidic conditions because a greater amount of NaOH (43–61 mL) was used compared to HCl (15–25 mL) to achieve a pH 10.0 and 3.0, respectively.

3.5. Mechanical Properties of Untreated and Treated Bamboo Strips

Table 7 shows the mechanical properties of the treated and untreated bamboo strips of four bamboo species. Treated strips appears to be much stronger compared to untreated strips. Treatment with boric acid increased the MOR, MOE, and maximum load in bending, as well as the compressive strength and the maximum load parallel to the grain. *G. scortechinii* showed superior bending properties, while *D. asper* was the least strong irrespective of the treatment. Similarly, concerning compression parallel to the grain, *G. scortechinii* tops other species in both compressive strength and maximum load. *D. asper* remained the weakest among the four species. The strength values for *G. scortechinii* and *G. levis* were rather close, which may be due to being in the same genus and having a similar structure of wall thickness among the bamboo species, and the internode lengths were longer than those for *D. asper* and *B. vulgaris*, even though the wall thickness for both bamboo species, from the bottom to the top portions, were the greatest [35]. The increased strength value can be correlated with the formation of boric acid (salt crystals) within the bamboo's microstructure, which may aid in accommodating the applied force during loading [44]. Nevertheless, some studies did report the negative effects of boric acid treatment towards the mechanical strength of bamboo [25,26]. It could be because water-based preservative formulations undergo hydrolytic reduction when they come into contact with wood sugars, causing a reaction with the cell wall components. Wood strength may be diminished due to fixation, a process that oxidizes the components of wood cell walls [45]. However, the goal of the boric acid treatment in this study is only to provide the bamboo with short-term protection against biodegradation agents. The concentration of boric acid used was relatively low, at 5%, and the WPG recorded was also low, ranging from 1.47 to 3.61% (Table 3). Therefore, the increase in MOR and MOE could be attributed, primarily, to an increase in the density of the treated bamboo strips.

Table 6. The pH and buffering capacity values of untreated and treated strips from four bamboo species.

Treatment	Species	pH	Acid * 0.1 N (HCl) (mL)	Alkali ** 0.1 N (NaOH) (mL)
Untreated	*Dendrocalamus asper*	6.48	39 [A]	22 [D]
	Bambusa vulgaris	5.92	21 [D]	27 [D]
	Gigantochloa scortechinii	6.02	19 [D]	19 [D]
	Gigantochloa levis	6.12	29 [B]	12 [E]
Treated with 5% Boric Acid	*Dendrocalamus asper*	6.24	24 [C]	51 [B]
	Bambusa vulgaris	6.32	21 [D]	61 [A]
	Gigantochloa scortechinii	6.26	14 [E]	43 [C]
	Gigantochloa levis	5.47	21 [C]	56 [B]

Notes: Mean value of 3 samples. * The amount of acid to reach pH 3.0. ** The amount of alkali to reach pH 10.0 Mean followed by the same letters ([A–E]) in the same column were not significantly different at $p \leq 0.05$.

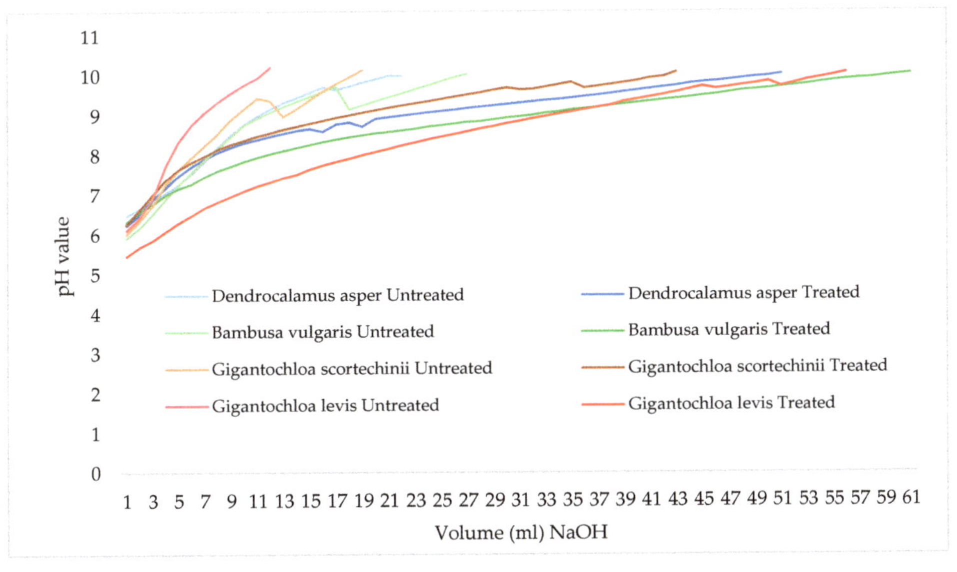

Figure 7. Stability of untreated and treated bamboo strips of four selected species in alkaline conditions.

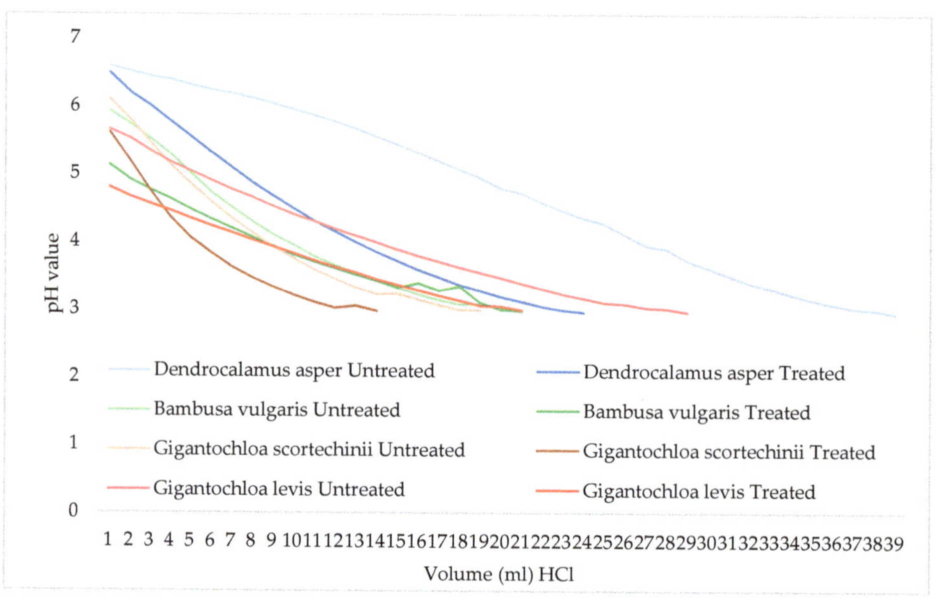

Figure 8. Stability of untreated and treated bamboo strips of four selected species in acidic conditions.

Table 7. Mean values for the mechanical properties of untreated and treated bamboo strips from different species.

Treatment	Species	Mean Values				
		Static Bending			Compression Parallel to the Grain	
		MOR (N/mm^2)	MOE (N/mm^2)	Max Load (N)	Compressive Strength (N/mm^2)	Max Load (N)
Untreated	Dendrocalamus asper	129.01 [B]	16,555.33 [D]	183.19 [B]	41.85 [C]	4754.66 [C]
	Bambusa vulgaris	130.23 [B]	15,402.88 [D]	169.23 [B]	42.67 [B]	4781.19 [B]
	Gigantochloa scortechinii	152.94 [A]	20,776.79 [B]	196.78 [A]	42.70 [B]	4432.90 [B]
	Gigantochloa levis	137.89 [B]	16,464.49 [D]	181.13 [B]	46.99 [B]	4998.34 [B]
Treated	Dendrocalamus asper	133.37 [B]	17,405.28 [C]	185.27 [B]	39.95 [C]	4574.63 [C]
	Bambusa vulgaris	156.88 [A]	17,503.49 [C]	201.10 [A]	46.71 [B]	5278.52 [B]
	Gigantochloa scortechinii	169.72 [A]	22,486.64 [A]	222.93 [A]	62.75 [A]	7112.48 [A]
	Gigantochloa levis	161.53 [A]	18,700.80 [C]	211.41 [A]	49.41 [B]	5402.29 [B]

Note: Means followed with the same letters [A,B,C,D] in the same column were not significantly different at $p \leq 0.05$.

3.6. Failure Behavior of Bamboo Strips in Mechanical Testing

The failure of bamboo strips loaded in static bending, for untreated and treated bamboo, is depicted in Figure 9. The bamboo fails in in two ways: (a) brittle shear tension mode, and (b) splintering tension mode. Brittle shear tension failure was caused primarily by extensive longitudinal shear in the node section and occurred in all portions of the node. Meanwhile, splintering tension failure was caused by splintering failure at the radial edge of the lower layer of the specimen and was common in the internodes of all portions. This finding is consistent with previous research using *Gigantochloa scortechinii* bamboo strips [46]. The bending failure for both the untreated and treated bamboo strips appeared in the node of the strips and in the lower layer, for which the failure in both the parenchyma and vascular bundles regions was due to the mechanical load applied to the specimens. The parenchyma and vascular bundles were crushed and split extensively in the node, whereas in the internodes, the failure only occurred in the parenchyma without any failure in the vascular bundles.

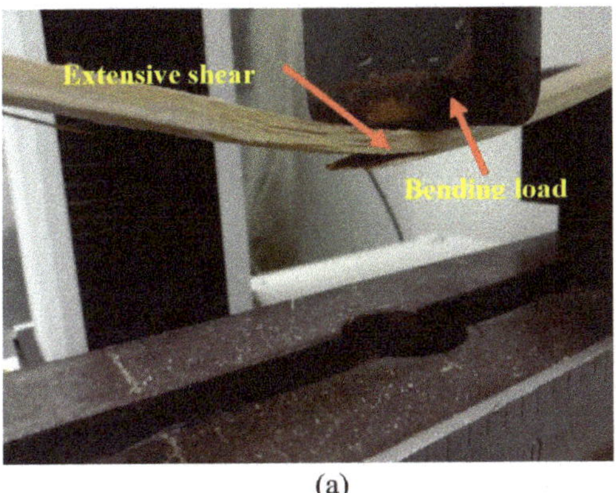

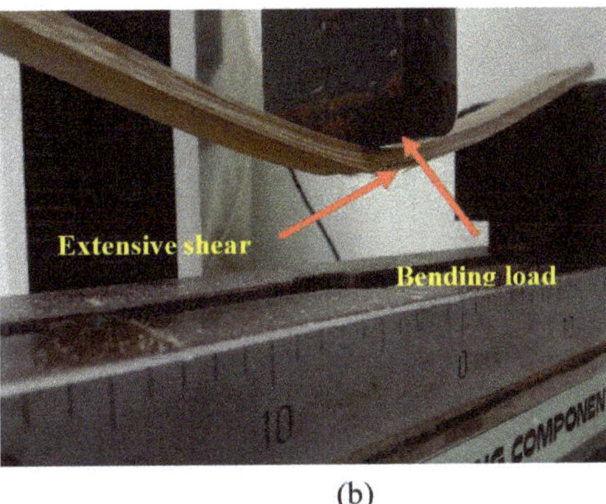

(a)   (b)

Figure 9. Failure mode of strips from four bamboo species loaded in static bending: (**a**) brittle shear tension mode, (**b**) splintering tension mode.

Figure 10 depicts the failure in compression of the bamboo strips, with various failure behaviors observed in both the untreated and treated strips. The failure is illustrated in Figure 10a, showing crushing from the bottom end and shear splitting in the middle and upwards; the shear splitting propagated above the crushed area and upwards, due to a weakness in compression. Figure 10b shows, on the other hand, that the crack tended to form from the bottom end and from the middle, upwards, and sheared perpendicularly to the maximum shear plane along 45° but terminated below the nodal area. At the node, the crack occurred from the bottom end and sheared perpendicularly, while for the internode, the crushing started from the bottom end and the shear splitting occurred in the middle, upwards. After about 80% of the stress load, creases appeared on the compression side of the bamboo. This could be attributed to the vascular bundle's thick polylamellated layer of fibre and the high percentage of parenchyma found in stress-resistant bamboo complements. The frequency of occurrence of these failures in bending and compression was nearly identical for both treated and untreated bamboo strips, indicating that the boric acid treatment had no effect on bamboo strip failure behavior. Table 7 shows that, despite the increased bending and compression of the treated bamboo strips, the values are not statistically different from the untreated bamboo strips, with the exception of compression in treated *G. scortechinii*.

(a)

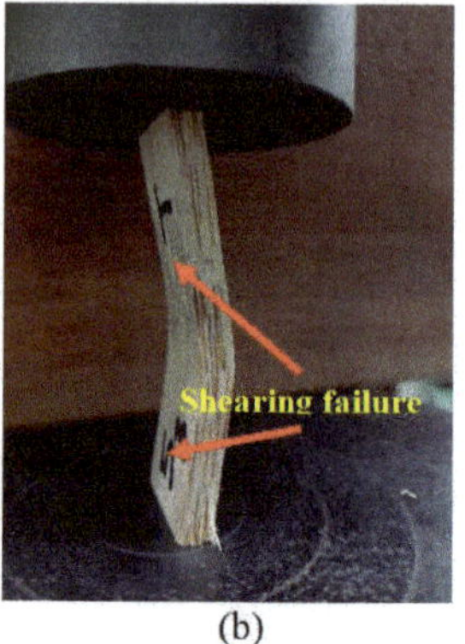

(b)

Figure 10. Failure mode of strips from four bamboo species for compression parallel to the grain: (**a**) failure due to crushing, (**b**) shearing failure in the bamboo strip.

4. Conclusions

The present study determines the effect of boric acid treatment on the physical, mechanical, adhesion and morphological properties of strips from four bamboo species *(Dendrocalamus asper, Bambusa vulgaris, Gigantochloa scortechinii* and *Gigantochloa levis)*. All four species of bamboo can be treated easily with boric acid, with a reasonable DSR and WPG. Based on the DSR and WPG values, the treatability of these bamboo species ranked (in descending order) as follows: *B. vulgaris, D. asper, G. levis* and *G. scortechinii*. In general, treatment with boric acid increased both the density and specific gravity slightly. *D. asper* was the most affected by the treatment. Treatment with boric acid reduced the WA and TS of the bamboo strips significantly. Based on the WA and TS results, *D. asper* was the most stable (high WA, low TS) while the most unstable was *B. vulgaris* (high WA, high TS). *G. scortechinii* showed superior bending properties, while *D. asper* was the least strong, irrespective of treatment. Similarly, concerning compression parallel to the grain, *G. scortechinii* tops other species in both compressive strength and maximum load. *D. asper* remained the weakest among the four species. Compared to *B. vulgaris* and *D. asper*, *G. levis* possessed acceptable strength next to *G. scortechinii*. Based on the overall properties, and despite having relatively poorer treatability and stability, both *G. scortechinii* and *G. levis* had greatly superior mechanical properties, which warrants these species being considered for the production of laminated products. Boric acid treatment was found to provide several benefits to bamboo strips intended for lamination, including increased wettability, dimensional stability, and mechanical strength.

Author Contributions: Conceptualization, L.S.H. and P.M.T.; methodology, P.M.T.; software, S.S.O.A.-E.; validation, A.S.M.R., L.K. and R.R.; formal analysis, M.A.R.L. and W.F.; investigation, N.M.Y., R.M.S.J. and R.D.; resources, P.M.T.; data curation, N.M.Y., R.M.S.J. and R.D.; writing—original draft preparation, N.M.Y.; writing—review and editing, L.S.H., L.K., M.A.R.L., W.F. and R.R.; visualization, S.S.O.A.-E.; supervision, L.S.H. and P.M.T.; project administration, L.S.H. and P.M.T.; funding acquisition, L.S.H. and P.M.T. All authors have read and agreed to the published version of the manuscript.

Funding: The presented work is supported by the Fundamental Research Grant Scheme (FRGS 2019-1), Reference code: FRGS/1/2019/WAB07/UPM/02/3; by the Ministry of Higher Education, Malaysia; a Higher Institution Centre of Excellence (HICoE) grant; and an RMK11 collaboration with the Malaysian Timber Industry Board (MTIB).

Data Availability Statement: Not applicable.

Acknowledgments: The authors also wish to express gratitude to the Faculty of Forestry and Environment, Universiti Putra Malaysia, for providing the mechanical testing facilities. This work was also supported by the Slovak Research and Development Agency under contracts No. APVV-19-0269 and No. SK-CZ-RD-21-0100.

Conflicts of Interest: The authors declare no conflict of interest.

References

1. Zhang, H.; Qian, Z.; Zhuang, S. Effects of soil temperature, water content, species, and fertilization on soil respiration in bamboo forest in subtropical China. *Forests* **2020**, *11*, 99. [CrossRef]
2. Karimah, A.; Ridho, M.R.; Munawar, S.S.; Amin, Y.I.; Damayanti, R.; Lubis, M.A.R.; Wulandari, A.P.; Iswanto, A.H.N.; Fudholi, A.; Asrofi, M.; et al. A Comprehensive Review on Natural Fibres: Technological and Socio-Economical Aspects. *Polymers* **2021**, *13*, 4280. [CrossRef] [PubMed]
3. Fatriasari, W.; Syafii, W.; Wistara, N.J.; Syamsu, K.; Prasetya, B. The Characteristic Changes of Betung Bamboo (Dendrocalamus asper) Pretreated by Fungal Pretreatment. *Int. J. Renew. Energy Dev.* **2014**, *3*, 133–143. [CrossRef]
4. Sharma, B.; Gatoo, A.; Bock, M.; Ramage, M. Engineered bamboo for structural applications. *Constr. Build. Mater.* **2015**, *81*, 66–73. [CrossRef]
5. Xiao, Y.; Yang, R.Z.; Shan, B. Production, environmental impact and mechanical properties of glubam. *Constr. Build. Mater.* **2013**, *44*, 765–773. [CrossRef]
6. Wang, X.; Cheng, K.J. Effect of glow-discharge plasma treatment on contact angle and micromorphology of bamboo green surface. *Forests* **2020**, *11*, 1293. [CrossRef]
7. Huang, X.Y.; Qi, J.Q.; Xie, J.L.; Hao, J.F.; Qin, B.D.; Chen, S.M. Variation in anatomical characteristics of bamboo, Bambusa rigida. *Sains Malays.* **2015**, *44*, 17–23.
8. Hanim, A.R.; Zaidon, A.; Anwar, U.M.K.; Rafidah, S. Effects of chemical treatments on durability properties of Gigantochloa scortechinii strips and ply-bamboo. *Asian J. Biol. Sci.* **2013**, *6*, 153–160. [CrossRef]
9. Hamdan, H.; Anwar, U.M.K.; Zaidon, A.; Tamizi, M.M. Mechanical properties and failure behaviour of *Gigantochloa scortechinii*. *J. Trop. For. Sci.* **2009**, *21*, 336–344.
10. Yang, S.; Li, H.; Fei, B.; Zhang, X.; Wang, X. Bond quality and durability of cross-laminated flattened bamboo and timber (CLBT). *Forests* **2022**, *13*, 1271. [CrossRef]
11. Bakar, E.S.; Nazip, M.N.M.; Anokye, R.; Lee, S.H. Comparison of three processing methods for laminated bamboo timber production. *J. For. Res.* **2019**, *30*, 363–369. [CrossRef]
12. Ashaari, Z.; Lee, S.H.; Zahali, M.R. Performance of compreg laminated bamboo/wood hybrid using phenolic-resin-treated strips as core layer. *Eur. J. Wood Wood Prod.* **2016**, *74*, 621–624. [CrossRef]
13. Mohmod, A.L.; Ariffin, W.T.W.; Ahmad, F. Anatomical features and mechanical properties of three Malaysian bamboos. *J. Trop. For. Sci.* **1990**, *2*, 227–234.
14. Mohmod, A.L.; Amin, A.H.; Kasim, J.; Jusuh, M.Z. Effects of anatomical characteristics on the physical and mechanical properties of Bambusa blumeana. *J. Trop. For. Sci.* **1993**, *6*, 159–170.
15. Wahab, R.; Mustafa, M.T.; Salam, M.A.; Sudin, M.; Samsi, H.W.; Rasat, M.S.M. Chemical composition of four cultivated tropical bamboo in genus Gigantochloa. *J. Agric. Sci.* **2013**, *5*, 66. [CrossRef]
16. Wahab, R.; Mohammed, A.S.; Aminuddin, M.; Samsi, H.W.; Shafiqur, R.; Mohd, S.; Izyan, K. Evaluation on strength of chemically treated 2 and 4 year-old bamboo *Bambusa vulgaris* through pressurized treatment. *Res. J. Pharm. Biol. Chem. Sci.* **2015**, *6*, 1282–1288.
17. Kasim, J.; Ahmad, A.J.; Mohamod, A.L.; Khoo, K.C. A note on the proximate chemical composition and fibre morphology of *Bambusa vulgaris*. *J. Trop. For. Sci.* **1994**, *6*, 356–358.
18. Nurhazwani, O.; Jawaid, M.; Tahir, P.M.; Abdul, J.H.; Hamid, S.A. Hybrid particleboard made from bamboo (Dendrocalamus asper) veneer waste and rubberwood (*Hevea brasilienses*). *BioResources* **2016**, *11*, 306–323. [CrossRef]
19. Wong, J.W.; Yeo, W.S.; Paridah, M.T. Sound absorption properties of kenaf bamboo particleboard at various mixing ratio and density. *AIP Conf. Proc.* **2017**, *1902*, 020001.
20. Siam, N.A.; Uyup, M.K.A.; Husain, H.; Awalludin, M.F. Anatomical, physical, and mechanical properties of thirteen Malaysian bamboo species. *BioResources* **2019**, *14*, 3925–3943. [CrossRef]
21. Luo, C.; Li, Y.; Chen, Y.; Fu, C.; Nong, X.; Yang, Y. Degradation of bamboo lignocellulose by bamboo snout beetle Cyrtotrachelus buqueti in vivo and vitro: Efficiency and mechanism. *Biotechnol. Biofuels* **2019**, *12*, 75. [CrossRef]
22. Satish, K.; Dobriyal, P.B. Preservative treatment of bamboo for structural uses. In Proceedings of the Bamboos. Current Research. International Bamboo Workshop, Cochin, India, 14–18 November 1990; Kerala Forest Research Institute: Kerala, India, 1990; pp. 199–206.
23. Baharuddin, N.; Lee, S.H.; Uyup, M.K.A.; Tahir, P.M. Effect of Preservative Treatment on Physical and Mechanical Properties of Bamboo (*Gigantochloa scortechinii*) Strips. *BioResources* **2022**, *17*, 5129–5145. [CrossRef]
24. Pizzi, A.; Baecker, A. A new boron fixation mechanism for environment friendly wood preservatives. *Holzforschung* **1996**, *50*, 507–510. [CrossRef]

25. Colakoglu, G.; Colak, S.; Aydin, I.; Yildiz, U.C.; Yildiz, S. Effects of boric acid treatment on mechanical properties of laminated beech veneer lumber. *Silva Fenn.* **2003**, *37*, 505–510. [CrossRef]
26. Ayrilmis, N.; Kartal, S.N.; Laufenberg, T.L.; Winandy, J.E.; White, R.H. Physical and mechanical properties and fire, decay, and termite resistance of treated oriented strandboard. *For. Prod. J.* **2005**, *55*, 74–81.
27. Simsek, H.; Baysal, E. Some physical and mechanical properties of borate-treated oriental beech wood. *Drv. Ind.* **2015**, *66*, 97–103. [CrossRef]
28. Paridah, M.T.; Chin, A.M.E.; Zaidon, A. Bonding properties of *Azadirachta excelsa*. *J. Trop. For. Prod.* **2001**, *7*, 161–171.
29. Nkeuwa, W.N.; Zhang, J.; Semple, K.E.; Chen, M.; Xia, Y.; Dai, C. Bamboo-based composites: A review on fundamentals and processes of bamboo bonding. *Compos. Part B Eng.* **2022**, *2022*, 109776. [CrossRef]
30. Li, J.; Zheng, H.; Sun, Q.; Han, S.; Fan, B.; Yao, Q.; Jin, C. Fabrication of superhydrophobic bamboo timber based on an anatase TiO_2 film for acid rain protection and flame retardancy. *RSC Adv.* **2015**, *5*, 62265–62272. [CrossRef]
31. Redzuan, M.S.J.; Paridah, M.T.; Anwar, U.M.K.; Juliana, A.H.; Lee, S.H.; Norwahyuni, M.Y. Effects of Surface Pretreatment on Wettability of *Acacia Mangium* Wood. *J. Trop. For. Sci.* **2019**, *31*, 249–258. [CrossRef]
32. *BS EN 373*; Methods of Testing Small Clear Specimens of Timber. BSI: London, UK, 1957.
33. Yamauchi, S. Raman spectroscopic study on the behavior of boric acid in wood. *J. Wood Sci.* **2003**, *49*, 227–234. [CrossRef]
34. Yamauchi, S.; Sakai, Y.; Watanabe, Y.; Kubo, M.K.; Matsue, H. Distribution of boron in wood treated with aqueous and methanolic boric acid solutions. *J. Wood Sci.* **2007**, *53*, 324–331. [CrossRef]
35. Zakikhani, P.; Zahari, R.; Sultan, M.T.B.H.H.; Majid, D.L.A.A. Morphological, mechanical, and physical properties of four bamboo species. *BioResources* **2017**, *12*, 2479–2495. [CrossRef]
36. Darwis, A.; Iswanto, A.H. Morphological Characteristics of *Bambusa vulgaris* and the Distribution and Shape of Vascular Bundles therein. *J. Korean Wood Sci. Technol.* **2018**, *46*, 315–322. [CrossRef]
37. Lee, S.H.; Lum, W.C.; Zaidon, A.; Maminski, M. Microstructural, mechanical and physical properties of post heat-treated melamine-fortified urea formaldehyde-bonded particleboard. *Eur. J. Wood Wood Prod.* **2015**, *73*, 607–616. [CrossRef]
38. Biswas, D.; Bose, S.K.; Hossain, M.M. Physical and mechanical properties of urea formaldehyde-bonded particleboard made from bamboo waste. *Int. J. Adhes. Adhes.* **2011**, *31*, 84–87. [CrossRef]
39. Borthakur, R.D.; Gogoi, P. Studies of Dimensional Stability, Thermal Stability and Biodegradation Resistance Capacity of Chemically Treated Bamboo. *Asian J. Biochem.* **2014**, *9*, 16–29. [CrossRef]
40. Priadi, T.; Lestari, M.D.; Cahyono, T.D. Posttreatment Effects of Castor Bean Oil and Heating in Treated Jabon Wood on Boron Leaching, Dimensional Stability, and Decay Fungi Inhibition. *J. Korean Wood Sci. Technol.* **2021**, *49*, 602–615. [CrossRef]
41. Paridah, M.T.; Hafizah, A.N.; Zaidon, A.; Azmi, I.; Nor, M.M.; Yuziah, M.N. Bonding properties and performance of multi-layered kenaf board. *J. Trop. For. Sci.* **2009**, *21*, 113–122.
42. He, G.; Yan, N. Effect of wood on the curing behavior of commercial phenolic resin systems. *J. Appl. Polym. Sci.* **2005**, *95*, 185–192. [CrossRef]
43. Lesar, B.; Straže, A.; Humar, M. Sorption properties of wood impregnated with aqueous solution of boric acid and montan wax emulsion. *J. Appl. Polym. Sci.* **2011**, *120*, 1337–1345. [CrossRef]
44. Gauss, C.; Kadivar, M.; Savastano, H., Jr. Effect of disodium octaborate tetrahydrate on the mechanical properties of Dendrocalamus asper bamboo treated by vacuum/pressure method. *J. Wood Sci.* **2019**, *65*, 27. [CrossRef]
45. Winandy, J.E.; Margaret, E.H. *Wood Protection Techniques and the Use of Treated Wood in Construction*; Forest Products Research: Madison, WI, USA, 1988; pp. 54–62.
46. Bahari, S.A.; Ahmad, M. Failure Behaviour of Semantan Bamboo Strips Loaded in Bending and Shear. *Key Eng. Mater. Trans Technol. Publ.* **2011**, *462–463*, 1176–1181.

forests

Article

Advisability-Selected Parameters of Woodworking with a CNC Machine as a Tool for Adaptive Control of the Cutting Process

Richard Kminiak [1], Miroslav Němec [2,*], Rastislav Igaz [2] and Miloš Gejdoš [3]

1. Department of Woodworking, Faculty of Wood Sciences and Technology, Technical University in Zvolen, Tomáš Garrigue Masaryka 24, 960 01 Zvolen, Slovakia
2. Department of Physics, Electrical Engineering and Applied Mechanics, Faculty of Wood Sciences and Technology, Technical University in Zvolen, Tomáš Garrigue Masaryka 24, 960 01 Zvolen, Slovakia
3. Department of Forest Harvesting, Logistics and Ameliorations, Faculty of Forestry, Technical University in Zvolen, Tomáš Garrigue Masaryka 24, 960 01 Zvolen, Slovakia
* Correspondence: nemec@tuzvo.sk; Tel.: +421-4-55206474

Abstract: The operation of CNC machining centers, despite their technological progress, can still be affected by undesirable events associated with the technological parameters of their operation. The minimization of these risks can be achieved via their adaptive control in the process of operation. Several input parameters for adaptive control are still the subject of research. The work aimed to find out the influence of the change in feed speed, revolutions, and radial depth of cut on the noise and temperature of the tool during the milling of wood-based composite material particleboard. At the same time, it was evaluated whether it is possible to use the measured values of these parameters in the future in the process of an adaptive control of the CNC machine with the minimization of their negative influence. The methods of measuring these parameters were chosen based on valid legislation and previous research. The results of the research show that all parameters influence both the noise and temperature of the tool, while the rate of the radial depth of cut has the greatest influence on the increase in temperature, and the noise is most affected by the revolutions. The effect of temperature during woodworking can also be characterized in terms of the potential long-term wear of the cutting tool. The setting of optimization algorithms of monitored parameters in the adaptive control of the CNC machining center will be the subject of further research.

Keywords: CNC machines; acoustics emission; wood-processing parameters; tool temperature; adaptive control

Citation: Kminiak, R.; Němec, M.; Igaz, R.; Gejdoš, M. Advisability-Selected Parameters of Woodworking with a CNC Machine as a Tool for Adaptive Control of the Cutting Process. *Forests* **2023**, *14*, 173. https://doi.org/10.3390/f14020173

Academic Editor: Stefano Grigolato

Received: 22 December 2022
Revised: 13 January 2023
Accepted: 14 January 2023
Published: 17 January 2023

1. Introduction

Universality, high precision, and product quality achieved through Computerized Numerical Control machining centers (CNC machines) have caused their dynamic development in practically all production spheres of industrial production [1–3]. These devices are also financially available for small- and medium-sized enterprises. The producers of these machines are looking for modern approaches to increase productivity and efficiency [4]. Modern CNC control units are designed to deal with obvious threats such as overload (via various sensor systems) and collision (by checking the syntax of the NC code and simulating it on the machine). However, despite these safety systems, various undesirable events still occur during the machining with CNC machining centers. This means that there are still risks that can seriously affect the operation of CNC machining centers [5]. Minimization of risks in the operation of CNC machines can be achieved by several approaches. A modern approach in this direction also represents the adaptive control of CNC machines based on achieved parameters in the process of operation, using neural networks and applying fuzzy logic approaches [6–10]. Since the machines work in different operating conditions with different materials, the risks of their operation can differ significantly. However, the basic risk factors from the point of view of machine operators include noise, vibrations, and from the point of view of their operation, the temperature of the cutting tool and the machine [11,12]. When the CNC machine is turned on, the working spindle with the tool, which rotates at a high cutting speed, produces noise. When in contact with the material, this noise increases significantly. It is also important to how the machine works and what material it processes. For example, in climb milling, the noise is greater than in conventional milling. The feed speed also has a great influence on the noise of the CNC machine [13–15].

The implementation of the Industry 4.0 philosophy, as well as the need to increase the efficiency of production processes, creates pressure for an increasingly greater degree of automation and leads to the minimization of the worker's role in the working process. It is becoming standard that one operator serves several production facilities and they perform production operations autonomously. The worker performs only supervision, and his main role has become interoperation control. Here, space is created for the application of control and safety mechanisms that will replace the original role of the operator-to eliminate the occurrence of unfavorable situations during the autonomous operation of the production equipment itself.

One of the ways to solve this problem is the collection of all available continuously measurable physical values and their subsequent interpretation using cyber-physical systems. The physical value that reacts more sensitively and faster to changes in the given process has a higher informative value. Our work aimed to assess the response of the tool temperature and the noise of the machining process to the change in the parameters of the cut layer.

Problems with the operation and damage of tools, or in the case of wood and wooden materials, the deterioration of the machined material also, can also be caused by high temperature, which arises as a result of high revolutions, cutting speed, feed, tool and workpiece materials, etc. In this direction, the temperature distribution on individual parts of the machine, tool, and workpiece, or the use of coolant, is important. Keeping the tool temperature below the critical level consequently extends its life [16–18]. Part of the mechanical energy consumed during cutting and milling is converted into thermal energy due to friction [19]. Unlike metals, the thermal conductivity of wood is low. Therefore, most of the heat generated during milling is transferred to the tool, which leads to its increased temperature [20–22]. The influence of the temperature of the tool during the cutting process on the service life of the tool was dealt with by many authors who discovered the influence of several parameters of the machining process, the properties of the tool, and the material being machined. Some dealt with the issue experimentally, others analytically or numerically, e.g., [23–25].

The current trend of setting and programming algorithms for CNC machines is for the machine to be able to evaluate the critical values of the given parameter and accordingly adjust the production cycle by adjusting the operating conditions. These conditions are met using the process of adaptive control [26,27]. The first step for the implementation and creation of an adaptive control system is the interpretation of the measured critical parameters into the creation of a structural model [28]. Such a system also helps in optimizing the time schedule of preventive maintenance of the machine with an effective cost allocation [29,30].

Published works in this area focus mainly on research into the factors of wear of cutting tools with selected cutting parameters. At the same time, adaptive control currently relies mainly on mathematical models and not on direct evaluation of critical parameters of the cutting process.

The aim of the work was to determine the influence of process parameters affecting the size of the cut layer (revolutions, feed speed, and radial depth of cut) on the temperature of the cutting tool and sound pressure level (A)-SPL (A) during the milling of the wood-composite particleboard (PB). At the same time, it was evaluated whether the obtained measured values of these parameters can be used in the future in the process of adaptive control of the CNC machine with the minimization of their negative impact.

2. Materials and Methods

2.1. Used Material and CNC Machine

Particleboards (PB) supplied by the company Kronospan Ltd., Zvolen, Slovakia, were used in the experiment. The density of the material given by the manufacturer is 600 kg·m^{-3} – 640 kg·m^{-3} (deciduous 10%, coniferous 90%), and urea formaldehyde glue with paraffin admixture is used. The manufacturer declares that the material complies with the EN 14322 standard, EN 312-2, and emission class E1 (EN ISO 12460-5). These PB had a thickness $h = 18$ mm, width $w = 2800$ mm, length $l = 2070$ mm. Within the experiment, a modified particleboard format of 1000 mm × 1000 mm was used. For each combination of considered parameters, 5 samples were used [31–33].

CNC machine specification:

The experiment took place on a 5-axis CNC machining center SCM Tech Z5 (Figure 1). Table 1 provides the basic technical–technological parameters given by the manufacturer of this machine.

Figure 1. CNC machining center SCM Tech Z5.

Table 1. Technical parameters of CNC machining center SCM Tech Z5.

Technical Parameters of CNC Machining Center SCM Tech Z5	
Useful desktop (mm)	X = 3050, y = 1300, z = 3000
Speed X axis (m·min^{-1})	$0 \div 70$
Speed Y axis (m·min^{-1})	$0 \div 40$
Speed Z axis (m·min^{-1})	$0 \div 15$
Vector rate (m·min^{-1})	$0 \div 83$
Technical Parameters of the Main Spindle–Electric Spindle with HSK F63 Connection	
Rotation axis C	640°
Rotation axis B	320°
Revolutions (rpm)	$600 \div 24{,}000$
Power (kW)	11
Maximum tool dimensions (mm)	D = 160 L = 180

Tool Characteristics:

The experiment used a diamond-end mill IGM Fachmann Ekonomik Z2+1 from the manufacturer of IGM tools from the Czech Republic (Figure 2). *Milling-Cutter parameters:* working diameter 16 mm, working length 26 mm, the diameter of the clamping head 16 mm, the direction of rotation-right, maximum permissible revolutions 24,000 rpm^{-1}. This milling cutter was installed in hydraulic clamp SOBO 302680291 GM 300 HSK 63F, manufactured by Gühring KG from Germany.

Figure 2. Diamond shank-type cutter IGM Fachmann Ekonomik Z2+1.

Parameters of milling process:

We carried out the high-speed milling process under the technological conditions listed in Table 2. As a basis for choosing the technological parameters of the experiment, the parameters recommended by the tool manufacturer were used. The revolutions were used from 75% to 100% of the recommended, the radial depth of cut values were in the range of $\frac{1}{2}$ and $\frac{1}{4}$ of the tool diameter, and the feed speed was in the range of $\pm33\%$ of the value recommended by the manufacturer.

Table 2. Parameters of milling process.

Parameter	Value
Feed speed (v_f) (m·min^{-1})	4 6 8
Radial depth of cut (a_e) (mm)	4 6 8
Axial depth of cut (ap) (mm)	18
Tool revolutions (n) (rpm·min^{-1})	15,000 17,500 20,000
Milling direction	Conventional milling
Machining strategy	Finishing
Material orientation	Technological direction of production PB in X-axis

2.2. Noise Measurement Methodology

The measurement was carried out in the enclosed places of the machining hall of the Technical University in Zvolen and was carried out under the same conditions that workers are exposed to during their daily work. A Brüel & Kjær type 2270 sound level meter (Denmark) was used for the measurement, which made it possible to measure and indicate sound pressure levels. It is a two-channel hand-held sound analyzer classified as accuracy class 1 with a Brüel & Kjær 4189 microphone. Its calibration was performed with a Brüel & Kjær 4231 calibrator. The sound level meter was placed on a tripod at a height of 1.50 m behind the yellow line on the ground, which is at 1 m from the CNC machine and serves as a safe distance for the worker, which he must not exceed during operation. The distance of the sound-level meter from the place of machining was 2 m [34]. The main descriptor was the equivalent sound pressure level *A* emitted by the CNC machine tool during milling, depending on the revolutions, the radial depth of cut, and the feed speed (Table 2).

During the measurement, the equivalent *A* sound pressure levels were recorded when the CNC machine was switched on, including the time when it was not under workload. During subsequent processing with the software Measurement Partner Suite (MPS–BZ 5503) (Brüel & Kjær, Nærum, Denmark), transitions were selected under the same operating conditions, which were repeated 3 times. The energy mean of these three values was calculated according to Formula (1) [34]:

$$L = 10log\frac{1}{n}\sum_1^n 10^{0,1 \times L_i},$$ (1)

where:

L is the energy mean of the equivalent sound pressure level A (dB),
n—number of measurement repetitions (3),
L_i—sound pressure level (A) of the i measurement (dB).

2.3. Temperature Measurement Methodology

The temperature measurement in the process of tool–workpiece interaction was carried out non-contact by means of two pieces of Raytek MI3 LTF pyrosensors (Raytek, Berlin, Germany) connected to the computer through the RAYMI3-MCOMMN communication interface (Raytek, Berlin, Germany). The layout of the measuring equipment is in Figure 3. The sensors allow non-contact temperature measurement with an optical resolution of 10:1 in the range −40 to 1000 °C in the wavelength range of 8 μm to 14 μm. The sensors were placed at 100 mm from the surface of the workpiece, perpendicular to the machined surface at a mutual axial distance of 50 mm in the direction of movement of the tool. The diameter of the scanned area on the surface of the workpiece is 10 mm in this case. This arrangement allows two tool-temperature measurements to be made in one tool pass. Data recording was performed at a frequency of 30 Hz, which allowed the recording of 5 images of the tool in the field of view of the sensor at the largest feeding speed. The emissivity of the scanned surface was set to 0.9. The tool temperature was measured after milling the 900 mm track, where the temperature conditions in the tool–workpiece environment system were stabilized. The tool temperature was determined as the maximum recorded temperature in the field of view of the sensors. The resulting temperature was determined as an average of 10 measurements.

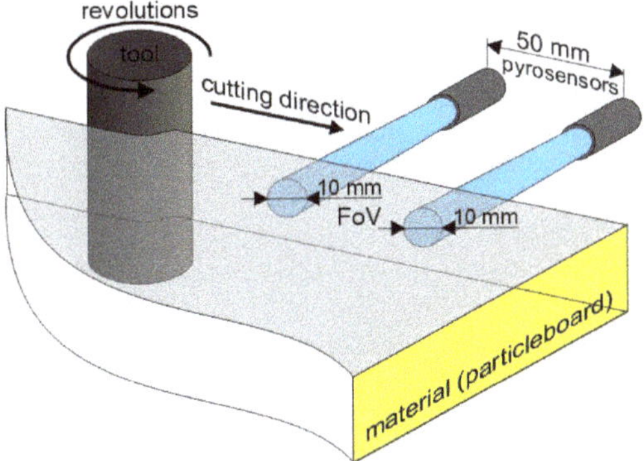

Figure 3. Layout diagram of pyro sensors for temperature measurement.

3. Results

3.1. Noise

In Figure 4 is the dependence of sound pressure level (A)–SPL (A) on revolutions at different feed speeds (gray 8 m·min^{-1}, orange 6 m·min^{-1}, blue 4 m·min^{-1}) and different radial depth of cuts.

From Figure 4, the sound pressure level (A) of the CNC machine increases with increasing feed speed (in the range of 15,000 min^{-1} to 20,000 min^{-1}) during all radial depth of cuts. This is since the dominant source of noise is the aerodynamic noise caused by the rotation of the CNC machine spindle. As the thickness of the removed chips only slightly decreases with the increase in revolutions, this phenomenon does not significantly affect the noise level. As expected, the highest noise is at the highest feed speed of 8 m·min^{-1}, when the tool removes the largest amount of material per unit of time (significant increase in chip thickness) and, on the contrary, the smallest at a feed speed of 4 m·min^{-1}.

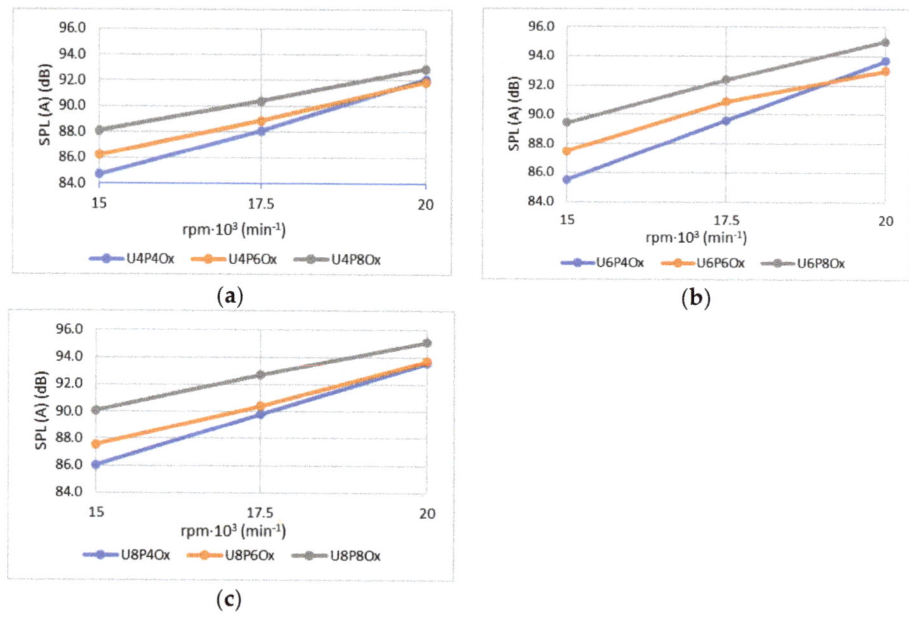

Figure 4. Dependence of sound pressure level (SPL) on change in revolutions, feed speed, and radial depth of cut ((**a**)—radial depth of cut 4 mm, (**b**)—radial depth of cut 6 mm, (**c**)—radial depth of cut 8 mm).

On Figure 5 is the dependence of the sound pressure level (A) on the revolutions at different material radial depth of cuts (grey 8 mm, orange 6 mm, blue 4 mm) and different feed speeds. Even in this case, it is possible to deduce from the graph that the noise of the CNC machine increases with increasing revolutions at all feed speeds. The highest sound pressure level (A) was measured at the highest-used radial depth of cut of 8 mm when there was the most pronounced interaction between the CNC milling cutter and the workpiece. This is caused by the fact that the active length of contact between the tool and the workpiece increases significantly when the material radial depth of cut increases. From the analysis of the processed data, it is clear the SPL (A) value when the cutter is not in engagement is at least 12 dB lower than it is in engagement for all set parameters. Its value ranges from 68 dB to 78 dB. Therefore, this value does not significantly affect the measurement results. The value of the extended measurement uncertainty ranges from 1.6 dB to 1.8 dB.

The results of the SPL (A) measurement at individual operating conditions show that the tonal components are, as expected, integer multiples of revolutions. At 15,000 revolutions per minute, the frequencies are 250 Hz, 500 Hz, and 1,000 Hz, and at 17,5000 and 20,000 revolutions, the frequencies are 315 Hz, 630 Hz, and 1,250 Hz. Considering that it is a milling cutter with two cutting edges, the highest tonal components are at frequencies of 500 Hz and 630 Hz (Figure 6).

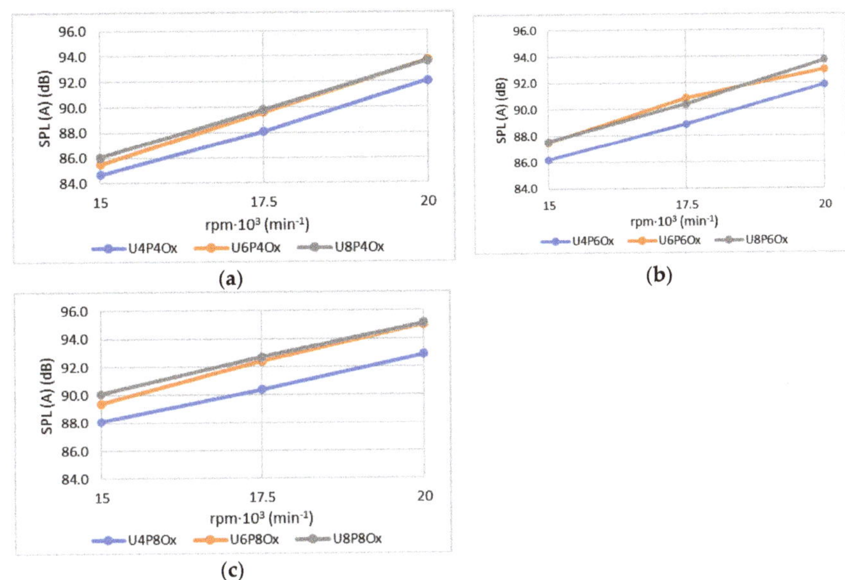

Figure 5. Dependence of sound pressure level (SPL) on revolutions at different radial depth of cuts and different feed speeds ((**a**)—feed speed 4 m·min^{-1}, (**b**)—feed speed 6 m·min^{-1}, (**c**)—feed speed 8 m·min^{-1}).

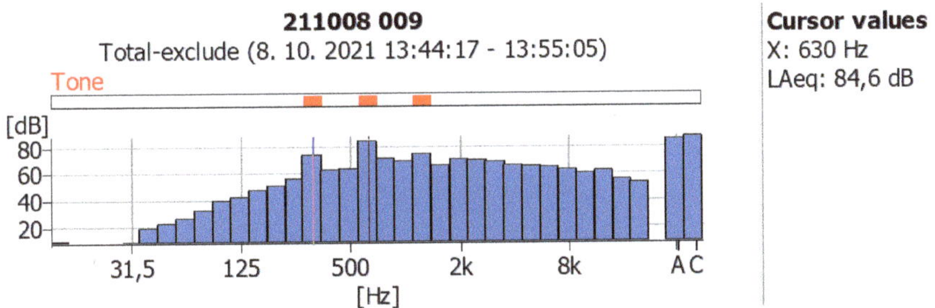

Figure 6. FFT analysis for 20,000 revolutions, feed speed 8 m·min^{-1}, and radial depth of cut 8 mm. The tone components are 315 Hz, 630 Hz, and 1,250 Hz, and the SPL (A) value at 630 Hz is 84,6 dB (processed using Measurement Partner Suite software–Brüel & Kjær, Denmark).

3.2. Temperature

On the graphs of the dependence of the temperature of the tool on the revolutions at a constant value of radial depth of cut and feed speed (Figure 7), the influence of the feed speed on temperature increase is relatively small. The temperature of the tool changes only slightly with the increase in revolutions, which is since the thickness of the removed chip (from 4 mm to 8 mm) only slightly decreases with the increase in revolutions. Wood has a low value of thermal conductivity (0.10–0.25 W·m^{-1}·K^{-1}) and thermal diffusivity (0.15–0.25·10^{-6} m^2·s^{-1}) in the direction perpendicular to the fibers [35–37], which causes slow heat transfer in the wood. Increasing the revolutions of the tool causes the thickness of the chip to decrease, and therefore the tool separates a thinner layer at the next contact of the blade with the material, which, due to thermal conductivity and diffusivity, has a higher temperature. The result is a slight increase in the temperature of the tool with increasing revolutions. However, this influence is not significant, because the thickness of the chip changes only slightly. When the tool speed changes from n = 15,000 min^{-1} to

$n = 20{,}000\ \mathrm{min}^{-1}$, the mean thickness of the chip decreases by approximately 25% (e.g., at feed speed $v_f = 8\ \mathrm{m\cdot s}^{-1}$ and radial depth of cut $a_e = 8$ mm, it is from 0.189 mm to 0.141 mm).

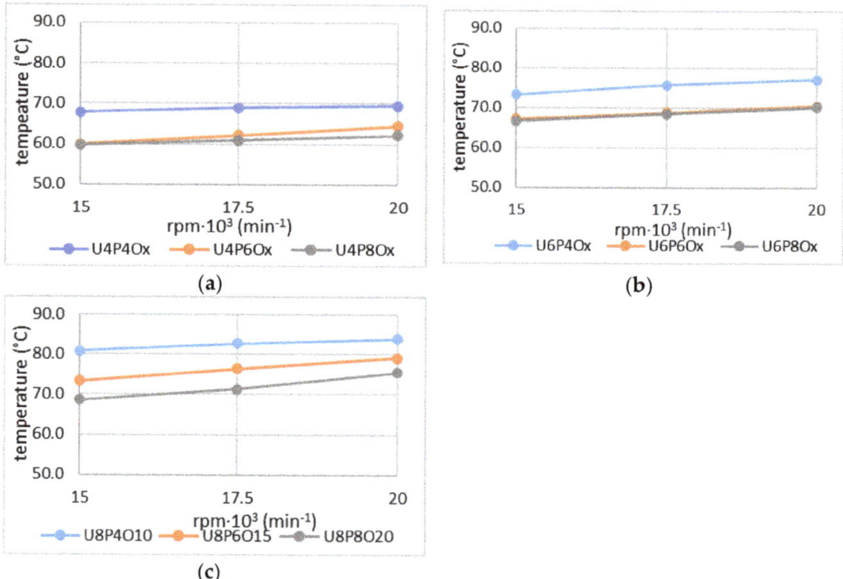

Figure 7. Dependence of temperature on change in revolutions, feed speed, and radial depth of cut ((**a**)—radial depth of cut 4 mm, (**b**)—radial depth of cut 6 mm, (**c**)—radial depth of cut 8 mm).

A more significant trend that can be seen from the graphs is that the temperature of the tool decreases with the increasing value of the feed speed. This trend is caused by a more significant increase in chip thickness (when the feed speed changes from $v_f = 4\ \mathrm{m\cdot s}^{-1}$ to $v_f = 8\ \mathrm{m\cdot s}^{-1}$, the average chip thickness increases from 0.071 mm to 0.141 mm, i.e., by 100%), which, due to the small value of thermal conductivity and diffusivity, causes the cutting edge to enter the colder material during the next cut of the tool. The length of the friction surface, when the tool separates the chip, remains unchanged when the feed speed is changed. The result is a temperature decrease in the tool, which is more effectively cooled by the "cold" material than heated by the friction of the tool against the material. This trend is best observed at the largest value of the radial depth of cut.

Other graphs show the dependence of the tool temperature change on the radial depth of cut (Figure 8). It is clear from the trends that increasing the radial depth of cut significantly affects the increase in tool temperature. The reason is the fact that as the radial depth of cut increases, the contact length of the tool with the material increases significantly, which results in a longer friction surface and a significant increase in the produced heat (at a radial depth of cut of 4 mm, the length of contact between the tool knife and the workpiece is 8.38 mm, and at a radial depth of cut of 8 mm increases to 12.57 mm, i.e., by 50%). This trend is clearly observable in all cases and is further synergistically supported by increasing the revolutions, which causes a decrease in the thickness of the chip and thus an increase in the temperature at the point of the cut due to the conduction of heat generated during the previous cut in the material. Confirmation of the radial depth of cut as the most important parameter affecting the temperature increase is also confirmed using statistics (Table 3). The standard deviation was less than 2.0 °C in all measurements.

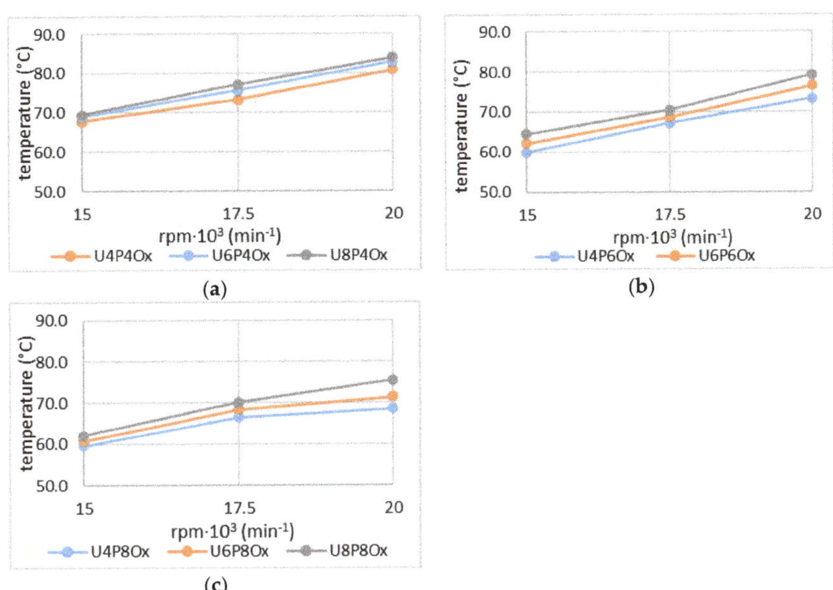

Figure 8. Dependence of the temperature change in the tool on the radial depth of cut and revolutions ((**a**)—feed speed-4 m·min^{-1}, (**b**)—feed speed-6 m·min^{-1}, (**c**)—feed speed 8 m·min^{-1}).

Table 3. The value of the correlation coefficient r_{xy} between the machine parameters and the measured temperature and noise parameters.

Parameter	Noise	Temperature
Revolutions	0.8739	0.2378
Feed Speed	0.3585	−0.5201
Radial depth of cut	0.2457	0.7908

Table 3 shows the correlation coefficients of noise and temperature depending on the change in the operating parameters of the CNC machine. Sound pressure level (A) showed the highest dependence on revolutions, while the extent of the radial depth of cut did not show a significant effect. In adaptive control, it is, therefore, necessary to focus mainly on the dependence of the sound pressure level (A) on the level of the revolutions for the selected type of tool and workpiece.

The tool–workpiece interaction temperature showed the greatest dependence on the radial depth of cut setting. As part of the adaptive control, it is, therefore, necessary to find a balanced ratio between the set radial depth of cut and revolutions, where the optimal value for both measured parameters appear to be between 15–17 thousand revolutions and radial depth of cut of 6 to 8 mm for the selected type of tool and workpiece. The dependence of temperature on feed speed is statistically confirmed only in an indirect plane. Thus, feed speed affects the temperature partly indirectly (negatively).

4. Discussion

The methods of measuring and evaluating tool temperature in CNC machines were objectified by Xie et al. (2009) [38]. Based on these principles, an objectified methodology for measuring the temperature of the tool was chosen. The same trends of the dependence of the tool temperature on the change in monitored parameters were identified in the work [16]. However, tool temperature does not appear to be a major problem in woodworking. The problem is rather in the machining of metallic materials due to the energy management of the machine, where the design of cooling mechanisms is also necessary [17]. In woodworking, the use of coolants and lubricants has proven impractical. Gisip et al. [39,40] demonstrated that the application of cold air or heat treatment of the tool before processing wood-agglomerated materials can have an impact on the overall life of the tool. Our results thus indirectly confirm that even if the temperature of the cutting tool during woodworking does not move within critical values, the application of this parameter in the adaptive control of CNC machines has significance. As part of the conducted research, the measured temperatures do not approach the "critical values". The critical temperature is considered to be approx. 400 °C, when there are visible changes in the color of the wood surface.

The parameter sound pressure level (A) has not been shown to be fundamental in terms of wear of the cutting tool [41]. In woodworking, noise as a more fundamental problem from a hygienic point of view was manifested earlier in sawmill wood processing and on technologically older machines than in CNC woodworking [42,43]. Monitoring the sound pressure level parameter (A) has, however, the importance in terms of its changes with a connection to the change in the cutting parameters of the machine and the tool used. Fluctuations in noise when changing cutting parameters can have a negative impact on the machining process in a long-term context. Therefore, it is advisable to incorporate the identified changes in this parameter into the adaptive control algorithm.

The presented results point to the fact that the noise of the tool as well as its temperature interacts with the thickness of the cut layer. When determining the thickness of the cut layer as an optimization parameter, both signals have the potential to be used for the identification of limit states as well as the optimization of process parameters.

The obtained results confirm the assumption that unfavorable machining conditions will be manifested by an increase in the tool temperature and the noise of the machining process. The common practice of operators, adjusting the parameters of the cut layer based on a subjective assessment of the noise level of the machining process or visual assessment and identification of thermal traces on the machined surface, is also based on this assumption.

5. Conclusions

The results of this study show that operating parameters affect both noise and tool temperature. As expected, the lowest SPL (A) = 84.7 dB was achieved at 15,000 rpm, 4 mm clearance, and a feed speed of 4 m·min^{-1}, and the highest SPL (A) = 95.1 dB at 20,000 rpm, a radial depth of cut of 8 mm and a feed speed of 8 m·min^{-1}. The lowest temperature of 59.7 °C was reached at 15,000 rpm, a 4 mm radial depth of cut, and an 8 m·min^{-1} feed speed, and the highest temperature was 84.0 °C at 20,000 rpm, an 8 mm radial depth of cut, and an 8 mm feed speed. m·min^{-1}. SPL (A) is most dependent on revolutions and interaction temperature is most sensitive to the radial depth of cut setting.

The applied methodology objectifies the given method and by its nature is suitable for application within adaptive control systems. For real use, however, it is necessary to expand the spectrum of the obtained data, not only in the range of the combination of parameters modulating the size of the cut layer but especially by the development over time resulting from the wear of the tool. A real application would help not only in the optimization of machining parameters but would also represent a suitable economy-detection tool. Based on the development of the monitored physical values, it would be possible to determine exactly the level of wear of the tool, at which it is necessary to replace it, without the need to remove the tool from the machine and procure specialized, financially demanding measuring equipment.

Author Contributions: Conceptualization, R.K., R.I. and M.N.; methodology, R.K., R.I. and M.N.; software, R.I. and M.N.; validation, R.K., R.I. and M.N.; formal analysis, M.G., R.I. and M.N.; resources, M.N. and R.K.; data curation, M.N. and R.I.; writing—original draft preparation, M.G.; writing—review and editing, M.G., M.N., R.I. and R.K.; visualization, M.G.; supervision, R.K.; project administration, M.N.; funding acquisition, M.N. and R.K. All authors have read and agreed to the published version of the manuscript.

Funding: This research was funded by Ministry of Education, Science, Research and Sport of the Slovak Republic, Grant Numbers VEGA 1/0324/21, Analysis of the risks of changes in the material composition and technological background on the quality of the working environment in small and medium-sized wood processing companies. VEGA 1/0714/21, Research of selected properties of sustainable insulation materials with the potential for use in wooden buildings. Slovak Research and Development Agency, Grant Number APVV-20-0403, FMA analysis of potential signals suitable for adaptive control of nesting strategies for milling wood-based agglomerates.

Data Availability Statement: Not applicable.

Conflicts of Interest: The authors declare no conflict of interest.

References

1. Lin, R.S.; Chen, S.L.; Liao, J.H. A New Method for Five-Axis CNC Curve Machining. *Eng. Lett.* **2011**, *19*, 179–184.
2. Xu, X.W.; Newman, S.T. Making CNC machine tools more open, interoperable and intelligent—A review of the technologies. *Comput. Ind.* **2006**, *57*, 141–152. [CrossRef]
3. Shin, Y.C.; Chin, H.; Brink, M.J. Characterization of CNC Machining Centers. *J. Manuf. Syst.* **1991**, *10*, 407–421. [CrossRef]
4. Petak, T.; Kurekova, E. New Methodology for Calibration of CNC Machines. *Int. J. Eng. Res. Afr.* **2015**, *18*, 144–151. [CrossRef]
5. Racz, S.G.; Breaz, R.E.; Cioca, L.I. Hazards That Can Affect CNC Machine Tools during Operation-An AHP Approach. *Safety* **2020**, *6*, 10. [CrossRef]
6. Gai, H.D.; Li, X.W.; Jiao, F.R.; Cheng, X.; Yang, X.H.; Zheng, G.M. Application of a New Model Reference Adaptive Control Based on PID Control in CNC Machine Tools. *Machines* **2021**, *9*, 274. [CrossRef]
7. Gutakovskis, V.; Gerins, E.; Naginevicius, V.; Gudakovskis, V.; Styps, E.; Servytis, R. Adaptive Control for the Metal Cutting Process. *Int. J. Eng. Res. Afr.* **2020**, *51*, 1–7. [CrossRef]
8. Chiu, H.W.; Lee, C.H. Intelligent Machining System Based on CNC Controller Parameter Selection and Optimization. *IEEE Access* **2020**, *8*, 51062–51070. [CrossRef]
9. Kasirolvalad, Z.; Motlagh, M.R.J.; Shadmani, M.A. An intelligent modeling system to improve the machining process quality in CNC machine tools using adaptive fuzzy Petri nets. *Int. J. Adv. Manuf. Technol.* **2006**, *29*, 1050–1061. [CrossRef]
10. Liu, Y.M.; Wang, C.J. Neural network based adaptive control and optimisation in the milling process. *Int. J. Adv. Manuf. Technol.* **1999**, *15*, 791–795. [CrossRef]
11. Yaman, K.; Basaltin, M. Investigations on the cutting parameters and the tool wear of SAE 1030 forged steel material by acoustic emission in turning operation. *J. Fac. Eng. Archit. Gaz.* **2017**, *32*, 1077–1088.
12. Miao, E.M.; Liu, H.; Fan, K.C.; Lv, X.X.; Liu, Y.; Hu, Y. Analysis of CNC machining based on characteristics of thermal errors and optimal design of experimental programs during actual cutting process. *Int. J. Adv. Manuf. Technol.* **2017**, *88*, 1363–1371. [CrossRef]
13. Qin, Y.M.; Park, S.S. Robust adaptive control of machining operations. In Proceedings of the IEEE International Conference on Mechatronics Automation, Vols. 1–4 Conference Proceedings, Niagara Falls, ON, Canada, 29 July–1 August 2005; pp. 975–979.
14. Mika, D.; Jozwik, J. Normative measurements of noise at CNC machines work stations. *Adv. Sci. Technol.* **2016**, *10*, 138–143. [CrossRef]
15. Seemuang, N.; McLeay, T.; Slatter, T. Using spindle noise to monitor tool wear in a turning process. *Int. J. Adv. Manuf. Technol.* **2016**, *86*, 2781–2790. [CrossRef]
16. Igaz, R.; Kminiak, R.; Krišťák, Ľ.; Němec, M.; Gergeľ, T. Methodology of Temperature Monitoring in the Process of CNC Machining of Solid Wood. *Sustainability* **2019**, *11*, 95. [CrossRef]
17. Kasim, A.; Kuntjoro, W.; Kasolang, S. Thermal Management System of CNC Machine Tool Spindle for Consistent Machining Accuracy. *J. Teknol.* **2015**, *76*, 97–101. [CrossRef]
18. Smith, S.; Woody, B.; Barkman, W.; Tursky, D. Temperature control and machine dynamics in chip breaking using CNC toolpaths. *CIRP Ann-Manuf. Technol.* **2009**, *58*, 97–100. [CrossRef]
19. Kowaluk, G.; Pałubicki, B.; Marchal, R.; Frąckowiak, I.; Beer, P. Friction and cutting forces when machining particleboards produced from different lignocellulose raw materials. *Ann. WULS For. Wood Technol.* **2007**, *61*, 362–365.
20. Troppová, E.; Tippner, J.; Hrčka, R.; Halachan, P. Qasi-stationary measurements of lignamon thermal properties. *Bioresources* **2013**, *8*, 6288–6296. [CrossRef]
21. Hrčka, R.; Babiak, M. Some non-traditional factors influencing thermal properties of wood. *Wood Res.-Slovak.* **2012**, *57*, 367–374.
22. Hrčka, R.; Babiak, M.; Németh, R. High temperature effect on diffusion coefficient. *Wood Res.-Slovak.* **2008**, *53*, 37–46.

23. Ishihara, M.; Noda, N.; Ootao, Y. Analysis of dynamic characteristics of rotating circular saw subjected to thermal loading and tensioning. *J. Therm. Stress.* **2010**, *33*, 501–517. [CrossRef]
24. Ratnasingam, J.; Pew, M.; Ramasamy, G. Tool temperature and cutting forces during the machining of particleboard and solid wood. *J. Appl. Sci.* **2010**, *10*, 2881–2886. [CrossRef]
25. Nasir, V.; Kooshkbaghi, M.; Cool, J.; Sassani, F. Cutting tool temperature monitoring in circular sawing: Measurement and multi-sensor feature fusion-based prediction. *Int. J. Adv. Manuf. Technol.* **2021**, *112*, 2413–2424. [CrossRef]
26. Prasad, B.S.; Prasad, D.S.; Sandeep, A.; Veeraiah, G. Condition Monitoring of CNC Machining Using Adaptive Control. *Int. J. Automat. Comput.* **2013**, *10*, 202–209. [CrossRef]
27. Kavaratzis, Y.; Maiden, J.D. Real-Time process monitoring and adaptive-control during CNC Deep hole drilling. *Int. J. Prod. Res.* **1990**, *28*, 2201–2218. [CrossRef]
28. Ford, D.G.; Myers, A.; Haase, F.; Lockwood, S.; Longstaff, A. Active vibration control for a CNC milling machine. *P. I. Mech. Eng. C-J. Mech.* **2013**, *228*, 230–245. [CrossRef]
29. Iskra, P.; Hernández, R.E. Toward a Process Monitoring and control of a CNC Wood Router: Development of an Adaptive Control System For routing White Birch. *Wood Fiber Sci.* **2010**, *42*, 523–535.
30. Laszewicz, K.; Górski, J.; Wilkowski, J. Long-term accuracy of MDF milling process—Development of adaptive control system corresponding to progression of tool wear. *Eur. J. Wood Wood Prod.* **2013**, *71*, 383–385. [CrossRef]
31. *EN 14322*; Wood-Based Panels-Melamine Faced Board for Interior Uses—Definition, Requirements and Classification. European Committee for Standardization: Brussels, Belgium, 2017.
32. *EN 312-2*; Particleboards-Specifications-Part 2: Requirements for General Purpose Boards for Use in Dry Conditions. European Committee for Standardization: Brussels, Belgium, 1996.
33. *EN ISO 12460-5*; Determination of Formaldehyde Release-Part 5: Extraction Method (Called the Perforator Method). European Committee for Standardization: Brussels, Belgium, 2016.
34. *STN EN ISO 9612*; Acoustics—Determination of Occupational Noise Exposure—Engineering Method. European Committee for Standardization: Brussels, Belgium, 2009.
35. Suleiman, B.M.; Larfeldt, J.; Leckner, B.; Gustavsson, M. Thermal Conductivity and Diffusivity of Wood. *Wood Sci. Technol.* **1999**, *33*, 465–473. [CrossRef]
36. Hrčka, R.; Babiak, M. Wood thermal properties. In *Wood in Civil Engineering*, 1st ed.; Concu, G., Ed.; Intech Open: Rijeka, Croatia, 2017; pp. 25–43. [CrossRef]
37. Krišťák, Ľ.; Igaz, R.; Ružiak, I. Applying the EDPS Method to the Research into Thermophysical Properties of Solid Wood of Coniferous Trees. *Adv. Mater. Sci. Eng.* **2019**, *2019*, 2303720. [CrossRef]
38. Xie, C.; Roddeck, W.; Liu, C.S.; Zhang, W.M. The Analysis and Research about Temperature and Thermal Error Measurement Technology of CNC Machine Tool. *Key Eng. Mater.* **2009**, *392–394*, 40–44. [CrossRef]
39. Gisip, J.; Gazo, R.; Stewart, H.A. Effects of cryogenic treatment and refrigerated air on tool wear when machining medium density fiberboard. *J. Mater. Process. Technol.* **2009**, *209*, 5117–5122. [CrossRef]
40. Gisip, J.; Gazor, R.; Stewart, H.A. Effects of refrigerated air on tool wear. *Wood Fiber Sci.* **2007**, *39*, 443–449.
41. Gorski, J.; Szymanovwski, K.; Podziewski, P.; Smietanska, K.; Czarniak, P.; Cyrankowski, M. Use of Cutting Force and Vibro-acoustic Signals in Tool Wear Monitoring Based on Multiple Regression Technique for Compreg Milling. *Bioresources* **2019**, *14*, 3379–3388. [CrossRef]
42. Osmolovsky, D.; Asminin, V.; Druzhinina, E. Reducing Noise from round woodworking machines by applying vibration damping friction pads between the saw blade and the clampping flange. *Akustika* **2019**, *32*, 138–140. [CrossRef]
43. Jungai, P.; Travnicek, P.; Ruzbarsky, J.; Kopecky, Z.; Solar, A. Noise emissions of older woodworking machines at parallel operation process. *MM Sci. J.* **2019**, *2019*, 2832–2838. [CrossRef]

Article

Life Cycle Assessment of a Road Transverse Prestressed Wooden–Concrete Bridge

Jozef Mitterpach *, Roman Fojtík, Eva Machovčáková and Lenka Kubíncová

Department of Wood Processing and Biomaterials, Faculty of Forestry and Wood Sciences, Czech University of Life Sciences Prague, Kamýcká 129, 16500 Prague, Czech Republic
* Correspondence: mitterpach@fld.czu.cz; Tel.: +420-224-383-789

Abstract: Through its anthropogenic activities in construction, human society is increasingly burdening the environment with a predominantly adverse impact. It is essential to try to use building materials that allow us to build environmentally friendly buildings. Therefore, this article deals with the determination of the environmental performance of a cross-prestressed timber-reinforced concrete bridge using life cycle assessment (LCA) compared with a reinforced concrete road bridge with a similar span and load. The positive environmental performance of the wooden concrete bridge was proved, with a relatively small (22.9 Pt) total environmental damage. The most significant impact on the environment is made by the wood–concrete bridge materials in three categories of impacts: Respiratory inorganics (7.89 Pt, 79.94 kg PM2.5 eq), Global warming (7.35 Pt, 7.28×10^4 kg CO_2 eq), and Non-renewable energy (3.96 Pt, 6.01×10^5 MJ primary). When comparing the wood–concrete and steel concrete road bridge, a higher environmental performance of 28% per m^2 for the wood--concrete bridge was demonstrated. Based on this environmental assessment, it can be stated that knowledge of all phases of the life cycle of building materials and structures is a necessary step for obtaining objective findings of environmental damage or environmental benefits of building materials or structures.

Keywords: LCA; wood; concrete; road; bridge; environmental impact

Citation: Mitterpach, J.; Fojtík, R.; Machovčáková, E.; Kubíncová, L. Life Cycle Assessment of a Road Transverse Prestressed Wooden–Concrete Bridge. *Forests* **2023**, *14*, 16. https://doi.org/10.3390/f14010016

Academic Editors: Petar Antov, Muhammad Adly Rahandi Lubis and Seng Hua Lee

Received: 16 November 2022
Revised: 16 December 2022
Accepted: 17 December 2022
Published: 21 December 2022

1. Introduction

With the world's growing population, it is necessary to travel and overcome longer distances. This is also connected with the broader use of roads and bridge structures. The bridges are becoming more and more loaded, and which necessitates their reconstruction or complete replacement. In the Czech Republic (Central Europe, total area 78,866 km^2, population 10.52 mil.), there a total of 17,850 bridges, 13.76% of which are in poor condition, 5.03% are in very poor condition, and 0.45% are in critical state condition. For bridges located on the class 2 roads (the connection between the districts) and class 3 roads (the connection between the municipalities or their connection to the other roads), 16.18% of bridges are in poor condition, 6.02% are in very poor condition, and 0.58% of the bridges are in a critical state condition. A significant part of the bridges in poor to critical state conditions are bridges placed on class 2 and class 3 roads [1]. From this information, almost 23% of class 2 and class 3 road bridges need to be renovated or completely replaced.

According to several world studies, infrastructure is the main cause of greenhouse gas (GHG) emissions, as infrastructure designs influence more than half of global climate change emissions [2–4]. Therefore, it is necessary to evaluate the time in which structural systems and the materials used for construction have the potential to minimize the impact on the environment [5–7]. Several authors, such as Balogun et al. [7], Horvath and Hendrickson [8], Zhang et al. [9], Du and Karoumi [10], and Pedneault et al. [11], conducted detailed studies comparing basic material groups in the construction of bridges, for example, steel, concrete, aluminum, etc., in their environmental impact assessment. Similar publications focused on the environmental impact of bridges [4,12–16] proved that there is a clear need for environmental information at various stages of the life cycle of the bridge structures.

The evaluation of the publication by Niu and Fink [17] confirms the fact that the phase of production of structural materials has the largest share of the environmental impact of bridges. Habert et al. [18] proved that two phases mainly contribute to the environmental impacts: the production of the materials and the maintenance of the bridge. Du et al. [19] found that regardless of the bridge type, the material manufacture phase dominates the whole life cycle in all indicators. The material production stage shall be considered carefully, since it is usually identified to contribute the largest share of environmental impact [17,20].

The bridges where the main structural material is concrete or various material variants of its reinforcement are dealt with by studies [16,21] considering the environmental impact. In publications, there are evaluations of the environmental impact of individual alternative solutions to the change in the material base of the reinforcement. In the case where concrete is used as the material for the main load-bearing structures of the bridge deck, it has been found that in the area of CO_2 emissions, the use of high-performance concrete caused 66% lower emissions than in the case of normal concrete [22]. For example, in the case of steel–concrete main load-bearing structures [23], these bridge decks are more expensive than concrete in terms of price but have a lower environmental impact. As part of material reuse, structural steel in reinforced concrete structures has significantly higher recyclability than concrete—up to 98% recyclability [10,24]. A comparison of composite bridge structures, for example, steel beams connected with an aluminum bridge deck or steel beams connected with a concrete bridge deck, is dealt with in [11]. It is obvious that the material composition of the structure has an important role in the life cycle stages of bridge structures.

However, in connection with structural materials, the construction design phase is also a very important phase of the life cycle of construction. This phase mainly affects the method of use, the amount of material used, and the impact on CO_2 emissions [25]. The SEGRO algorithm was used to design prefabricated bridges made of two isostatic beams with a double U-shaped cross-section and isostatic spans. [26]. It has been shown that using the algorithm in the bridge design phase reduces CO_2 emissions. The combination of bridge length with material and construction solutions also has an impact on the environment [27,28]. It is obvious that the design solution can affect not only the choice of the bridge material used but especially its environmental impacts during the life cycle.

Therefore, this article deals with determining the environmental performance of material composition on the classification and quantification of environmental impacts of a transverse prestressed Wooden–Concrete Bridge (WCB) for road transport (with a small bridge up to 10 m). In the next step, the design level and its environmental impacts are evaluated, and the WCB is compared with a similar Steel Concrete Bridge (SCB) per m^2 of the bridge area.

2. Materials and Methods

2.1. Characteristics of the Road Transverse Prestressed Wooden–Concrete Bridge (WCB)

The transverse prestressed wooden–concrete bridge (WCB) for road transport is in the village of Bohunice in the Czech Republic (Central Europe) (Table 1, Figures 1–3). The bridge is located on road 2 class, and it was designed as a single-field bridge (one field between two supports) to short span; the length is 13.650 m with a total width of 6.230 m. The reconstruction of the bridge was focused on the part of the substructure and on the superstructure. The foundations of abutments and piers, of which the originals were retained, were designed as spread footing. Bridge abutments, capping beams, and abutment walls are made of concrete with strength class C30/37 with a resistance to corrosion induced by chlorides and a resistance to alternate freeze and thaw attacks. The description of the concretes used in the construction is specified in European standard [29]. The concrete is reinforced with steel-bar reinforcement. The shaft width of concrete abutments is 2.725 m. The capping beams and the abutment wall are reinforced with concrete reinforcement with a diameter of 12 mm and the binder bars with a diameter of 8 mm by 200 mm. The linear elastomer bridge bearing system was chosen and placed on the capping beam mainly due to its low height, with dimensions of the elastomer system 200 × 52 × 300 mm. The structure was also anchored in the horizontal and transverse directions by means of steel loop with a bolt. The superstructure of the bridge was placed on the bridge bearings. The horizontal bearing structure is a patented system, coupling load-bearing wooden parts and reinforced concrete. The wood–steel part is made of transversely prestressed beams made of spruce glued laminated timber (GLT) of strength class GL32h. The height of the wooden beams is 600 mm, and the width is 100 mm. The prestress is realized by means of high-strength steel bars. The prestressing bars are equipped with large steel backplates and nuts on both sides. The beams are protected by polyurea insulation spray. The wood–steel load-bearing structure was placed directly on the elastomer bridge bearings and anchored with an anchoring loop. The roadway was reinforced and cast on the self-supporting structure. The bridge deck is reinforced concrete with strength class C30/37 with resistance to corrosion induced by chlorides and with the resistance to alternate freeze and thaw attacks with a maximum height of 300 mm. The transverse prestressing bars are 14 mm in diameter at an axial distance of 100 mm. Coupling between the wood–steel part and the concrete deck was performed using special coupling screws VB-48-7.5x165. All steel elements were heat galvanized. The roadway of WCB is directly movable with short parapets made of reinforced concrete C30/37. Both parapets are fitted with 1.100 m high steel railings made of structural steel with a yield strength of 235 MPa (S235). The parameters of the steel used are given in [30].

Table 1. Structural units and materials composition of WCB and SCB bridges.

Structural Elements	Material	WCB m^3	WCB t	SCB m^3	SCB t
Abutments and piers	Concrete C30/37	65.160	162.900	31.009	77.523
	Reinforcement B500B	0.917	7.200	0.480	3.765
Capping beam	Concrete C30/37	3.507	8.767	13.166	32.915
	Reinforcement B500B	0.022	0.172	0.019	0.149
Abutment wall	Concrete C30/37	3.163	7.908	3.163	7.908
	Reinforcement B500B	0.006	0.047	0.006	0.047
Main construction and bridge deck	Concrete C30/37	17.100	43.180	16.801	42.003
	Glued laminated timber GL32h	33.000	20.483	-	-
	Steel	-	-	1.670	13.128
	Reinforcement B500B	0.080	0.626	0.803	6.302
	Shear connections—engineering steel	0.036	0.279	0.041	0.320
	Prestressing elements	0.220	0.600	-	-
Bridge bearing	Elastomer	0.178	0.570	0.031	0.100
Bridge top	Polyurea	0.082	0.004	-	-
	Asphaltic concrete	-	-	2.898	6.376
	Connecting spray PS	-	-	0.090	0.019
	Medium grain aggregates (16+)	-	-	2.898	4.637
	Mastic asphalt	-	-	1.932	2.125
	Concrete	-	-	1.932	4.250
Bridge equipment	Railings structural steel	27.300	0.493	0.139	1.090
	Concrete monolithic parapet C30/37	4.100	10.430	5.148	12.870
	Reinforcement B500B	0.038	0.301	0.147	1.151

Figure 1. Digital picture of WCB bridge.

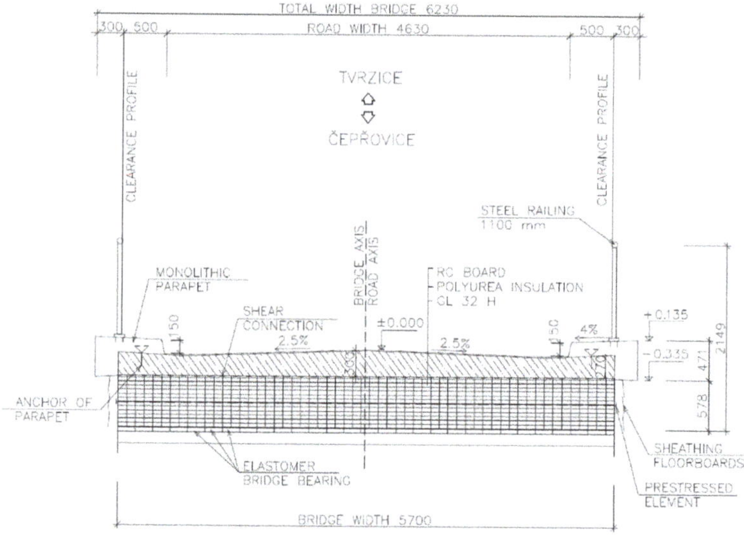

Figure 2. Transverse section of WCB bridge (scale 1:50).

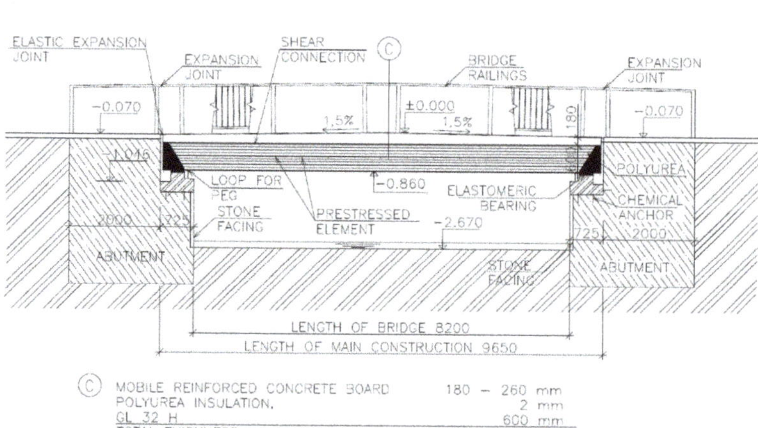

Figure 3. Longitudinal section of WCB bridge (scale 1:100).

2.2. Characteristics of the Road Steel Concrete Bridge (SCB)

The composite steel-reinforced concrete bridge, Steel Concrete Bridge (SCB), for road transport is in the village in the municipality of Glovčín (Czech Republic, Central Europe) (Table 1, Figures 4–6). The SCB was designed as part of road class 2. It is a single-field (one field between two supports) bridge construction with a total bridge length of 11.064 m. The total width of the bridge is 5.1 m. The reconstruction of the bridge was focused on the part of the substructure and on the superstructure. The bridge abutments were founded at strip foundation, which remains as the original, made of reinforced concrete with strength class C25/30. The substructure of the bridge includes two monolithic reinforced concrete abutments. The shaft of abutments, capping beams, and abutment walls were designed from the reinforced concrete with strength class C30/37 with resistance to corrosion induced by chlorides and a resistance to alternate freeze and thaw attacks. The description of the concretes used in the construction is specified in the European standard [29]. The width of the abutments shaft is 1.300 m. The load-bearing structure was mounted on two rectangular reinforced elastomeric bridge bearings on each abutment. The elastomer bearings were located on capping beams in the axes of the main beams. Elastomer bearings were anchored, i.e., structurally secured against movements in the open joint between the bearings and the subsequent construction. One omnidirectionally movable elastomeric bearing and one longitudinally movable elastomeric bearing were designed on abutment 1. One transversely movable elastomer bearing, and one fixed elastomer bearing were designed on abutment 2. The SCB bridge was designed as a full-beam girder with a lower steel–concrete bridge deck. The load-bearing structure of the bridge includes two main longitudinal beams with a span of 12.0 m.

The main beams were designed as welded I profiles with a height of 1.47 m with both-side column bracing. The mutual axial distance of the beams is 4.75 m. The beams are interconnected in the bottom part by an oblique end and perpendicular intermediate crossbeams at a distance of 1.5 m. The bottom bridge deck is made of a system of longitudinal beams and cross beams. Crossbeams with a span of 4.75 m were designed from cold-rolled profile HEB 260 and carry internal longitudinal beams, which are also designed from cold-rolled profiles HEB 260 at a mutual distance of 1.50 m. For the structural coupling of the reinforced concrete slab of the bridge deck with steel structure, perforated coupling strips made of the structural steel with a yield strength of 355 MPa were welded to the upper flanges of the bridge deck steel crossbeams. The properties of the steel used are specified in the European standard [30]. A railing with vertical infill made of structural steel with a yield strength of 235 MPa (S235) was welded to the upper flange of the main load-bearing beams. The load-bearing steel beams were composed of structural steel with a yield strength of 355 MPa (S355). The bridge deck slab was designed as monolithic reinforced concrete, 4.736 m wide and 13.06 long. The thickness of the slab is variable. The minimum slab thickness is 200 mm. The slab is designed from reinforced concrete C30/37, and it supports the roadway. The crossbeams and reinforced concrete slab are connected in the roadway part. Coupling was realized by means of the coupling strip made of 12 mm thick sheet steel, which is perforated with a diameter of 30 mm holes and welded to the upper flange of the crossbeams by a T-filled weld. The bottom primary load-bearing reinforcement of the concrete slab was passed through the holes at the coupling strip. The roadway structure on the bridge comprises asphalt concrete with a thickness of 40 mm, connecting spray, coated medium grain aggregates, 35 mm thick mastic asphalt, and NAIP insulation on the sealing layer. On both sides of the bridge, monolithic parapets are composed of reinforced concrete with strength class C30/37 with resistance to corrosion induced by chlorides and with the resistance to alternate freeze and thaw attacks. The monolithic parapets are 618 mm wide. On both sides of the bridge, a 1.30 m high railing is designed with a vertical infill made of structural steel, which is anchored to the parapet via anchor plates by means of glued anchors.

Figure 4. Digital picture of SCB bridge.

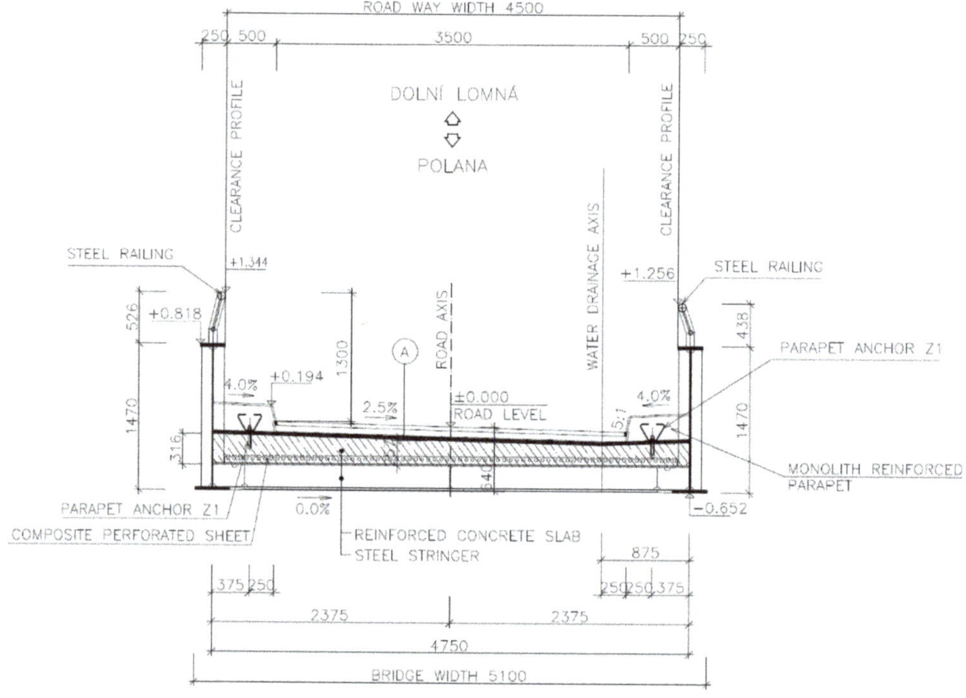

Figure 5. Transverse section of SCB bridge (scale 1:50).

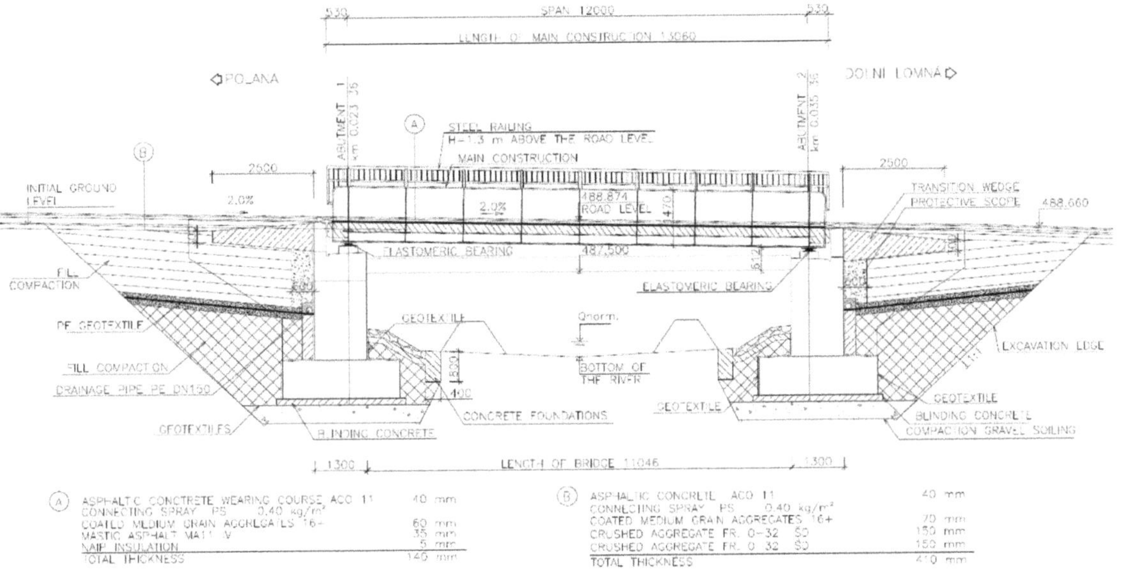

Figure 6. Longitudinal section of SCB bridge (scale 1:100).

2.3. Life Cycle Assessment Methodology

The Life Cycle Assessment (LCA) method was chosen for the environmental impact assessment [31,32]. The system boundaries for LCA included in the assessment were divided into construction materials (product stage for raw material supply, transport to the manufacturer, and manufacturing). The materials in Table 1 were included in the inventory analysis. The functional unit was the whole WCB selected for the WCB environmental performance calculation. For the comparison of WCB and SCB, a functional square meter (1 m^2) of bridging was chosen to consider the different bridging lengths and different widths of bridge structures, and thus the results can be used for other environmental comparisons for low-span road bridges.

The SimaPro 9 database software [33] and the IMPACT 2002+ method [34] were used for life cycle impact assessment (LCIA). LCIA was evaluated at midpoints of environmental damage: Respiratory inorganics (kg PM2.5 eq), Global warming (kg CO$_2$ eq), Non-renewable energy (MJ primary), Land occupation (m^{2}org.arable), Terrestrial ecotoxicity (kg TEG soil), Non-carcinogens (kg C$_2$H$_3$Cl eq), Carcinogens (kg C$_2$H$_3$Cl eq), Mineral extraction (MJ surplus), Terrestrial acid/nutria (kg SO$_2$ eq), Aquatic ecotoxicity (kg TEG water), Respiratory organics (kg C$_2$H$_4$ eq), Ionizing radiation (Bq C-14 eq), Ozone layer depletion (kg CFC-11 eq), Aquatic acidification (kg SO$_2$ eq), and Aquatic eutrophication (kg PO$_4$ P$_{-lim}$). The dataset covers all relevant process steps and technologies over the supply chain of the represented cradle-to-gate inventory with good overall data quality [35].

3. Results and Discussion

3.1. Life Cycle Impact Assessment of WCB

The largest overall impact has the WCB structural elements: Abutments and piers (56.7%, 13 Pt), followed by the Main construction and bridge deck (33.4%, 7.66 Pt), and Bridge equipment (3.41%, 0.78 Pt) (Figure 7). A Monte Carlo simulation on a single score of the whole WCB showed a mean of 22.27, median of 22.26, standard error of the mean (SEM) of 0.23, standard deviation (SD) of 3.06, and coefficient of variability (CV) of 13.75%. In the environmental impact categories at the midpoints (Tables 2 and 3), the greatest negative impact is found on Respiratory inorganics (34.4%, 7.89 Pt, 79.935 kg PM2.5 eq), followed by

Global warming (32%, 7.35 Pt, 7.28 × 10⁴ kg CO_2 eq), and Non-renewable energy (17.2%, 3.96 Pt, 6.01 × 10⁵ MJ primary). Additionally, according to the results of Du et al. [19], Global Warming, Human Toxicity, and Particulate Matter Formation are the most important environmental damage categories. This finding points to a significant negative impact of the construction materials used on the quality of human health, as PM2.5 particles are known to penetrate the pulmonary alveoli of the lower respiratory tract.

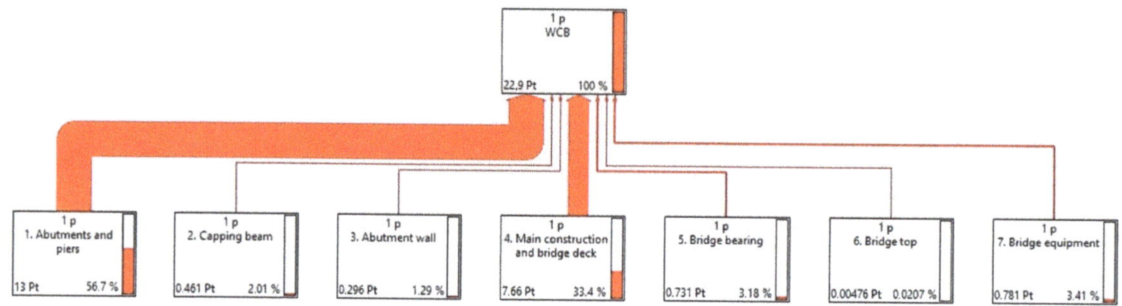

Figure 7. WCB structural damage tree (SimaPro, IMPACT 2002+).

Concrete and steel are the biggest contributors to the overall negative impacts. The negative effects of the use of other materials due to their quantity are less significant compared to concrete and steel. The findings of this research are similar to the results of other authors, such as Liu et al. [4], Du et al. [19], and Hettinger et al. [36]. According to Du et al. [19], steel and reinforcement are key materials that affect GWP (global warming potential) and CED (cumulative energy demand), and this means the bridge, which uses steel as the main load-bearing structure, has a much higher potential in the GWP impact mitigation. When looking at the initial construction of equivalent bridge designs, steel-reinforced concrete girders appear to have lower overall environmental effects than steel girders. However, steel girders are reusable and recyclable at the end of their useful life [8]. Habert et al. [18] found that, within concrete structures, the concrete used for the bridge deck clearly dominates in the impact on a conventional bridge. This effect is significantly reduced for a high-performance bridge. The carbon emission of cement is the greatest in the material production stage, which has the greatest impact on the environment in the whole life cycle of the bridge. Therefore, the decarbonization of the cement industry will have a significant impact on the carbon reduction of the infrastructure industry [4].

In the case of the alternative use of concrete reinforcement in bridge construction, namely variants of bars made of different composite materials [16], the net difference between the variants has been shown to be approximately 35.500 kg in CO_2 production. Composite materials in bridge structures [37–39] may be an alternative to the currently most widely used steel. Therefore, WCB is trying to replace concrete and steel and instead use composite materials based on renewable raw materials, i.e., wood. In their study, Jena and Kaewunruen [38] compared the use of LCA two bridge systems for footbridges made from modern composite materials. The results showed the importance of composite materials in reducing the environmental demands of bridge infrastructure.

The use of the materials itself seems important, but it is also necessary to point out the structural level of the bridge and its environmental design. Therefore, in the following sections of this article, in addition to examining the uncertainty analysis, the research focuses on comparing the environmental properties of materials for two similar bridges, but on a different design basis and materials basis.

Table 2. WCB environmental damage (Pt), midpoint categories, SimaPro, IMPACT 2002+: 1. Abutments and piers, 2. Capping beam, 3. Abutment wall, 4. Main construction and bridge deck, 5. Bridge bearing, 6. Bridge top, 7. Bridge equipment.

Impact Category	Unit	Total	1	2	3	4	5	6	7
Total	Pt	22.9	13.0	0.461	0.296	7.66	0.731	4.76×10^{-3}	0.781
Respiratory inorganics	Pt	7.89	4.54	0.152	9.09×10^{-2}	2.64	0.206	1.34×10^{-3}	0.254
Global warming	Pt	7.35	4.63	0.175	0.121	2.00	0.142	9.23×10^{-4}	0.289
Non-renewable energy	Pt	3.96	2.17	7.93×10^{-2}	5.27×10^{-2}	1.18	0.310	2.01×10^{-3}	0.155
Land occupation	Pt	1.22	0.202	7.26×10^{-3}	4.73×10^{-3}	0.988	3.04×10^{-3}	1.98×10^{-5}	1.13×10^{-2}
Terrestrial ecotoxicity	Pt	1.14	0.659	2.18×10^{-2}	1.28×10^{-2}	0.387	2.71×10^{-2}	1.76×10^{-4}	3.31×10^{-2}
Non-carcinogens	Pt	0.729	0.455	1.44×10^{-2}	7.98×10^{-3}	0.221	9.36×10^{-3}	6.09×10^{-5}	2.13×10^{-2}
Carcinogens	Pt	0.539	0.285	8.65×10^{-3}	4.46×10^{-3}	0.197	2.95×10^{-2}	1.92×10^{-4}	1.33×10^{-2}
Mineral extraction	Pt	4.57×10^{-2}	2.75×10^{-2}	7.69×10^{-4}	3.39×10^{-4}	1.48×10^{-2}	9.55×10^{-4}	6.22×10^{-6}	1.41×10^{-3}
Terrestrial acid/nutri	Pt	3.86×10^{-2}	1.93×10^{-2}	1.06×10^{-3}	9.60×10^{-4}	1.38×10^{-2}	1.63×10^{-3}	1.06×10^{-5}	1.77×10^{-3}
Aquatic ecotoxicity	Pt	2.36×10^{-2}	1.47×10^{-2}	4.58×10^{-4}	2.46×10^{-4}	7.20×10^{-3}	6.15×10^{-4}	4.00×10^{-6}	4.09×10^{-4}
Respiratory organics	Pt	1.27×10^{-2}	7.57×10^{-3}	2.31×10^{-4}	1.21×10^{-4}	3.79×10^{-3}	6.20×10^{-4}	4.04×10^{-6}	3.68×10^{-4}
Ionizing radiation	Pt	3.87×10^{-3}	1.65×10^{-3}	7.30×10^{-5}	5.79×10^{-5}	1.47×10^{-3}	4.29×10^{-4}	2.79×10^{-6}	1.82×10^{-4}
Ozone layer depletion	Pt	6.47×10^{-4}	3.64×10^{-4}	1.34×10^{-5}	8.94×10^{-6}	1.97×10^{-4}	4.52×10^{-5}	2.94×10^{-7}	1.85×10^{-5}
Aquatic eutrophication	Pt	-	-	-	-	-	-	-	-
Aquatic acidification	Pt	-	-	-	-	-	-	-	-

Table 3. WCB characterization, midpoint categories, SimaPro, IMPACT 2002+: 1. Abutments and piers, 2. Capping beam, 3. Abutment wall, 4. Main construction and bridge deck, 5. Bridge bearing, 6. Bridge top, 7. Bridge equipment.

Impact Category	Unit	Total	1	2	3	4	5	6	7
Respiratory inorganics	kg PM2.5 eq	79.935	46.018	1.540	0.921	26.778	2.088	0.014	2.577
Global warming	kg CO_2 eq	7.28×10^4	4.58×10^4	1.73×10^3	1.20×10^3	1.98×10^4	1.40×10^3	9.14	2.86×10^3
Non-renewable energy	MJ primary	6.01×10^5	3.30×10^5	1.21×10^4	8.00×10^3	1.80×10^5	4.70×10^4	3.06×10^2	2.36×10^4
Land occupation	m^2org.arable	15,293.188	2544.069	91.185	59.383	12418.011	38.155	0.248	142.137
Terrestrial ecotoxicity	kg TEG soil	1.98×10^6	1.14×10^6	3.77×10^4	2.22×10^4	6.70×10^5	4.69×10^4	3.05×10^2	5.73×10^4
Non-carcinogens	kg C_2H_3Cl eq	1846.830	1151.506	36.526	20.225	560.850	23.717	0.154	53.852
Carcinogens	kg C_2H_3Cl eq	1364.350	722.252	21.919	11.296	499.937	74.840	0.487	33.619
Mineral extraction	MJ surplus	6950.275	4176.453	116.894	51.509	2245.212	145.210	0.945	214.052
Terrestrial acid/nutri	kg SO_2 eq	508.365	254.687	13.911	12.648	182.140	21.524	0.140	23.315
Aquatic ecotoxicity	kg TEG water	6.45×10^6	4.01×10^6	1.25×10^5	6.72×10^4	1.97×10^6	1.68×10^5	1.09×10^3	1.12×10^5
Respiratory organics	kg C_2H_4 eq	42.308	25.220	0.771	0.402	12.610	2.065	0.013	1.226
Ionizing radiation	Bq C-14 eq	1.31×10^5	5.58×10^4	2.47×10^3	1.96×10^3	4.97×10^4	1.45×10^4	9.43	6.15×10^3
Ozone layer depletion	kg CFC-11 eq	0.004	0.002	0.000	0.000	0.001	0.000	0.000	0.000
Aquatic acidification	kg SO_2 eq	44.843	10.394	1.814	2.263	20.710	6.258	0.041	3.365
Aquatic eutrophication	kg PO_4 P-lim	8.873	5.089	0.196	0.137	2.706	0.471	0.003	0.271

3.2. Comparison of WCB and SCB Environmental Performance

The design level and its environmental impacts are evaluated via a comparison of WCB and SCB (Table 1, Figures 4–6) per m^2 of the bridge area. Ideally, a WCB and an SCB design for exactly the same situation would be the only comparable bridges but using m^2 is the best solution for bridge structures of different lengths and different widths. LCIA was evaluated with the IMPACT 2002+ method, which contains 15 midpoints of environmental damage (Table 4). LCIA for bridge comparison thus provides a multicriteria analysis, using not only inputs (Table 1) but also outputs (Table 4, Figure 8), where each bridge has its own different and unique material impact on the environment. Similar conclusions were introduced by Hettinger et al. [36], who state that the environmental profile of the bridges, defined by eleven indicators, is strongly connected to the bill of quantities. Du et al.'s [19] interpretation of their environmental analysis indicates that the bridge's environmental performance is governed by multiple indicators. Each bridge has a unique performance among different indicators.

Table 4. Characterization and environmental damage of WCB and SCB per m^2 (SimaPro, IMPACT 2002+, midpoint categories).

Impact Category	Unit	WCB	SCB	Unit	WCB	SCB
Respiratory inorganics	kg PM2.5 eq	54.936	88.599	mPt	92.782	128.194
Global warming	kg CO$_2$ eq	57,215.27	81,722.027	mPt	86.419	130.184
Non-renewable energy	MJ primary	468,549.53	701,965.683	mPt	46.540	75.441
Land occupation	m^{2}org.arable	14,194.63	4420.413	mPt	14.310	5.211
Terrestrial ecotoxicity	kg TEG soil	1,333,088.65	2,042,480.86	mPt	13.415	16.439
Non-carcinogens	kg C$_2$H$_3$Cl eq	1126.85	2041.858	mPt	8.575	10.757
Carcinogens	kg C$_2$H$_3$Cl eq	866.241	1339.417	mPt	6.334	6.985
Mineral extraction	MJ surplus	3607.60	8748.035	mPt	0.538	0.773
Terrestrial acid/nutri	kg SO$_2$ eq	622.813	382.403	mPt	0.454	0.650
Aquatic ecotoxicity	kg TEG water	3,830,468.60	5,662,764.03	mPt	0.278	0.225
Respiratory organics	kg C$_2$H$_4$ eq	25.169	49.209	mPt	0.149	0.203
Ionizing radiation	Bq C-14 eq	128,415.22	135,768.087	mPt	0.046	0.077
Ozone layer depletion	kg CFC-11 eq	0.003	0.004	mPt	0.008	0.009
Aquatic acidification	kg SO$_2$ eq	107.981	2.688	mPt	0.000	0.000
Aquatic eutrophication	kg PO$_4$ P-lim	7.292	7.98	mPt	0.000	0.000
Total	-	-	-	mPt	269.847	375.147

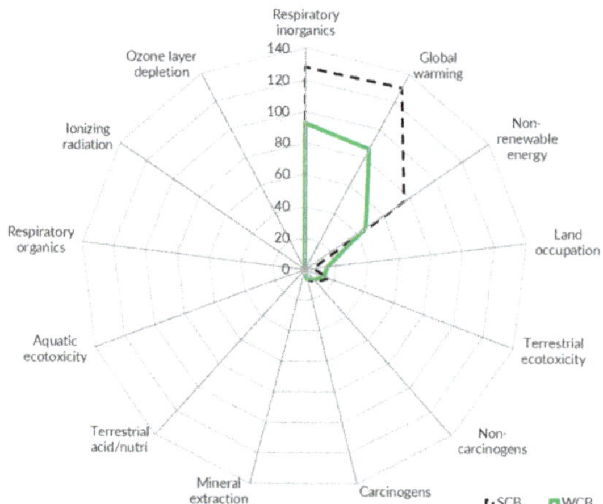

Figure 8. Comparison of WCB and SCB environmental damage (mPt) per m^2 (SimaPro, IMPACT 2002+, midpoint categories).

From Table 4, where the specific values of emissions to the environment are given at the midpoints (characterizations), and from Figure 9, where the total environmental damage is presented in standard eco point units (mPt, mili eco point), it can be seen that the most significant damage from both bridges is within three categories of environmental damage: Respiratory inorganics (WCB = 54.9, SCB = 88.6, kg PM2.5 eq), Global warming (WCB = 57,215, SCB = 81,722, kg CO_2 eq), and Non-renewable energy (WCB = 468,549, SCB = 701,965, MJ primary). These three categories form the triangle of the most significant environmental damage, accounting for 84% of the total negative environmental impacts for WCB and 89% for SCB (Figure 9).

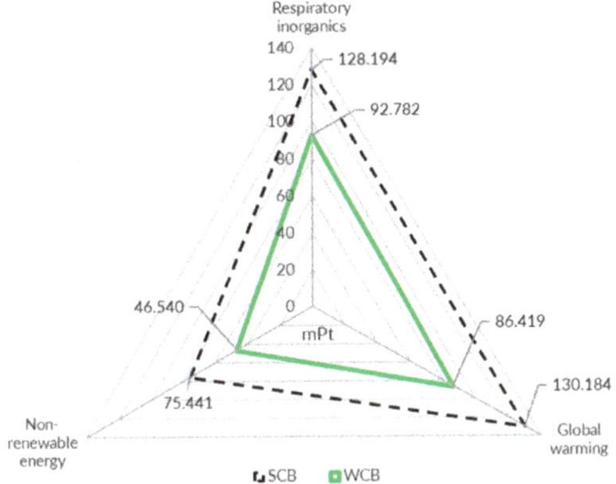

Figure 9. Comparison of three most significant WCB and SCB environmental damage (mPt) per m^2 (SimaPro, IMPACT 2002+, midpoint categories).

Pedneault et al. [11] analysis has shown a carbon footprint of 8960 kg CO_2 eq/m^2 for a concrete deck on steel beams (CD) and 4870 kg CO_2 eq/m^2 for the aluminum deck (AD) on steel beams according to the scope 2 (complete life cycle without traffic diversion), it is slightly higher than the steel–concrete bridge with 4090 kg CO_2 eq/m^2 [40] and a few times higher the concrete slab frame bridges with 1370 kg CO_2 eq/m^2 [19]. The study from Hammervold et al. [12] demonstrates a ten times lower carbon footprint per m^2 compared to the study from Pedneault et al. [11], but they considered only routine repair actions and studied different bridge designs and lengths.

Compared to this case, the values are about 10 times smaller; for example, the amount of steel in the study of Pedneault et al. [11] is 136.82 kg/m^2 for CD and 78.08 kg/m^2 for AD, compared to 114.276 kg/m^2 for WCB and 418.824 kg/m^2 for SCB in this study. Pedneault et al. [11] used the same evaluation method but different software to calculate the damage. In addition, this research deals purely with the material level of the bridges with different structural systems and a different scope of inventory analysis. The reserves in the calculations using the database program are also confirmed by Habert et al. [18], which state that for the prestressed steel, it is commonly accepted that this steel is essentially produced using the blast furnace method. However, because the efficiency of a blast furnace plant can be different from one plant to another, it was assumed that the environmental load could be 20% higher than the load calculated with Ecoinvent data. Therefore, trends may be more interesting than absolute numbers. Material damage trends are the same. This means that the greatest damage to the road bridges that were compared is caused by concrete and steel in the construction. The other materials used for WCB and SCB have a significantly smaller impact on the environment than concrete and steel.

Several authors [7,8,41,42] concluded that concrete is a more environmentally friendly material than steel, but steel is used in lower quantities for the same design conditions. It may be similarly advantageous in terms of environmental impact in certain design solutions. By comparing composite constructions, coupling steel beams with an aluminum bridge deck, and steel beams coupled with a concrete bridge deck [11], it is shown that all categories of environmental impact are in favor of aluminum bridge decks. One of the reasons is the environmental contribution of the recycling of steel and aluminum. Their use as a substitute for original material (concrete) can reduce environmental impact by up to 40%. Martínez-Muñoz et al. [28] state that prestressed concrete is the best alternative for bridge lengths less than 17 m. The prestressed cellular concrete deck is the best alternative for bridge lengths between 17 m and 25 m because no box girder solution is used. For lengths between 25 m and 40 m, the best solution depends on the percentage of recycled structural steel. If this value is more than 90%, then the best alternative is a composite box-girder bridge deck. However, if the value of recycled structural steel is lower than 90%, then the most environmentally friendly alternative is a prestressed concrete bridge deck.

It is desirable to minimize these two main materials (concrete and steel) in bridge structures and to reasonably replace them with a suitable composite material incorporating its design advantages and possibilities.

When comparing the environmental damages of WCB and SCB, it is clear that a difference of 28% per m^2 is in favor of WCB. The difference in the use of steel in the compared bridges is 304.547 kg/m^2, and concrete 157.742 kg/m^2. This means that the use of a composite based on renewable raw material, i.e., wood, and its proper integration into the bridge construction system can significantly affect the overall environmental impacts as well as impacts on the individual components of the environment. The main difference is in the categories of environmental impacts: Global warming (43.765 mPt, 14,506.757 kg CO_2 eq), followed by Respiratory inorganics (35.412 mPt, 33.663 kg PM2.5 eq), and Non-renewable energy (28.901 mPt, 233,416.153 MJ primary). The only two categories of environmental impacts where the WCB has worse results are the Land occupation (9.099 mPt, 9774.217 m^2org.arable) and the Aquatic ecotoxicity (0.053 mPt, 183.229 $\times$ 10^4 kg TEG water) (Table 4, Figure 8). This is understandable in context with the use of wood as a construction material. Given their overall impact (Figure 9, Table 4), they appear less significant. Hettinger et al. [36] dealt with a similar topic. The result of this research is that the composite bridge generates significantly less environmental impacts than its equivalent made of prestressed concrete.

As already mentioned, wood as a material is considered one of the main renewable resources. From the point of view of ecology and environmental sustainability, the use of wood and wood-based composite materials is also very advantageous in construction because it is one of the major contributors to greenhouse gas emissions. When comparing different wood products, bricks, concrete, and steel products in terms of global warming impact, it has been shown that concrete and brick are at the lower end of the wood GWP range, and the cement and steel are at the top of the wood GWP range [43]. Wooden beams have become an ecological and cost-effective variant for steel and concrete elements. In terms of the structural system in bridge engineering, wood most often appears as a glued laminated timber (GLT) element in arched, girder, or hybrid structural systems. GLT is, in many cases, combined with steel or concrete in the form of slabs to increase the overall load-bearing capacity of the structural systems [44,45]. The planned timber is most often used for the construction of truss, hanging, and strut-framed constructional systems. The advantage of these systems is not only the use of renewable material but also the use of a lower volume of materials for a larger span than other constructional systems. In terms of mechanical parameters, wood can fully replace other commonly used materials and at the same time contribute to improving the quality of the environment.

4. Conclusions

Human society is increasingly burdening the environment with its anthropogenic activities in construction. It is therefore essential to try to use building materials and construct buildings that are environmentally friendly. Therefore, this article deals with determining the environmental performance of a road wooden–concrete bridge using Life Cycle Assessment (LCA). The research was a comparison of the wood–concrete bridge with the steel–concrete bridge, where both serve as a road bridge for short spans.

When comparing the wood–concrete and steel–concrete road bridges, higher environmental performance for the wood–concrete bridge was demonstrated. The comparison showed that the wood–concrete bridge has more favorable environmental properties. The advantage of WCB is, for example, the use of renewable resources, especially wood, as a material for the main load-bearing structure. WCB is an adequate alternative to steel, concrete, and other composite structural systems with similar load and dimensional parameters. It can be observed that the involvement of renewable bio-material resources in the construction industry in road infrastructure can have a positive impact on the environment. At the same time, it turned out that the environmental properties of wood-based composite materials will have a significant positive effect with the help of a suitable construction solution.

The results can be used for a good comparison of the environmental characteristics of bridge structures, roads, or urban infrastructure. WCB is an innovative design system that has proven to be suitable for environmental impact due to its LCA method evaluation. Based on this environmental assessment, it can be stated that clear knowledge of all phases of the life cycle of building materials and structures is a necessary step to obtain objective findings about environmental damage or the environmental benefits of building materials or structures.

Author Contributions: Conceptualization, J.M.; methodology, J.M.; software, J.M.; validation, J.M., R.F., E.M. and L.K.; formal analysis, J.M.; investigation, J.M., E.M. and L.K.; resources, R.F.; data curation, J.M. and R.F.; writing—original draft preparation, J.M., E.M. and L.K.; writing—review and editing, J.M., E.M. and L.K.; visualization, J.M., E.M. and L.K; supervision, J.M. and L.K.; project administration, J.M. All authors have read and agreed to the published version of the manuscript.

Funding: This research was funded by Faculty of Forestry and Wood Sciences, Czech University of Life Sciences Prague.

Institutional Review Board Statement: Not applicable.

Informed Consent Statement: Not applicable.

Data Availability Statement: Data are available in article.

Acknowledgments: Research was carried out thanks to the Department of Wood Processing and Biomaterials, Faculty of Forestry and Wood Sciences, Czech University of Life Sciences Prague.

Conflicts of Interest: The authors declare no conflict of interest.

References

1. Czech Republic Supreme Audit Office. Stav Mostů v ČR k 1.1.2020. 2020. Available online: https://www.nku.cz/cz/kontrola/analyzy/stav-mostu-id11407/ (accessed on 1 December 2022).
2. Creutzig, F.; Agoston, P.; Minx, J.; Canedell, J.G.; Andrew, R.M.; Quéré, C.L.; Peters, G.P.; Sharifi, A.; Yamagata, Y.; Dhakal, S. Nature, Urban infrastructure choice's structure climate solutions. *Clim. Chang.* **2016**, *6*, 1054–1056. [CrossRef]
3. Thacker, S.; Adshead, D.; Fay, M.; Hallegatte, S.; Harvey, M.; Meller, H.; O'Regan, N.; Rozenberg, J.; Watkins, G.; Hall, J.W. Infrastructure for sustainable development. *Nat. Sustain.* **2019**, *2*, 324–331. [CrossRef]
4. Liu, Y.; Wang, Y.; Shi, C.; Zhang, W.; Luo, W.; Wang, J.; Li, K.; Yeung, N.; Kite, S. Assessing the CO_2 reduction target gap and sustainability for bridges in China by 2040. *Renew. Sustain. Energy Rev.* **2022**, *154*, 111811. [CrossRef]
5. *ISO/TS 21929-2*; Sustainability in Building Construction—Sustainability Indicators—Part 2: Framework for the Development of Indicators for Civil Engineering Works. ISO: Geneva, Switzerland, 2015.
6. Onat, N.C.; Kucukvar, M. Carbon footprint of construction industry: A global review and supply chain analysis. *Renew. Sustain. Energy Rev.* **2020**, *124*, 109783. [CrossRef]
7. Balogun, T.B.; Tomor, A.; Lamond, J.; Gouda, H.; Booth, C.A. Life-cycle assessment environmental sustainability in bridge design and maintenance. *Proc. ICE Eng. Sustain.* **2020**, *173*, 365–375. [CrossRef]
8. Horvath, A.; Hendrickson, C.T. Steel vs. Steel-Reinforced Concrete Bridges: Environmental Assessment. Journal of Infrastructure Systems. *ASCE* **1998**, *4*, 111–117. [CrossRef]
9. Zhang, C.; Amaduddin, M.; Canning, L. Carbon dioxide evaluation in a typical bridge deck replacement project. *Proc. ICE Energy* **2011**, *164*, 183–194. [CrossRef]
10. Du, G.; Karoumi, R. Environmental life cycle assessment comparison between two bridge types: Reinforced concrete bridge and steel composite bridge. In Proceedings of the Sustainable Construction Materials and Technologies, Kyoto, Japan, 8 February 2013.

11. Pedneault, J.; Desjardins, V.; Margni, M.; Conciatori, D.; Fafard, M.; Sorelli, L. Economic and environmental life cycle assessment of a short-span aluminium composite bridge deck in Canada. *J. Clean. Prod.* **2021**, *310*, 127405. [CrossRef]
12. Hammervold, J.; Reenaas, M.; Brattebø, H. Environmental Life Cycle Assessment of Bridges. *J. Bridge Eng.* **2013**, *18*, 153–161. [CrossRef]
13. Patel, J. Bridging the gap: Enabling lower carbon footprint and creating economic value from application of modern high strength niobium steels. In Proceedings of the IABSE Conference, Kuala Lumpur 2018: Engineering the Developing World—Report, Kuala Lumpur, Malaysia, 25–27 April 2018; Volume 2018, pp. 867–874. [CrossRef]
14. Zheng, Y.; Zhou, L.; Xia, L.; Luo, Y.; Taylor, S. Investigation of the behaviour of SCC bridge deck slabs reinforced with BFRP bars under concentrated loads. *J. Eng. Struct.* **2018**, *171*, 500–515. [CrossRef]
15. Hajiesmaeili, A.; Pittau, F.; Denarié, E.; Habert, G. Life Cycle Analysis of Strengthening Existing RC Structures with R-PE-UHPFRC. *Sustainability* **2019**, *11*, 6923. [CrossRef]
16. Cadenazzi, T.; Dotelli, G.; Rossini, M.; Nolan, S.; Nanni, A. Cost and environmental analyses of reinforcement alternatives for a concrete bridge. *Struct. Infrastruct. Eng.* **2020**, *16*, 787–802. [CrossRef]
17. Niu, Y.; Fink, G. Life Cycle Assessment on modern timber bridges. *Wood Mater. Sci. Eng.* **2019**, *14*, 212–225. [CrossRef]
18. Habert, G.; Arribe, D.; Dehove, T.; Espinasse, L.; Le Roy, R. Reducing environmental impact by increasing the strength of concrete: Quantification of the improvement to concrete bridges. *J. Clean. Prod.* **2012**, *35*, 250–262. [CrossRef]
19. Du, G.; Pettersson, L.; Karoumi, R. Soil-steel composite bridge: An alternative design solution for short spans considering LCA. *J. Clean. Prod.* **2018**, *189*, 647–661. [CrossRef]
20. Bouhaya, L.; Le Roy, R.; Feraille-Fresent, A. Simplified environmental study on innovative bridge structure. *Environ. Sci. Technol.* **2009**, *43*, 2066–2071. [CrossRef]
21. Steele, K.; Cole, G.; Parke, G.; Clarke, B.; Harding, J. Highway bridges and environment-sustainable perspectives. *Proc. Inst. Civ. Eng.* **2003**, *156*, 176–182. [CrossRef]
22. Lounis, Z.; Daigle, L. Environmental Benefits of Life Cycle Design of Concrete Bridges. In Proceedings of the 3rd International Conference on Life Cycle Management, Zurich, Switzerland, 27–29 August 2007; pp. 1–6.
23. Gervasio, H.; Da Silva, L.S.O.E. Comparative life-cycle analysis of steel-concrete composite bridges. *Struct. Infrastruct. Eng.* **2008**, *4*, 251–269. [CrossRef]
24. World Steel Association, Steel Statistical Yearbook 2018. Brussels, Belgium. 2018. Available online: https://worldsteel.org/wp-content/uploads/Steel-Statistical-Yearbook-2018.pdf (accessed on 1 December 2022).
25. Kim, K.J.; Yun, W.G.; Cho, N.; Ha, J. Life cycle assessment based environmental impact estimation model for pre-stressed concrete beam bridge in the early design phase. *Environ. Impact Assess. Rev.* **2017**, *95*, 47–56. [CrossRef]
26. Yepes, V.; Martí, J.V.; García-Segura, T. Cost and CO_2 emission optimization of precast–prestressed concrete U-beam road bridges by a hybrid glowworm swarm algorithm. *Autom. Constr.* **2015**, *49*, 123–134. [CrossRef]
27. Pang, B.; Yang, P.; Wang, Y.; Kendall, A.; Xie, H.; Zhang, Y. Life cycle environmental impact assessment of a bridge with different strengthening schemes. *Int. J. Life Cycle Assess.* **2015**, *20*, 1300–1311. [CrossRef]
28. Martínez-Muñoz, D.; Martí, J.V.; Yepes, V. Comparative Life Cycle Analysis of Concrete and Composite Bridges Varying Steel Recycling Ratio. *Materials* **2021**, *14*, 4218. [CrossRef] [PubMed]
29. *EN 1992-1-1*; Eurocode 2: Design of Concrete Structures—Part 1-1: General Rules and Rules for Buildings. CEN: Bruxelles, Belgium, 2004.
30. *EN 1993-1-1*; Eurocode 3: Design of Steel Structures—Part 1-1: General Rules and Rules for Buildings. CEN: Bruxelles, Belgium, 2005.
31. *ISO 14040*; Environmental Management, Life Cycle Assessment-Principles and Framework. ISO: Geneva, Switzerland, 2006.
32. *ISO 14044*; Environmental Management, Life Cycle Assessment-Requirements and Guidelines. ISO: Geneva, Switzerland, 2006.
33. PRé Consultants, SimaPro Sustainability Software for Fact-Based Decisions, Introduction to LCA with SimaPro 2019. Available online: https://pre-sustainability.com/solutions/tools/simapro/ (accessed on 4 June 2021).
34. Jolliet, O.; Margni, M.; Charles, R.; Humbert, S.; Payet, J.; Rebitzer, G.; Rosenbaum, R. IMPACT 2002+: A new life cycle assessment methodology. *Int. J. Life Cycle Assess.* **2003**, *8*, 324–330. [CrossRef]
35. Ecoinvent.org., Ecoinvent Database. 2021. Available online: https://www.ecoinvent.org/database/database.html (accessed on 23 May 2021).
36. Hettinger, A.; Birat, J.; Hechler, O.; Braun, M. Sustainable bridges—LCA for a composite and a concrete bridge. In *Economical Bridge Solutions Based on Innovative Composite Dowels and Integrated Abutments*; Petzek, E., Băncilă, R., Eds.; Springer Vieweg: Wiesbaden, Germany, 2015. [CrossRef]
37. Sonnenschein, R.; Gajdosova, K.; Holly, I. FRP Composites and their Using in the Construction of Bridges. *Procedia Eng.* **2016**, *161*, 477–482. [CrossRef]
38. Jena, T.; Kaewunruen, S. Life Cycle Sustainability Assessments of an Innovative FRP Composite Footbridge. *Sustainability* **2021**, *13*, 13000. [CrossRef]
39. Ali, H.T.; Akrami, R.; Fotouhi, S.; Bodaghi, M.; Saeedifar, M.; Yusuf, M.; Fotouhi, M. Fiber reinforced polymer composites in bridge industry. *Structures* **2021**, *30*, 774–785. [CrossRef]

40. Mara, V.; Haghani, R.; Sagemo, A.; Storck, L.; Nilsson, D. Comparative study of different bridge concepts based on life-cycle cost analyses and life-cycle assessment. In Proceedings of the Asia-Pacific Conference on FRP in Structures, Melbourne, Australia, 11–13 December 2013.
41. Widman, J. Environmental impact assessment of steel bridges. *J. Constr. Steel Res.* **1998**, *46*, 291–293. [CrossRef]
42. Collings, D. An environmental comparison of bridge forms. *Bridge Eng.* **2006**, *159*, 163–168. [CrossRef]
43. Hill, C.A.S.; Dibdiakova, J. The environmental impact of wood compared to other building materials. *Int. Wood Prod. J.* **2016**, *4*, 215–219. [CrossRef]
44. Dias, A.M.P.G.; Ferreira, M.C.P.; Jorge, L.F.C.; Martins, H.M.G. Timber–concrete practical applications–bridge case study. *Struct. Build.* **2011**, *164*, 131–141. [CrossRef]
45. Rodrigues, J.N.; Dias, A.M.P.G.; Providência, P. Timber-Concrete Composite Bridges: State-of-the-Art Review. *BioResources* **2013**, *8*, 6630–6649. [CrossRef]

Article

Management of Forest Residues as a Raw Material for the Production of Particleboards

Marta Pędzik [1,2], Karol Tomczak [1,3], Dominika Janiszewska-Latterini [1], Arkadiusz Tomczak [3] and Tomasz Rogoziński [2,*]

1   Center of Wood Technology, Łukasiewicz Research Network—Poznan Institute of Technology, 60-654 Poznań, Poland
2   Department of Furniture Design, Faculty of Forestry and Wood Technology, Poznań University of Life Sciences, 60-625 Poznań, Poland
3   Department of Forest Utilization, Faculty of Forestry and Wood Technology, Poznań University of Life Sciences, 60-625 Poznań, Poland
*   Correspondence: tomasz.rogozinski@up.poznan.pl

Abstract: Expanding the base of raw materials for use in the production of wood-based materials, researchers and panel manufacturers around the world are increasingly trying to produce panel prototypes from raw materials available in a given area and climate, or by managing waste from wood industry processing. The aim of the study was therefore to test the hypothesis that forest residues de-rived from Scots pine roundwood harvesting have the same suitability for the production of three-layer particleboard as the wood of the most valuable part of the Scots pine stem, by comparing selected properties of raw wood material and final product—particleboard. Characterization of both the raw material and the physical-mechanical and hygienic properties of the produced panels was carried out. For these panels from the tree trunk, MOR was 14.6 N/mm^2, MOE 1960 N/mm^2 and IB 0.46 N/mm^2. The MOR and IB values turned out to be higher for the panel from the branch and are 16.5 and 0.72 N/mm^2, respectively. Excessive swelling of the panels resulted in all manufactured particleboards meeting the standardized performance requirements of EN 312 for interior furnishing panels (including furniture) for use in dry conditions (type P2).

Keywords: wood-based materials properties; formaldehyde; alternative raw materials; forest residues

Citation: Pędzik, M.; Tomczak, K.; Janiszewska-Latterini, D.; Tomczak, A.; Rogoziński, T. Management of Forest Residues as a Raw Material for the Production of Particleboards. *Forests* **2022**, *13*, 1933. https://doi.org/10.3390/f13111933

Academic Editor: Nadir Ayrilmis

Received: 27 October 2022
Accepted: 14 November 2022
Published: 16 November 2022

Publisher's Note: MDPI stays neutral with regard to jurisdictional claims in published maps and institutional affiliations.

1. Introduction

According to the Food and Agricultural Organization of the United Nations report [1], global roundwood production in 2020 (including wood fuel (WF) and industrial roundwood (IR)) was estimated at 3966 million m^3 (WF—1945 million m^3 and IR 2021 million m^3). Compared to 2000 [2], global timber production has increased by about 24%. Available models and calculations show that if the world's human population reaches 10 billion, the demand for wood will be higher than the world's supply of this raw material, which may lead to increased wood prices and uncontrolled deforestation of protected forest areas for illegal wood trading. The importance of forests and the need to protect its resources is the reason to take steps toward environmental sustainability as one of the main parameters when selecting raw materials for industrial purposes. Therefore, to maintain stability in the production of wood and protect the most valuable wood resources, measures should be taken to increase the production of wood-based materials that can replace raw wood.

Forests residues (FR) represent a raw wood material which occurs during logging operations in both mature stands and thinning interventions [3,4]. Forest residues that are eligible for further use include branches, with needles and leaves, and tree tops, as well as undersized or damaged stem parts [5]. FR is usually left to decompose naturally or for the local population, for domestic heating purposes [5–7]. However, FR has the potential to be a suitable renewable source of energy also at the industrial scale [8]. While searching for alternative uses of a given raw material, it is necessary to take into account the three pillars of sustainable development, which are environmental, social, and economic aspects [9]. According to estimates, the amount of forest residues available for further use in Polish forests ranges from a few to several million cubic meters [10,11]. The average share of FR in the total mass of harvested wood is about 11% and depends mainly on the fertility of the habitat [6,12]. The average wood volume of forest residues (mainly pine wood residues) in the total harvest was estimated at 37.7 m^3/ha [6].

Besides utilizing FR for energy purposes [5,13–15], one of the possible alternative use of forest residues is to exploit them as material for wood-based boards, i.e., particleboards [16,17]. Currently, forest wood, lignocellulosic biomass [18,19], recycled wood [20,21], and industrial wood (residues in the form of cuttings and sawdust) [22–24] are processed for the production of wood-based materials including particleboards. Combinations of lignocellulosic raw materials, recycled wood and industrial particles [25], and even wood bark [26,27] can also be used. By expanding the base of raw materials for use in the production of wood materials, researchers and panels producers from all around the world are more and more often trying to produce prototypes of boards from raw materials available in a given area and climate, or using waste from the processing of the wood industry for added-value applications. The search for new alternative raw materials is one of the key issues in line with the bioeconomy approach, especially for particleboard production, which is in high demand [28].

In addition, it is possible to produce boards from raw materials of other origins, such as annual plant residues, e.g., straws and grasses [28,29], crop residues [30,31], post-production residues from the food industry, e.g., seed husks [32,33], or residues from garden tree pruning [34,35], sugarcane bagasse [29] and other lignocellulosic wastes [18,19]. When cereal straw is used, to obtain a satisfactory board quality, binders other than those used commonly are often employed to improve the adhesion of straw particles [36,37]. When using alternative raw materials other than wood, it is most advantageous to use mixtures of wood and non-wood raw materials, as this allows to control of the strength parameters of the manufactured boards and reduces the radicality of adjusting technological parameters. At the same time, the use of alternative raw materials additives contributes to the reduction of significant amounts of waste (of various origins) and also reduces the costs of purchasing raw materials, which has positive environmental and economic aspects.

Several studies have focused on the evaluation of FR for particleboard production. Iwakiri et al. 2017 reported that it is possible to produce particleboards based on 50% of forest residues from branches, tree tops, stumps, and roots in a mixture with industrial pine particles highlighting that the obtained panels meet the requirements of EN 312 standard [38]. Panels produced from a mixture of industrial particles and forest residues in a 50/50% ratio were considered the maximum possible, as they showed a statistically equal average to panels produced with a predominance of industrial particles and tree tops and branches (in a 75/25% ratio) and also a mixture of three other types of materials. Wronka and Kowaluk (2022) proved that there is a possibility to use different content (up to 50%) of Scots pine (*Pinus sylvestris* L.) branches to produce three-layer particleboard. Their results indicate that the higher bulk density of branch particles than the industrial material, along with an increase in the content of branch particles, results in a significant increase in internal bonding. On the other hand, the values of flexural strength and modulus of elasticity decreased with an increase in the content of branch particles up to 100%. Therefore, a 50% content in the panel of branch particles characterized by a maximum diameter of 40 mm was also considered maximum [39]. Nurek et al. 2019 showed that the mean density of unfractionated forest residues was app. 800 kg/m^3 [8]. They also noticed the correlation between density and practicle size—the smaller the fraction the greater the density. Rahman et al. (2013) showed that a practical board from stem wood has better properties than particleboard made from branches and twigs of surian tree (*Toona sinensis* Roem) [40]. In addition, wood from garden resources (citrus branches) and beech twigs (*Fagus orientalis* Lipsky) and poplar wood trunks (*Populus alba* L.) can be successfully used to produce particleboards with good mechanical properties [41]. On the other hand, Nemli et al. 2004 identified the presence of branch wood as a parameter which is negatively influencing the mechanical properties of particleboards produced from black locust biomass [42]. The same phenomenon was observed by Kowaluk et al. (2011). Therefore, further research is needed to test the possibility of applying FR in the sustainable production of particleboards and converting them into value-added products [43]. Most of the published studies are focusing on the properties of wood-based boards produced from mixing wood and branches wood [30,39] or other materials [19,33,35,41]. Moreover, in particular, knowledge about the properties of particleboards produced from FR derived from logging interventions in Scots pine (*Pinus sylvestris* L.) stands is still limited.

The aim of the study was therefore to test the hypothesis that forest residues derived from Scots pine roundwood harvesting have the same suitability for the production of three-layer particleboard as the wood of the most valuable part of the Scots pine stem, by comparing selected properties of raw wood material and the final product—particleboard.

2. Materials and Methods

2.1. Raw Wood Material

The wood used in the study was harvested as part of a clearcut in a 94-year-old Scots pine stand in the State Forests in Poland. Wood raw material from 3 trees was taken and debarked for laboratory analyses and particular board preparation. Figure 1 presents graphically the elements of the raw material obtained at individual stages.

Figure 1. The process of obtaining individual elements of raw materials.

A 1.0 m long log was taken from each tree, measured from the base of the trunk, and pieces of branches were cut from each crown. Immediately after the tree was cut, approximately 4 cm thick discs were cut at the breast height from each tree for green (GD) and basic (BD) density estimation. Respectively the same was done with branches. Each disc was weighed with an accuracy of 0.001 g using a Steinberg laboratory scale (Steinberg Systems SBS-LW-200A, Berlin, Germany). After weighing, the thickness of each sample was measured to an accuracy of 0.01 mm using a certified Vogel calliper in five different places, and the diameter of each disc was measured by using the cross-over method according to the minimum and maximum size of the sample (Vogel Germany GmbH & Co. KG, Kevelaer, Germany). Then, the volume of each disc was calculated by Equation (1) according to Pérez-Harguindeguy et al. dimensional methodology [44]:

$$V = \pi \times (0.5r)^2 \times h \qquad (m^3) \qquad (1)$$

where V is the volume of disc, r is the average diameter of the disc, and h is the average height of each sample.

In the next stage, discs were transported to the laboratory where were dried for 48 h at 105 °C in a laboratory oven. After drying the samples were placed in a desiccator until cooled. Next, each sample was weighed and measured by the same procedure that had been used for samples in fresh condition. The green density was calculated using Equation (2), while basic density was calculated using Equation (3):

$$GD = \frac{m_f}{v_m} \ (kg/m^3) \tag{2}$$

$$BD = \frac{m_s}{v_m} \ (kg/m^3) \tag{3}$$

where m_f is the mass of fresh felled wood, m_s is the mass of an oven-dried sample and v_m is the volume of the fresh sample.

The remaining parts of both raw materials were first cut into smaller elements on a format saw (Felder, Hall in Tirol, Austria) and then shredded in a cutting mill Condux (Mankato, MN, United States). Obtained particles were divided into fractions on a vibrating sorter (Allgaier, Uhingen, Germany) equipped with sieves with mesh sizes: 8.0, 2.0, 1.0, and 0.5 mm. The core layer fraction consisted of particles retained on a 2.0 mm sieve, and the surface layers were particles from a sieve 1.0 mm in diameter and smaller. The fractional composition of the tree stem and branches stem intended for core layers are determined on a laboratory vibrating screen AS 200 tap (Fritsch, Idar-Oberstein, Germany) with the following sets of mesh sieves: 8.0, 4.0, 2.0, 1.0, 0.50, 0.25, and <0.25 mm. Their bulk density (ρ) was also determined according to Equation (4):

$$\rho = \frac{m_c - m_n}{V} \ (kg/m^3) \tag{4}$$

where m_c is the weight of the measuring vessel with the raw material (kg), m_n is the mass of the measuring vessel (kg) and V is the capacity measuring vessel (m^3).

2.2. Adhesives

A melamine-urea-formaldehyde (MUF) adhesive (Swiss Krono Sp. z o.o., Żary, Poland) was used for the production of particleboards. Table 1 presents the selected properties of the adhesive. The resination was 11% for the surface layers and 9% for the core layer particles of dry resin content referred to dry particles (*w/w*). The hardener is a 40% water solution of NH_4NO_3 which was used in an amount of 2 wt.% for a core layer and 3 wt.% for the surface layers. A paraffin emulsion (0.8 wt.% for the core layer and 0.2 wt.% for the surface layers) was also added to the resin to protect the manufactured boards from exposure to water.

Table 1. Properties of MUF adhesive.

Properties	Value
Dry mass	67.5%
Relative density	1.27 g/cm^3
pH	8.4
Dynamic viscosity	326 mPa s
Gel time	95 s

2.3. Particleboard Production

Two three-layer particleboards with a nominal density of 670 kg/m^3 and dimensions 500 × 700 × 16 mm were produced the share of surface layers in the panels was 35 wt.%. The raw materials were dried at 100 °C to a moisture content of 2–3% in a chamber dryer with forced air circulation (Ashaki Kagaku Co., Ltd., Tokyo, Japan) and then glued in a rotary laboratory sealer with pneumatic spraying (Lodige, Paderborn, Germany).

The boards were pressed on a hydraulic single-level press (Simpelkamp, Krefeld, Germany) using the pressing parameters: unit pressure 2.5 MPa, temperature 200 °C, pressing factor 7 s per one mm of nominal board thickness After manufacturing, respectively, the boards were conditioned in an air-conditioning chamber at a relative humidity of $65 \pm 5\%$ and a temperature of 20 ± 2 °C. Then parameters were determined, such as density, modulus of elasticity in bending (MOE), modulus of rupture (MOR), internal bond (IB), thickness swelling (TS) after 24 h water immersion, and water absorption (WA), formaldehyde content by perforator method and formaldehyde emission by the chamber method according to procedures defined in the European standards: EN 120 [45], EN 310 [46], EN 317 [47], EN 319 [48], EN 323 [49], EN 717-1 [47].

2.4. Statistical Analysis

In the first step, the Shapiro–Wilk test was performed to verify the normal distribution of data. The result of the test rejected the normal distribution hypothesis. To compare data between samples the non-parametric Mann–Whitney U test was performed. Statistical inference was performed at significance level $\alpha = 0.05$. The program Statistica 13.1 (TIBCO Software Inc., Palo Alto, CA, USA) and RStudio and the R package (R Core Team 2022) were used for the calculations.

3. Results

3.1. Characteristics of Raw Material

Wood discs collected from branches were characterized by a greater green density of wood than wood collected from tree stems. The difference between these two wood locations was app. 11 kg/m^3. The maximum GD of branches was over 1000 kg/m^3 and was higher than the maximum stem wood GD by 61.83 kg/m^3 (Table 2).

Table 2. Basic statistics of green density (kg/m^3) of raw wood material collected from tree stems and branches of wood.

Raw Wood Material	Mean	Sd. Dev.	Minimum	Maximum	Median	U-W Test
Tree stem	982.26	6.69	976.21	989.44	981.13	Ns *
Tree branches	993.20	40.18	938.75	1051.27	995.52	

* Ns—no significant differences.

When in the case of basic density wood from branches was characterized by lower density. The differences between obtained results from the stems and branches were app. 160 kg/m^3 and was statistically significant (Table 3).

Table 3. Basic statistics of basic density (kg/m^3) of raw wood material collected from tree stems and branches wood.

Raw Wood Material	Mean	Sd. Dev.	Minimum	Maximum	Median	U-W Test
Tree stem	555.07	14.02	545.95	571.22	548.05	0.03
Tree branches	395.41	35.54	340.40	436.53	393.88	

The bulk density of the material is closely related to the density of the raw material from which it was obtained. Already shredded to particle form, the material from the tree trunk showed a bulk density of about 90 kg/m^3, while that from tree branches was about 70 kg/m^3.

Figure 2 presents the fractional composition of particles from tree stems and branches. The analysis shows that the fractions with the size of 4.0, 2.0, and 1.0 mm represent the largest mass share of particles obtained from the tree stem, respectively 30, 36, and 28%. In turn, for the particles from the branches, the largest fraction came from the 2.0 and 1.0 mm sieves—a total of 72% of the entire particle mixture. Very small amounts of fine particles from the sieve with a mesh smaller than 1.0 mm were observed for the trunk particles it is approx. 7% and for the branches particles approx. 4%.

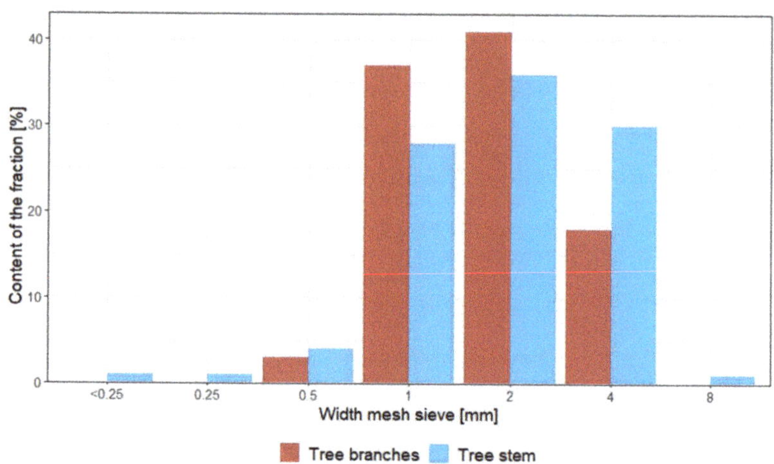

Figure 2. Fractional composition of particles from tree stem and branches.

3.2. Properties of Particleboards Depending on Raw Wood Material

Physical-mechanical properties of the investigated particleboards are reported in Table 4 and the hygienic properties are given in Table 5. No statistically significant differences were found regarding MOR. All the mechanical properties of the manufactured boards have met the requirements of P3 type 16 mm thick boards—non-load-bearing boards for use in humid conditions. For these panels, MOR 14.6 N/mm^2, MOE 1960 $N/mm^{2,}$ and IB 0.46 N/mm^2 were obtained. On the other hand, the MOR and IB values turned out to be higher for the branch panel and are respectively 16.5 and 0.72 N/mm^2. Such high values made it possible to qualify the board to the P5 type—load-bearing boards for use in humid conditions. Comparing the achieved results of internal bond tests to the EN 312 standard requirements it should be noted, that the IB values of tree branches panels exceed the minimal requirements for P7 type—heavy duty load-bearing boards for use in humid conditions, which is 0.70 N/mm^2. In the case of TS and WA, the stem panels showed better performance, but were still insufficient to achieve adequate water resistance.

Table 4. The mean value and standard deviation of examined properties of 3-layer particleboards from tree stem and branches wood.

Raw Wood Material	MOR	MOE	IB	TS	WA
	[N/mm²]			[%]	
Tree stem	14.6 ± 1.9	2960 ± 190	0.46 ± 0.08	31.2 ± 1.49	78.4 ± 3
Tree branches	16.5 ± 0.9	2640 ± 120	0.72 ± 0.04	37.6 ± 2.07	97.7 ± 1
U-W test result	Ns *	0.0031	0.0009	0.0006	0.0008

* Ns—no significant differences.

Table 5. Hygienic properties of 3-layer particleboards from tree stems and branches of wood.

Raw Wood Material	Perforator Value [mg/100 g Oven Dry Board]	Formaldehyde Emission [ppm]
Tree stem	1.2	0.025
Tree branches	1.5	0.021

The formaldehyde content was determined using the EN 120 perforator method, and it was found that boards manufactured with formaldehyde-containing resin achieved very low formaldehyde content. Many factors can affect the formaldehyde content of particleboard, and certainly, these are the pressing parameters, the raw material, the type and amount of hardener, and the properties of the resin itself: the type of resin and the molar ratio (of urea and formaldehyde) and the formaldehyde content in the resin [28]. A lower formaldehyde content was observed in panels made from the stem portion of 1.2 mg/100 g oven dry board. The obtained formaldehyde emission level is also satisfactorily low. For the E1 emission class, acceptable values for raw wood-based boards are $\leq$8mg/100 g oven dry board content and release $\leq$0.124 mg/m^3 air.

4. Discussion

The main aim of this study was to estimate the suitability of forest residues (pine branches) for the production of three-layer particleboard. Moreover, particleboards from the stems of trees from which the forest residues were collected, were produced for comparison.

Raw wood material collected from pine stems was characterized by a mean green density of around 982 kg/m^3 when the basic density of the same material was 555 kg/m^3. Obtained green mean values results were similar to results noticed on freshly felled pine logs by Tomczak et al. (2020) and Tomczak et al. (2016) [50,51]. The wood of branches is characterized by a different anatomical structure and structure than stem wood [52]. Mainly due to the high occurrence of reaction wood, which, among other things, has an impact on wood density [53,54]. The mean value of wood branches green density was 993 kg/m^3 when the basic density was 395 kg/m^3. Findings about branches basic density concerning mostly tropical species [55–57]. When studies about wood branches properties of European species are very limited. The basic density of branches of wood was significantly lower than the density of samples collected from the steam. This result is not comparable with a study carried out by Dibdiakova and Vadla (2012) [58] and Gryc et al. (2011) [59], who compared the basic density of branches and wood basic density of the steam of spruce from different sites. In both presented studies obtained branch wood density was higher than in our study.

Currently, alternative lignocellulosic raw materials are high in high demand [19] which will be able to compete with other industries due to the lack of wood, such as the paper industry, or for energy purposes. Forest residues also could be a good alternative for high quality as a raw material for board production. Alamsyah et al. (2020) produced OSB and particleboard from branches and twigs of surian tree (*Toona sinensis* Roem) [60], Rahman et al. (2013) manufactured boards from branches and stem wood in three types: only from stem wood, only from branch wood and stem-branch mix of bahdi (*Lannea coromandelica* Merr.) [40]. According to the comparison they presented, stem particles were characterized by the highest mechanical properties, then branch-stem particles, and the lowest quality were obtained in branch particles. In the study by Jahan-Latibari and Roohnia (2010), two types of forest residues were used to make particleboard, they were poplar branches, small-diameter poplar wood (3–8 cm), and beech wood [61]. Their results showed that the characteristics of particleboard made from poplar branches and small-diameter wood were comparable to that made from mature beech wood. In our study panels made of branch particles were characterized by significantly higher quality than stem ones. Which led to classified branch panels as P7 type—heavy-duty load-bearing boards for use in humid conditions according to EN 312. Such differences between our and presented studies can be explained by species' wood properties. At the stage of production of wood-based materials, it is important to increase the durability of products, which extends their life cycle, thus influencing the extension of the period of carbon sequestration in the wood contained in them, which is a beneficial environmental aspect. The use of a board with such good mechanical parameters for the manufacture of furniture from them can allow reducing the amount of waste wood-based materials created by slower consumption of products and replacement with new ones. Due to the non-emissivity of wood-based panels, the control of free formaldehyde content and emissions is important. Sustainable production of low-emission materials raises ecological issues that are extremely important for the environment and subsequent use, without polluting the air and disposing of products.

In our study, the bulk density of raw material obtained from stem wood was higher than the bulk density of the wood from branches. The research of Kowaluk et al. 2019 used a mixture of industrial wood chips only with an admixture of coniferous wood (mainly *Pinus sylvestris* L.), thanks to which it obtained a bulk density of approx. 164 kg/m^3 [62]. In the same study, applewood particles (classified as moderately heavy species) were used, the density of which was approx. 245 kg/m^3. The higher the bulk density of the ligno-cellulosic material, the smaller the thickness of the mat prepared to be pressed to form a particleboard. Knowledge of the material density is especially important when preparing mats from alternative lignocellulosic raw materials, which are characterized by a much lower density than pine wood particles. An example is straw particles obtained from alternative lignocellulosic raw materials, such as rapeseed—approx. 42 kg/m^3 and rye approx. 25 kg/m^3 [36]. The lower bulk density of materials may cause some technical limitations in the industrial-scale manufacturing process. In addition, there may be some logistical problems in developing an optimal and sustainable transportation mechanism [28]. Due to the high volume of logged timber—around 40 mln m^3 per year, State Forests (SF) in Poland are a huge source of forest residues [63]. The exact amount of FR production is difficult to estimate. Due to many environmental factors that affect the amount of waste [6,12]. Gendek et al. [6], in reference to estimates in a paper by Zajączkowski (2013) [64], estimated that forest residues in Poland in 2031 could amount to 2.34 million m3, an increase of 12.5% over the 2021 estimate. In recent years SF in Poland offer to sell FR for energy purposes as M2E assortments [65]. The starting price of the auction starts around 4–5 EUR/m^3 (accessed on 1 September 2022). According to the results of our study, this raw material is also relevant and sustainable for high-value particleboard production. Due to the relatively low price, high accessibility, and quality of the final product. Unfortunately, the destabilization of the timber market in Poland, caused among others energy problems and war in Ukraine (stopped wood import). Moreover, EU sanctions have stopped the import of forest residues from Belarus. Therefore currently the final auction price of wood biomass sells by State Forest significantly increased even to over 100 EUR/m^3.

From an environmental and economic point of view, particleboard production could be based on the idea of using less valuable wood waste and residues. For example, those generated in the woodworking process like particles and residues from the production of wood-based panels, constitute a significant waste [66]. This would create a value-added product that contributes to improved waste management. Moreover, additions of other materials such as waste from the leather industry [67,68], tetrapak packaging [69,70], and construction materials [71,72] can be used. When material innovations are used to make panels, the hygienic properties of the panels made from them must be controlled, the standards of which must be strictly observed. When using wood material, the likelihood of high emission values and formaldehyde is much lower than for alternative raw materials. There are many additives to amine resins that are designed to reduce formaldehyde content in the board, but often their mechanical and physical properties are adversely affected. An example of the use of formaldehyde (FS) Scavenger Solution described in Basbog 2022, which was synthesized using a mixture of monoethanolamine (MEA), ammonium chloride (AC), and distilled water (DW) [73]. Adhesives that are considered biobased on a mixture of citric acid and glycerin [74], waste melamine impregnated paper [75], protein-based [76], and cornstarch and tannin-based wood adhesives [42,77] can be used.

5. Conclusions

One possible alternative use of forest residues is their use as an alternative material for wood-based panels, namely particleboard. Based on the research, it can be concluded that forest biomass in the form of pine branches can replace roundwood. The results obtained clearly show that the use of wood material in the form of forest biomass residues improves MOR, which is worth noting, and significantly increases the IB value from 0.46 N/mm^2 for boards made of stem wood particles to 0.72N/mm^2. Mechanically, the obtained panels meet the minimum requirements for P5-type boards, but nevertheless, the TS value is insufficient. All of the particleboards produced met standardized performance requirements, making them suitable for use as a board for interior fitments (including furniture) for use in dry conditions (type P2).

Future research should be directed toward improving the quality of particleboard produced from raw materials that are alternatives to roundwood. Consideration should be given to raw materials with lower market prices and the use of additives in the form of alternative adhesives, including those that increase water resistance. In addition, further research is needed to reduce competition for raw materials in the energy sector, which is currently a major obstacle to the cultivation of many industrial crops.

Author Contributions: Conceptualization, M.P. and K.T.; methodology, M.P., K.T. and A.T.; formal analysis, D.J.-L., A.T. and T.R.; investigation, M.P., K.T., D.J.-L.; data curation, M.P. and K.T.; writing— original draft preparation, M.P. and K.T.; writing—review and editing, D.J.-L., T.R.; visualization, M.P. and K.T. All authors have read and agreed to the published version of the manuscript.

Funding: This research received no external funding.

Institutional Review Board Statement: Not applicable.

Informed Consent Statement: Not applicable.

Data Availability Statement: Not applicable.

Conflicts of Interest: The authors declare no conflict of interest.

References

1. *FAO Global Forest Resources Assessment 2020 (FRA 2020)*; Food and Agricultural Organization of the United Nations: Rome, Italy, 2020.
2. *FAO Global Forest Resources Assessment 2000 (FRA 2000)*; Food and Agricultural Organization of the United Nations: Rome, Italy, 2000.
3. Stampfer, K.; Kanzian, C. Current State and Development Possibilities of Wood Chip Supply Chains in Austria. *Croat. J. For. Eng. J. Theory Appl. For. Eng.* **2006**, *27*, 135–145.
4. Eker, M. Assessment of Procurement Systems for Unutilized Logging Residues for Brutian Pine Forest of Turkey. *Afr. J. Biotechnol.* **2011**, *10*, 2455–2468.
5. Moskalik, T.; Gendek, A. Production of Chips from Logging Residues and Their Quality for Energy: A Review of European Literature. *Forests* **2019**, *10*, 262. [CrossRef]
6. Gendek, A.; Wezyk, P.; Moskalik, T. Share and Accuracy of Estimation of Logging Residues in the Total Volume of Harvested Timber. *Sylwan* **2018**, *162*, 679–687.
7. Giuntoli, J.; Caserini, S.; Marelli, L.; Baxter, D.; Agostini, A. Domestic Heating from Forest Logging Residues: Environmental Risks and Benefits. *J. Clean. Prod.* **2015**, *99*, 206–216. [CrossRef]
8. Nurek, T.; Gendek, A.; Roman, K. Forest Residues as a Renewable Source of Energy: Elemental Composition and Physical Properties. *Bioresources* **2019**, *14*, 6–20. [CrossRef]
9. Janeiro, L.; Patel, M.K. Choosing Sustainable Technologies. Implications of the Underlying Sustainability Paradigm in the Decision-Making Process. *J. Clean. Prod.* **2015**, *105*, 438–446. [CrossRef]
10. Płotkowski, L. Forest Biomass Balance, Current Status and Medium- and Long-Term Projection. In *Biomass for Electric Power and Heating. Opportunities and Problems*; Wieś jutra: Warszawa, Poland, 2007.
11. Gornowicz, R.; Gałązka, S.; Kuźmiński, R.; Kwaśna, H.; Łabędzki, A.; Łakomy, P.; Pilarek, Z.; Polowy, K.; Sierota, Z. Forest Biomass as a Source of Bioenergy and an Important Component of the Forest Ecosystem. In *Challenges and Opportunities of 21st Century Forestry*; Sierota, Z., Ed.; Instytut Badawczy Leśnictwa: Sękocin Stary, Poland, 2015; pp. 61–67.
12. Gałęzia, T. Economic Methods for the Utilisation of Logging Residues. *For. Res. Pap.* **2016**, *77*, 50–55. [CrossRef]

13. Liu, W.; Hou, Y.; Liu, W.; Yang, M.; Yan, Y.; Peng, C.; Yu, Z. Global Estimation of the Climate Change Impact of Logging Residue Utilization for Biofuels. *For. Ecol. Manag.* **2020**, *462*, 118000. [CrossRef]

14. Kanzian, C.; Holzleitner, F.; Stampfer, K.; Ashton, S. Regional Energy Wood Logistics-Optimizing Local Fuel Supply. *Silva Fenn.* **2009**, *43*, 113–128. [CrossRef]

15. Roman, K.; Roman, M.; Wojcieszak-Zbierska, M.; Roman, M. Obtaining Forest Biomass for Energy Purposes as an Enterprise Development Factor in Rural Areas. *Appl. Sci.* **2021**, *11*, 5753. [CrossRef]

16. Walker, J.F.C. Wood Panels: Particleboards and Fiberboards. In *Primary Wood Processing Principle and Practices*; Chapman & Hall: London, UK, 1993.

17. Sahin, H.T.; Arslan, M.B. Weathering Performance of Particleboards Manufactured from Blends of Forest Residues with Red Pine (*Pinus brutia*) Wood. *Maderas Cienc. Y Tecnol.* **2011**, *13*, 337–346. [CrossRef]

18. Lee, S.H.; Lum, W.C.; Geng, B.J.; Kristak, L.; Antov, P.; Rogoziński, T.; Pędzik, M.; Taghiyari, H.R.; Lubis, M.A.R.; Fatriasari, W.; et al. Particleboard from Agricultural Biomass and Recycled Wood Waste: A Review. *J. Mater. Res. Technol.* **2022**, *20*, 4630–4658. [CrossRef]

19. Pędzik, M.; Janiszewska, D.; Rogoziński, T. Alternative Lignocellulosic Raw Materials in Particleboard Production: A Review. *Ind. Crops Prod.* **2021**, *174*. [CrossRef]

20. Iždinský, J.; Vidholdová, Z.; Reinprecht, L. Particleboards from Recycled Wood. *Forests* **2020**, *11*, 1166. [CrossRef]

21. Iždinský, J.; Reinprecht, L.; Vidholdová, Z. Particleboards from Recycled Pallets. *Forests* **2021**, *12*, 1597. [CrossRef]

22. Simal Alves, L.; da Silva, S.A.M.; dos Anjos Azambuja, M.; Varanda, L.D.; Christofóro, A.L.; Lahr, F.A.R. Particleboard Produced with Sawmill Waste of Different Wood Species. *Adv. Mat. Res.* **2014**, *884–885*, 689–693. [CrossRef]

23. Yano, B.B.R.; Silva, S.A.M.; Almeida, D.H.; Aquino, V.B.M.; Christoforo, A.L.; Rodrigues, E.F.C.; Carvalho Junior, A.N.; Silva, A.P.; Lahr, F.A.R. Use of Sugarcane Bagasse and Industrial Timber Residue in Particleboard Production. *Bioresources* **2020**, *15*, 4753–4762. [CrossRef]

24. Gößwald, J.; Barbu, M.C.; Petutschnigg, A.; Krišťák, Ľ.; Tudor, E.M. Oversized Planer Shavings for the Core Layer of Lightweight Particleboard. *Polymers* **2021**, *13*, 1125. [CrossRef]

25. Sandak, A.; Sandak, J.; Janiszewska, D.; Hiziroglu, S.; Petrillo, M.; Grossi, P. Prototype of the Near-Infrared Spectroscopy Expert System for Particleboard Identification. *J. Spectrosc.* **2018**, 1–11. [CrossRef]

26. Tudor, E.M.; Dettendorfer, A.; Kain, G.; Barbu, M.C.; Réh, R.; Krišťák, Ľ. Sound-Absorption Coefficient of Bark-Based Insulation Panels. *Polymers* **2020**, *12*, 1012. [CrossRef] [PubMed]

27. Tudor, E.M.; Scheriau, C.; Barbu, M.C.; Réh, R.; Krišťák, Ľ.; Schnabel, T. Enhanced Resistance to Fire of the Bark-Based Panels Bonded with Clay. *Appl. Sci.* **2020**, *10*, 5594. [CrossRef]

28. Janiszewska, D.; Żurek, G.; Martyniak, D.; Bałęczny, W. Lignocellulosic Biomass of C3 and C4 Perennial Grasses as a Valuable Feedstock for Particleboard Manufacture. *Materials* **2022**, *15*, 6384. [CrossRef] [PubMed]

29. Hýsková, P.; Hýsek, Š.; Schönfelder, O.; Šedivka, P.; Lexa, M.; Jarský, V. Utilization of Agricultural Rests: Straw-Based Composite Panels Made from Enzymatic Modified Wheat and Rapeseed Straw. *Ind. Crops Prod.* **2020**, *144*, 112067. [CrossRef]

30. Kariuki, S.W.; Wachira, J.; Kawira, M.; Murithi, G. Crop Residues Used as Lignocellulose Materials for Particleboards Formulation. *Heliyon* **2020**, *6*, e05025. [CrossRef]

31. Gűler, C.; Ibrahim Sahin, H.; Yeniay, S. The Potential for Using Corn Stalks as a Raw Material for Production Particleboard with Industrial Wood Chips. *Wood Res.* **2016**, *61*, 299–306.

32. Kucuktuvek, M.; Kasal, A.; Kuskun, T.; Ziya Erdil, Y. Utilizing Poppy Husk-Based Particleboards as an Alternative Material in Case Furniture Construction. *Bioresources* **2017**, *12*, 839–852.

33. Borysiuk, P.; Auriga, R.; Bujak, M. Sunflower Hulls as Raw Material for Particleboard Production. *Biul. Inf. OB-RPPD* **2020**, *1–2*, 32–44.

34. Auriga, R.; Borysiuk, P.; Misiura, Z. Evaluation of the Physical and Mechanical Properties of Particle Boards Manufactured Containing Plum Pruning Waste. *Biul. Inf. OB-RPPD* **2021**, *1–2*, 5–11.

35. Pędzik, M.; Auriga, R.; Kristak, L.; Antov, P.; Rogoziński, T. Physical and Mechanical Properties of Particleboard Produced with Addition of Walnut (*Juglans regia* L.) Wood Residues. *Materials* **2022**, *15*, 1280. [CrossRef]

36. Dukarska, D.; Pędzik, M.; Rogozińska, W.; Rogoziński, T.; Czarnecki, R. Characteristics of Straw Particles of Selected Grain Species Purposed for the Production of Lignocellulose Particleboards. *Part. Sci. Technol.* **2021**, *39*, 213–222. [CrossRef]

37. Hafezi, S.M.; Enayati, A.; Hosseini, K.D.; Tarmian, A.; Mirshokraii, S.A. Use of Amino Silane Coupling Agent to Improve Physical and Mechanical Properties of UF-Bonded Wheat Straw (*Triticum aestivum* L.) Poplar Wood Particleboard. *J. For. Res.* **2016**, *27*, 427–431. [CrossRef]

38. Iwakiri, S.; Trianoski, R.; Chies, D.; Tavares, E.L.; França, M.C.; Lau, P.C.; Iwakiri, V.T. Use of Residues of Forestry Exploration of *Pinus taeda* for Particleboard. *Rev. Árvore* **2017**, *41*, e410304. [CrossRef]

39. Wronka, A.; Kowaluk, G. Upcycling Different Particle Sizes and Contents of Pine Branches into Particleboard. *Polymers* **2022**, *14*, 4559. [CrossRef] [PubMed]

40. Rahman, K.; Shaikh, A.A.; Rahman, M.; Alam, N.; Alam, R. The Potential for Using Stem and Branch of Bhadi (*Lannea coromandelica*) as a Lignocellulosic Raw Material for Particleboard. *Int. Res. J. Biol. Sci.* **2013**, *2*, 8–12.

41. Maraghi, M.M.R.; Tabei, A.; Madanipoor, M. Effect of Board Density, Resin Percentage and Pressing Temperature on Particleboard Properties Made from Mixing Poplar Wood Slab, Citrus Branches and Twigs of Beech. *Wood Res. J.* **2018**, *63*, 669–682.

42. Nemli, G.; Hiziroglu, S.; Usta, M.; Serin, Z.; Ozdemir, T.; Kalaycioglu, H. Effect of Residue Type and Tannin Content on Properties of Particleboard Manufactured from Black Locust. *For. Prod. J.* **2004**, *54*, 36–40.
43. Braghiroli, F.L.; Passarini, L. Valorization of Biomass Residues from Forest Operations and Wood Manufacturing Presents a Wide Range of Sustainable and Innovative Possibilities. *Curr. For. Rep.* **2020**, *6*, 172–183. [CrossRef]
44. Pérez-Harguindeguy, N.; Díaz, S.; Garnier, E.; Lavorel, S.; Poorter, H.; Jaureguiberry, P.; Bret-Harte, M.S.; Cornwell, W.K.; Craine, J.M.; Gurvich, D.E.; et al. New Handbook for Standardised Measurement of Plant Functional Traits Worldwide. *Aust. J. Bot.* **2013**, *61*, 167. [CrossRef]
45. EN 312; Particleboards–Specifications. European Commission for Standardization: Brussels, Belgium, 2010.
46. EN 310; Wood-Based Panels–Determination of Modulus of Elasticity in Bending and of Bending Strength. European Commission for Standardization: Brussels, Belgium, 1993.
47. EN 717-1; Wood-Based Panels–Determination of Formaldehyde Release–Part 1: Formaldehyde Emission by the Chamber Method. European Commission for Standardization: Brussels, Belgium, 2006.
48. EN 319; Particleboards and Fibreboards–Determination of Tensile Strength Perpendicular to the Plane of the Board. European Commission for Standardization: Brussels, Belgium, 1999.
49. EN 323; Wood-Based Panels–Determination of Density. European Committee for Standardization. European Commission for Standardization: Brussels, Belgium, 1993.
50. Tomczak, K.; Tomczak, A.; Jelonek, T. Effect of Natural Drying Methods on Moisture Content and Mass Change of Scots Pine Roundwood. *Forests* **2020**, *11*, 668. [CrossRef]
51. Tomczak, A.; Jakubowski, M.; Jelonek, T.; Wąsik, R.; Grzywiński, W. Mass and Density of Pine Pulpwood Harvested in Selected Stands from the Forest Experimental Station in Murowana Goślina. *Acta Sci. Pol. Silvarum Colendarum Ratio Ind. Lignaria* **2016**, *15*, 105–112. [CrossRef]
52. Dadzie, P.K.; Amoah, M.; Ebanyenle, E.; Frimpong-Mensah, K. Characterization of Density and Selected Anatomical Features of Stemwood and Branchwood of *E. cylindricum*, *E. angolense* and *K. ivorensis* from Natural Forests in Ghana. *Eur. J. Wood Wood Prod.* **2018**, *76*, 655–667. [CrossRef]
53. Shirai, T.; Yamamoto, H.; Yoshida, M.; Inatsugu, M.; Ko, C.; Fukushima, K.; Matsushita, Y.; Yagami, S.; Lahjie, A.M.; Sawada, M.; et al. Eccentric Growth and Growth Stress in Inclined Stems of Gnetum Gnemon. *IAWA J.* **2015**, *36*, 365–377. [CrossRef]
54. Jourez, B.; Riboux, A.; Leclercq, A. Comparison of Basic Density and Longitudinal Shrinkage in Tension Wood and Opposite Wood in Young Stems of *Populus euramericana* Cv. Ghoy When Subjected to a Gravitational Stimulus. *Can. J. For. Res.* **2001**, *31*, 1676–1683. [CrossRef]
55. Dadzie, P.K.; Amoah, M. Density, Some Anatomical Properties and Natural Durability of Stem and Branch Wood of Two Tropical Hardwood Species for Ground Applications. *Eur. J. Wood Wood Prod.* **2015**, *73*, 759–773. [CrossRef]
56. Dadzie, P.K.; Amoah, M.; Frimpong-Mensah, K.; Oheneba-Kwarteng, F. Some Physical, Mechanical and Anatomical Characteristics of Stemwood and Branchwood of Two Hardwood Species Used for Structural Applications. *Mater. Struct.* **2016**, *49*, 4947–4958. [CrossRef]
57. Moreira, L.d.S.; Andrade, F.W.C.; Balboni, B.M.; Moutinho, V.H.P. Wood from Forest Residues: Technological Properties and Potential Uses of Branches of Three Species from Brazilian Amazon. *Sustainability* **2022**, *14*, 11176. [CrossRef]
58. Dibdiakova, J.; Vadla, K. Basic Density and Moisture Content of Coniferous Branches and Wood in Northern Norway. *EPJ Web. Conf.* **2012**, *33*, 02005. [CrossRef]
59. Gryc, V.; Horacek, P.; Slezingerova, J.; Vavrcik, H. Basic Density of Spruce Wood, Wood with Bark, and Bark of Branches in Locations in the Czech Republic. *Wood Res.* **2011**, *56*, 23–32.
60. Alamsyah, E.M.; Sutrisno; Sumardi, I.; Darwis, A.; Suhaya, Y.; Hidayat, Y. The Possible Use of Surian Tree (*Toona sinensis* Roem) Branches as an Alternative Raw Material in the Production of Composite Boards. *J. Wood Sci.* **2020**, *66*, 25. [CrossRef]
61. Jahan-Latibari, A.; Roohnia, M. Potential of Utilization of the Residues from Poplar Plantation for Particleboard Production in Iran. *J. For. Res.* **2010**, *21*, 503–508. [CrossRef]
62. Kowaluk, G.; Szymanowski, K.; Kozlowski, P.; Kukula, W.; Sala, C.; Robles, E.; Czarniak, P. Functional Assessment of Particleboards Made of Apple and Plum Orchard Pruning. *Waste Biomass Valorization* **2020**, *11*, 2877–2886. [CrossRef]
63. *Statistics Poland Statistical Yearbook of Forestry in Poland*; Statistics Poland: Warsaw, Poland, 2021.
64. Zajączkowski, S. Forecasts of Timber Harvest in Poland in the 20-Year Perspective and the Possibility of Using Them to Estimate Timber Resources for Energy Purposes. In *Forest Biomass for Energy Purposes*; Gołos, P., Kaliszewski, A., Eds.; Instytut Badawczy Leśnictwa: Sękocin Stary, Poland, 2013; pp. 21–32.
65. Zarządzenie Dyrektora Generalnego Lasów Państwowych Nr 57, z Dnia 22 Września 2021 r., Ws. Zasad Sprzedaży Drewna w Państwowym Gospodarstwie Leśnym Lasy Państwowe Na Lata 2022–2023. (Ordinance of the Director General of the State Forests Rules for the Sale of Wood in the State Forests for 2022–2023.). Available online: https://drewno.zilp.lasy.gov.pl/drewno/file.2021-09-24.2553304359 (accessed on 1 September 2022).
66. Pędzik, M.; Kwidziński, Z.; Rogoziński, T. Particles from Residue Wood-Based Materials from Door Production as an Alternative Raw Material for Production of Particleboard. *Drv. Ind.* **2022**, *73*, 351–357. [CrossRef]
67. Kibet, T.; Tuigong, D.R.; Maube, O.; Mwasiagi, J.I. Mechanical Properties of Particleboard Made from Leather Shavings and Waste Papers. *Cogent Eng.* **2022**, *9*, 2076350. [CrossRef]

68. Oliveira, R.C.d.; Bispo, R.A.; Trevisan, M.F.; Gilio, C.G.; Rodrigues, F.R.; Silva, S.A.M.d. Influence of Leather Fiber on Modulus of Elasticity in Bending Test and of Bend Strength of Particleboards. *Mater. Res.* **2021**, *24*, e20210287. [CrossRef]

69. Bekhta, P.; Lyutyy, P.; Hiziroglu, S.; Ortynska, G. Properties of Composite Panels Made from Tetra-Pak and Polyethylene Waste Material. *J. Polym. Environ.* **2016**, *24*, 159–165. [CrossRef]

70. Auriga, R.; Borysiuk, P.; Auriga, A. An Attempt to Use „Tetra Pak" Waste Material in Particleboard Technology. *Ann. WULS For. Wood Technol.* **2021**, *114*, 70–75. [CrossRef]

71. Oriire, L.T.; Aina, K.S.; Olajide, O.B.; Aguda, L.O.; Adiji, A.O. Evaluation of Durability Performance of Rice Husk-Cement Bonded Particleboards. *J. Agric. For. Soc. Sci.* **2022**, *18*, 50–58. [CrossRef]

72. Nazerian, M.; Nanaii, H.A.; Gargarii, R.M. Silica (SiO$_2$) Content on Mechanical Properties of Cement-Bonded Particleboard Manufactured from Lignocellulosic Materials. *Drv. Ind.* **2018**, *69*, 317–328. [CrossRef]

73. Başboğa, İ.H. Particleboard Manufacturing with Low Formaldehyde Content for Indoor Applications. *Wood Mater. Sci. Eng.* **2022**, 1–7. [CrossRef]

74. Nitu, I.P.; Rahman, S.; Islam, M.d.N.; Ashaduzzaman, M.d.; Shams, M.d.I. Preparation and Properties of Jute Stick Particleboard Using Citric Acid–Glycerol Mixture as a Natural Binder. *J. Wood Sci.* **2022**, *68*, 30. [CrossRef]

75. Başboğa, İ.H.; Taşdemir, Ç.; Yüce, Ö.; Mengeloğlu, F. Utilization of Different Size Waste Melamine Impregnated Paper as an Adhesive in the Manufacturing of Particleboard. *Int. J. Adhes. Adhes.* **2022**, 103275, in press. [CrossRef]

76. Raydan, N.D.V.; Leroyer, L.; Charrier, B.; Robles, E. Recent Advances on the Development of Protein-Based Adhesives for Wood Composite Materials—A Review. *Molecules* **2021**, *26*, 7617. [CrossRef] [PubMed]

77. Oktay, S.; Kızılcan, N.; Bengü, B. Environment-Friendly Cornstarch and Tannin-Based Wood Adhesives for Interior Particleboard Production as an Alternative to Formaldehyde-Based Wood Adhesives. *Pigment Resin Technol.* **2022**, *ahead of print*. [CrossRef]

 *forests*

MDPI

Article

Ignition of Wood-Based Boards by Radiant Heat

Iveta Marková [1,*], Martina Ivaničová [2], Linda Makovická Osvaldová [1] , Jozef Harangózo [2] and Ivana Tureková [2]

[1] Department of Fire Engineering, Faculty of Security Engineering, University of Žilina, Univerzitná 1, 010 26 Zilina, Slovakia

[2] Constantine the Philosopher University in Nitra, Tr. A. Hlinku 1, 949 74 Nitra, Slovakia

* Correspondence: iveta.markova@uniza.sk; Tel.: +421-41-513-6799

Abstract: Particleboards (PB) and oriented strand boards (OSB) are commonly used materials in building structures or building interiors. The surface of boards may hence become directly exposed to fire or radiant heat. The aim of this paper is to evaluate the behaviour of uncoated particleboards and OSB exposed to radiant heat. The following ignition parameters were used to observe the process of particleboard and OSB ignition: heat flux intensity (from 43 to 50 kW.m^{-2}) and ignition temperature. The time-to-ignition and mass loss of particleboards and OSB with thicknesses of 12, 15 and 18 mm were monitored and compared. The experiments were conducted on a modified device in accordance with ISO 5657: 1997. Results confirmed thermal degradation of samples. Heat flux had a significant effect on mass loss (burning rate) and time-to-ignition. OSB had higher ignition time than particleboards and the thermal degradation of OSB started later, i.e., at a higher temperature than that of particleboards, but OSB also had higher mass loss than particleboards. The samples yielded the same results above 47 kW.m^{-2}. Thermal analysis also confirmed a higher thermal decomposition temperature of OSB (179 °C) compared to particleboards (146 °C). The difference in mass loss in both stages did not exceed 1%.

Keywords: particleboard; OSB; heat release; time-to-ignition; mass loss

Citation: Marková, I.; Ivaničová, M.; Osvaldová, L.M.; Harangózo, J.; Tureková, I. Ignition of Wood-Based Boards by Radiant Heat. *Forests* **2022**, *13*, 1738. https://doi.org/10.3390/f13101738

Academic Editors: Réh Roman, Petar Antov, Ľuboš Krišťák, Muhammad Adly Rahandi Lubis and Seng Hua Lee

Received: 23 September 2022
Accepted: 18 October 2022
Published: 20 October 2022

Publisher's Note: MDPI stays neutral with regard to jurisdictional claims in published maps and institutional affiliations.

1. Introduction

The production of wood-based boards encompasses the utilization of wood of lower quality classes [1–4] and obtaining suitable materials with improved physical and mechanical properties [5–10]. Properties of particleboards (PB) are described in detail in the work of [11,12]. The oriented strand boards (OSB) belong to this product group, but they are also considered an input material in the furniture and construction industries [9,13–15]. A description of OSB in terms of their preparation and properties is defined in the work of [16,17]. These materials are also analysed within the scope of insulation materials [18–21]. They are a part of sandwich panels in low-energy houses [22–25]. They are typically used as an interior sheathing material [26] or furniture [27–30]. Research on the fire resistance of the mentioned materials is also rich [31–35].

Large-size wood-based materials form the largest percentage of wood material in timber houses [36–38]. These materials can be directly exposed to fire [39–41] or the effect of radiant heat [42,43]. Thermal degradation and potentially even ignition of wood-based boards are caused by the action of the ignition source [41–48]. These processes are affected by both the combustible material and the environment in which it is located [49,50]. The ignition process cannot be characterized by a single property [51]. Rantuch et al. [52] used ignition parameters to define the term ignition. Two of these ignition parameters (critical heat flux and ignition temperature) are used here to compare OSB and PB with thicknesses of 12, 15 and 18 mm. This article presents the differences in the results of the research between PB and OSB due to the influence of external heat flux.

Ignition is the ability of a sample to ignite under the action of an external thermal initiator and under defined test conditions, according to [53]. According to ISO 3261 [54], it is the ability of a material to ignite. The process of ignition is characterized by the time-to-ignition of a sample, which depends on the ignition temperature, thermal properties of materials, sample conditions (size, humidity, orientation) and critical heat flux [55]. Definition of "ignition temperature" can be interpreted as the minimum temperature to which the air must be heated so that the sample placed in the heated air environment ignites, or the surface temperature of the sample just before the ignition point [56–58].

Separate attention is paid to the issue of simulating the ignition of wood under external heat flux from calculations of ignition parameters [59,60]. A prediction model presented in Chen et al.'s paper [61] studies the pyrolysis and ignition time of wood under external heat flux. The solution of the model provides the temperature at each point of the solid and the local solid conversion, and the time-to-ignition of the wood is predicted with the solution of surface temperature [62]. Chen et al. [61] obtained good agreement between experimental and theoretical results.

The aim of this paper is to evaluate the behaviour of uncoated particleboards and OSB exposed to radiant heat. The significant influence of board density and thickness on time-to-ignition and mass loss of PB and OSB samples is monitored and observed. At the same time, the difference in the thermal degradation of PB and OSB samples is sought by comparing the results between time-to ignition and mass loss of PB and OSB samples.

2. Materials and Methods

2.1. Experimental Samples

Particleboards (PB) and OSB with thicknesses of 12, 15 and 18 mm (Figure 1a) were used as samples. Selected thicknesses correspond to those typically used in the construction and insulation of houses, in the construction of ceilings, soffits, partitions, etc. The samples were sourced from the company BUČINA DDD, Zvolen, under the product name Particleboard raw unsanded (Table 1). These particleboards contain softwood strands, mainly spruce, and a urea–formaldehyde adhesive mixture [63].

The samples of oriented strand boards were obtained from the company Kronospan-Jihlava, under product name OSB/3 SUPERFINISH ECO (Figure 1b), without surface treatment. These OSB are multi-layered boards made of flat wood chips of a specific shape and thickness. The chips in the outer layers are oriented parallel to the length or width of the board, the chips in the middle layers may be oriented randomly or generally perpendicular to the lamellae of the outer layers. They are bonded with melamine formaldehyde resin and PMDI, and they are flat-pressed. The boards contain mainly a mixture of different softwood species [64].

Figure 1. Example of experimental samples prepared in accordance with ISO 5657 [65]: (**a**) particleboard (PB); (**b**) OSB.

The samples of OSB were cut to specific dimensions (165 × 165 mm) according to ISO 5657: 1997 [65]. Selected sheet board materials were stored at a specific temperature (23 °C ± 2 °C) and relative humidity (50 ± 5%).

There were tested air-conditioned samples, because the change in moisture will affect the thermal parameters of the samples and subsequently the thermal processes [16,66]. The density of samples (Table 1) was determined according to EN 323: 1996 [67]. The remaining parameters were obtained from the safety data sheets (Table 2).

Table 1. The density of PB and OSB samples according to EN 323: 1996 [67].

Samples	Designation	Density (kg.m^{-3}) for Thickness (mm)		
		12	15	18
Particleboard (PB)	PB	690 ± 9.8	713 ± 9.7	644 ± 10.1
Oriented Strand Boards (OSB)	OSB	562 ± 7.9	570 ± 12.1	569 ± 12.8

Table 2. Physical and chemical properties and fire-technical characteristics of particleboards and OSB with thicknesses of 10–18 mm.

Parameters	PB [64]	OSB [63]
Density (kg.m^{-3})	665	630
Moisture (%)	5	5
Swelling (%)	3.5	15
Thermal conductivity (W.m^{-2}.K^{-1})	0.10–0.14	0.13
Specific heat (J.kg^{-1}.K^{-1}) [67]	-	1460–1470
Formaldehyde content (mg.100 g^{-1})	6.5	8
Flame spread rating (mm.min^{-1})	-	83.8
Reaction to fire	D-s1, d0	

2.2. Methodology

2.2.1. Determination of Mass Loss and Time-to-Ignition

The measuring instrument was calibrated, and heat flux values used for selected samples were logged in Tureková et al. [68,69].

Time-to-ignition and mass loss were determined for the selected level of heat flux density and thickness of the sheet board materials according to a modified procedure based on ISO 5657: 1997 [65]. This modification included a change of the igniter. The ignition was caused only by heat flux, without the use of an open flame (Figure 2).

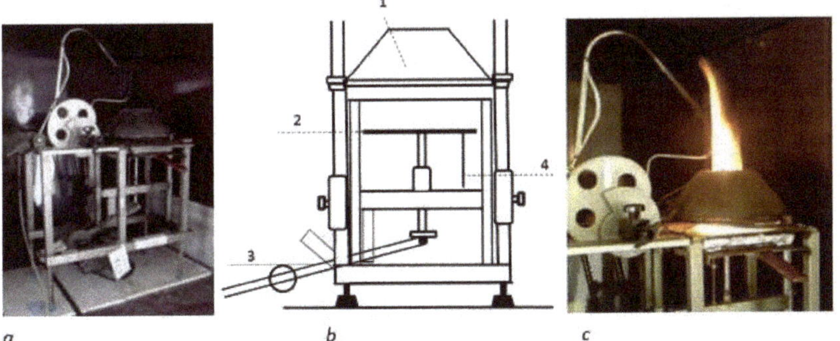

Figure 2. Scheme of the equipment for determination of flammability of materials at radiant heat flux of 10–50 kW.m^{-2} according to ISO 5657: 1997 [65]. (**a**) Real test equipment and equipment scheme. (**b**) Scheme of the used equipment with description of components: 1—heating cone, 2—board for sample, 3—movable arm, 4—connection point for recording experimental data. (**c**) Detailed look at the burning of the particleboard sample with 18 mm thickness in 100 s.

The samples were placed horizontally and exposed to a heat flux of 43 to 50 kW.m^{-2} by an electrically heated cone calorimeter. Orientation experiments determined the minimum heat flux required to maintain flame combustion. Time-to-ignition and mass loss were monitored in the interval of 43 to 50 kW.m^{-2} at each thickness of the sheet board material in a series of five repetitions.

The factors which affect time-to-ignition and mass loss are type of sample, board thickness and heat flux density. The obtained results of the ignition and mass loss temperatures were statistically evaluated by statistics. The following factors were used: mixture samples, board thickness (12, 15 and 18 mm) and heat flux density (from 43 to 50 kW.m^{-2}).

2.2.2. Thermal Analysis (Thermogravimetry TGA) of PB and OSB

This analytical method was chosen as the weight of the analysed samples in milligrams. These methods are used in observations and comparison of thermal decomposition of samples, and in the research of the changes and conditions of the chemical reaction course. Thermogravimetry (TGA) studies the course of both thermolysis and polymer burning and records the changes in the weight of the heated sample. The sample was stabilized for 24 h under standard conditions; the test was conducted on a Mettler TA 3000 with a TC 10A processor and TG 50 thermogravimetric weights module in the air and flow rate of 200 mL.min^{-1}. The heat increased at a rate of 10 °C.min^{-1}. The test was carried out up to a temperature of 700 °C. The samples for TGA were specifically prepared by disintegration, and the weight of the OSB sample was 10.225 mg and PB was 10.543 mg.

3. Results and Discussion

The minimum value of radiant heat flux for particleboards and OSB was approximately 43 kW.m^{-2}. This value represented the critical heat flux for the selected samples. The maximum value of the radiant heat flux, to which the selected sheet board materials were exposed, was 50 kW.m^{-2}. The heat flux was gradually increased by 1 kW.m^{-2} (Table 3).

Table 3. Time-to-ignition and mass loss of samples with different thicknesses using heat fluxes from 43 to 50 kW.m^{-2} at a distance of 20 mm.

Radiant Heat Flux (kW.m^{-2})	Thickness (mm)	PB		OSB	
		Time-to-Ignition (s)	Mass Loss Δm (%)	Time-to-Ignition (s)	Mass Loss Δm (%)
43	12	89.0 ± 5.215	17.108 ± 0.520	107.4 ± 32.920	19.018 ± 0.742
	15	92.6 ± 3.441	14.604 ± 0.375	172.8 ± 68.271	16.528 ± 1.103
	18	117.0 ± 5.513	13.198 ± 0.173	170.0 ± 19.279	12.436 ± 0.402
44	12	80.0 ± 5.366	17.594 ± 0.409	80.80 ± 14.372	20.188 ± 1.210
	15	86.4 ± 4.882	15.452 ± 0.355	108.0 ± 31.093	16.092 ± 0.885
	18	102.8 ± 4.308	13.754 ± 0.239	140.0 ± 31.698	13.256 ± 0.745
45	12	78.2 ± 0.748	17.96 ± 0.301	100.2 ± 21.673	20.870 ± 0.889
	15	84.4 ± 2.057	15.27 ± 0.294	86.4 ± 10.442	17.026 ± 0.541
	18	92.2 ± 2.481	13.87 ± 0.286	111.2 ± 24.235	13.716 ± 0.303
46	12	71.6 ± 1.624	18.406 ± 0.522	84.4 ± 9.002	21.868 ± 0.879
	15	76.0 ± 2.280	15.714 ± 0.290	93.4 ± 21.767	17.272 ± 0.647
	18	89.0 ± 7.974	13.776 ± 0.565	98.8 ± 12.592	13.504 ± 0.228
47	12	66.4 ± 2.870	18.91 ± 0.288	71.0 ± 8.671	22.026 ± 0.908
	15	73.8 ± 0.797	16.23 ± 0.363	67.08 ± 5.403	17.500 ± 0.455
	18	75.6 ± 3.720	14.48 ± 0.339	103.6 ± 18.391	13.818 ± 0.266
48	12	64.0 ± 1.490	19.11 ± 0.338	58.60 ± 5.953	23.206 ± 0.505
	15	69.4 ± 1.959	16.27 ± 0.373	63.40 ± 7.116	18.366 ± 0.910
	18	75.0 ± 2.000	14.65 ± 0.225	77.60 ± 25.881	14.222 ± 0.826
49	12	60.6 ± 2.241	19.75 ± 0.439	65.0 ± 11.436	23.578 ± 0.858
	15	66.0 ± 2.283	16.59 ± 0.333	62.20 ± 3.2497	18.764 ± 0.571
	18	67.2 ± 1.166	15.17 ± 0.131	63.20 ± 3.187	14.678 ± 0.899
50	12	59.8 ± 2.638	19.91 ± 0.415	56.80 ± 2.039	24.302 ± 0.814
	15	64.4 ± 2.497	16.5 ± 0.335	59.40 ± 5.607	19.402 ± 0.586
	18	66.8 ± 2.093	15.94 ± 0.945	60.20 ± 5.741	14.846 ± 1.033

The sample was placed horizontally under the cone calorimeter and exposed to selected heat fluxes which led to gradual thermal degradation and generation of flammable gases. Thermal degradation (Figure 3) is manifested by mass loss (Table 3). Ignition occurs when the critical temperature is reached [69]. Time-to-ignition was recorded, while considering only the permanent ignition of the surface of the analysed sample when exposed to a selected level of heat flux density. The carbonized residue (Figure 4) remained on the surface which has been exposed to radiant heat [70–73], which proves the thermal insulation properties of the particleboard and OSB [25].

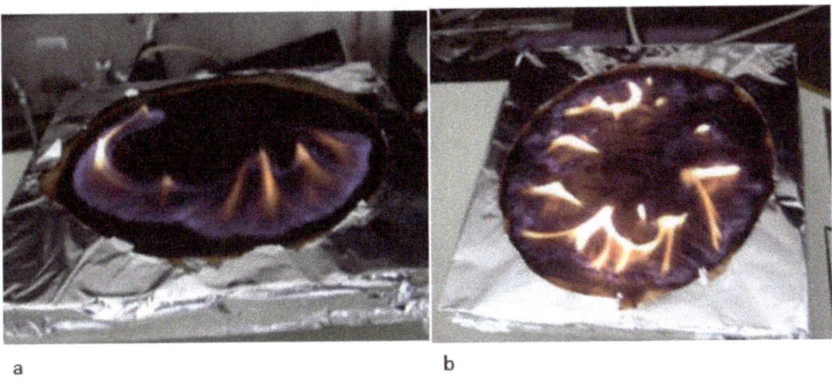

a   b

Figure 3. (**a**) Burning process of particleboard samples with 15 mm thickness after their ignition by radiant heat at 48 kW.m^{-2} in 80 s; (**b**) OSB sample with 15 mm thickness during experiment in 80 s at 48 kW.m^{-2}.

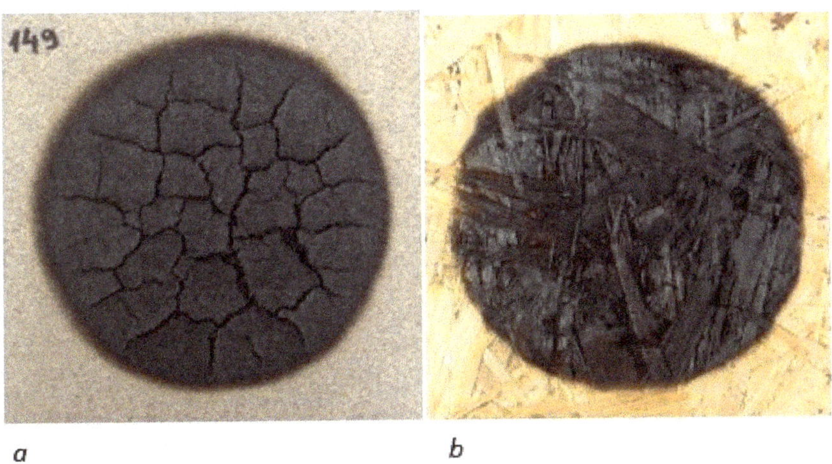

a   *b*

Figure 4. Cooled samples 10 min after the end of the experiment, sample thickness of 18 mm: (**a**) PB; (**b**) OSB.

Time-to-ignition of particleboard and OSB samples of the same thickness (Figure 4) differed in experiments with lower heat flux values, i.e., at 43 to 46 kW.m^{-2}. Particleboards and OSB with thicknesses of 12 and 15 mm had the same time-to-ignition values starting from 47 kW.m^{-2} (Figure 4 and Table 1). Samples of particleboard and OSB with a thickness of 18 mm showed the same time stamps starting from 48 kW.m^{-2} (Figure 5).

In comparison with OSB, particleboards generally showed lower time-to-ignition values. The cause can be found in the board structure. OSB consist of larger wood chips compared to particleboards.

The box plot graph for time-to-ignition OSB and PB samples shows the dispersion of the obtained data (Figure 5c). PB samples, in all thicknesses, have comparable results (in Figure 5c), marked with the numbers 2 as PB 12, 4 as PB 15 and 6 as PB 18. The above matrix presents the data obtained from heat flux 43 to 50 $kW.m^{-2}$. It confirms the fact that the thickness of the sample does not have a significant influence on time-to-ignition for PB samples. OSB samples show a significant dispersion of the obtained data and confirm the ratio with increasing heat flux; the ignition time is shortened (see also in Figure 5a).

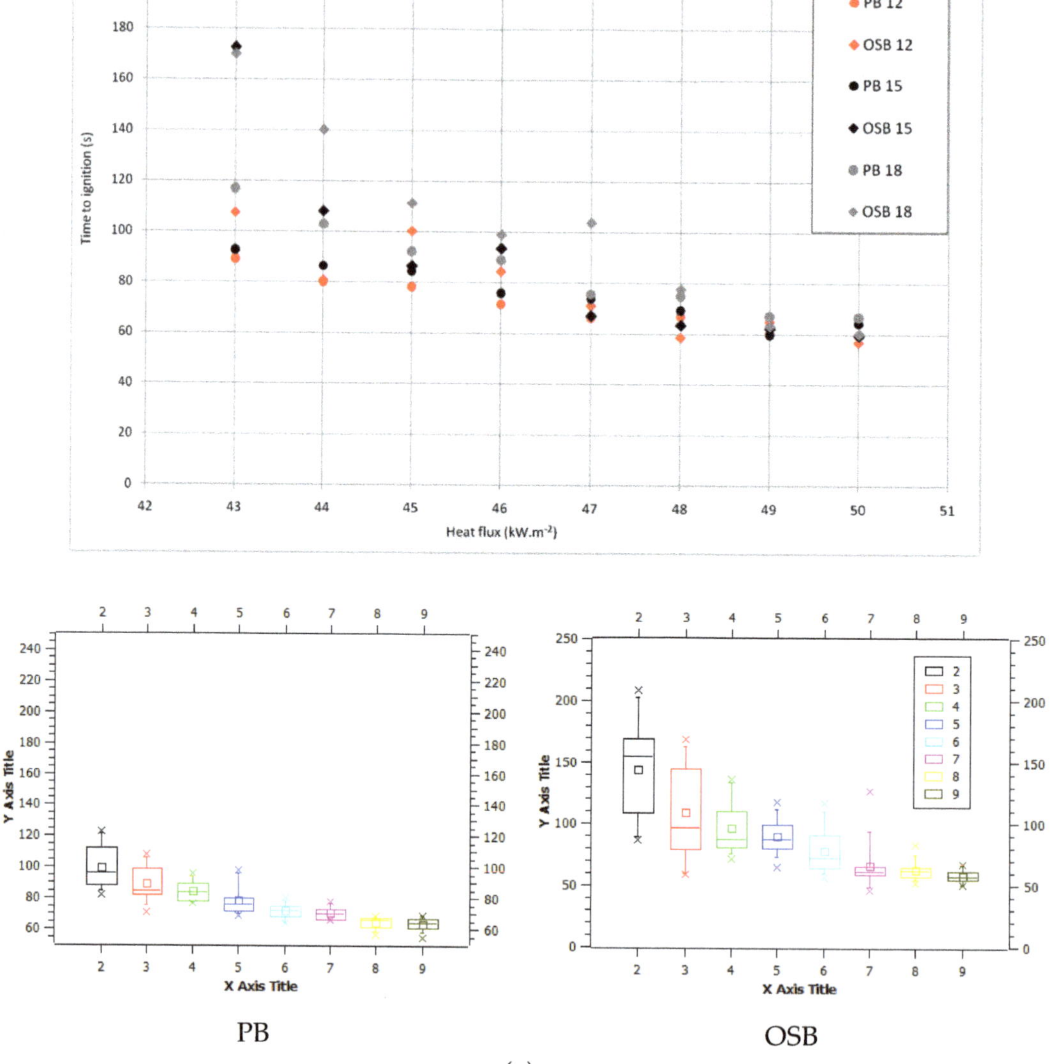

(a)

Figure 5. *Cont.*

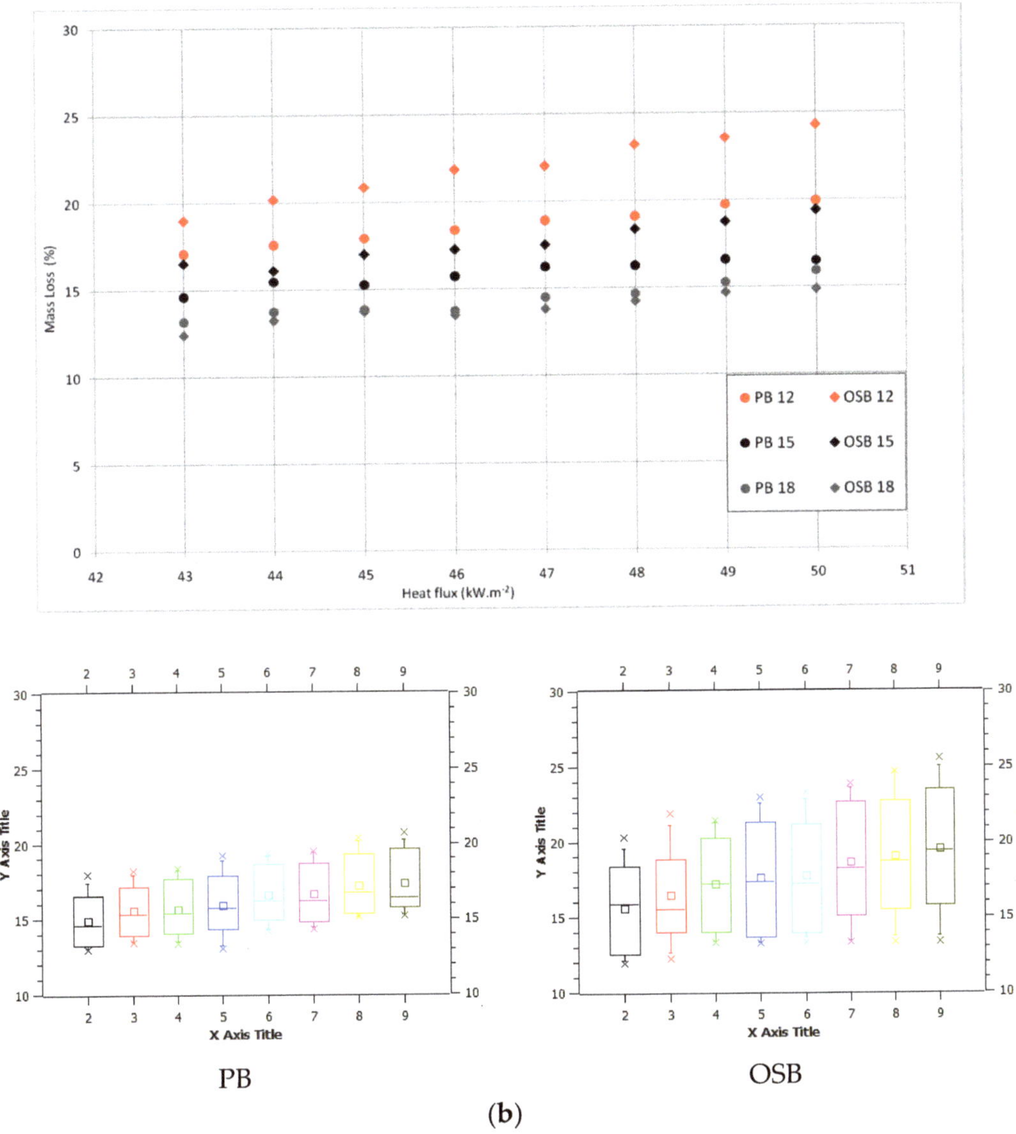

Figure 5. Comparison of time-to-ignition and mass loss values of particleboards and OSB depending on the heat flux values. (**a**) Comparison of time-to-ignition results with box graph, where X Axis is heat flux 43–50 $kW.m^{-2}$ and Y Axis is time-to-ignition for PB and OSB samples. (**b**) Comparison of mass loss results with box graph, where X Axis is heat flux 43–50 $kW.m^{-2}$ and Y Axis is mass loss for PB and OSB samples. Legends: PB 12—PB samples with 12 mm thickness, PB 15—PB samples with 15 mm thickness, PB 18—PB samples with 18 mm thickness, OSB 12—OSB samples with 12 mm thickness, OSB 15—OSB samples with 15 mm thickness, OSB 18—OSB samples with 18 mm thickness. Box graphs have X Axis marks as 2—43 $kW.m^{-2}$; 3—44 $kW.m^{-2}$; 4—45 $kW.m^{-2}$; 5—46 $kW.m^{-2}$; 6—47 $kW.m^{-2}$; 7—48 $kW.m^{-2}$; 8—49 $kW.m^{-2}$; and 9—50 $kW.m^{-2}$. Confidential interval 95%.

The values of time-to-ignition and mass loss of OSB have a greater dispersion of results, as evidenced by the created box graphs (Figure 5). The variability results from the nature of the board, which is composed of large-area wood particles from pressed flat chips that are pressed under the influence of high pressure and temperature (Figure 6). The binder is a formaldehyde-based resin [74]. Osvald et al. [75] do not assume the influence of the bonding material (glue as well as other additives) on the thermal degradation of the OSB surface.

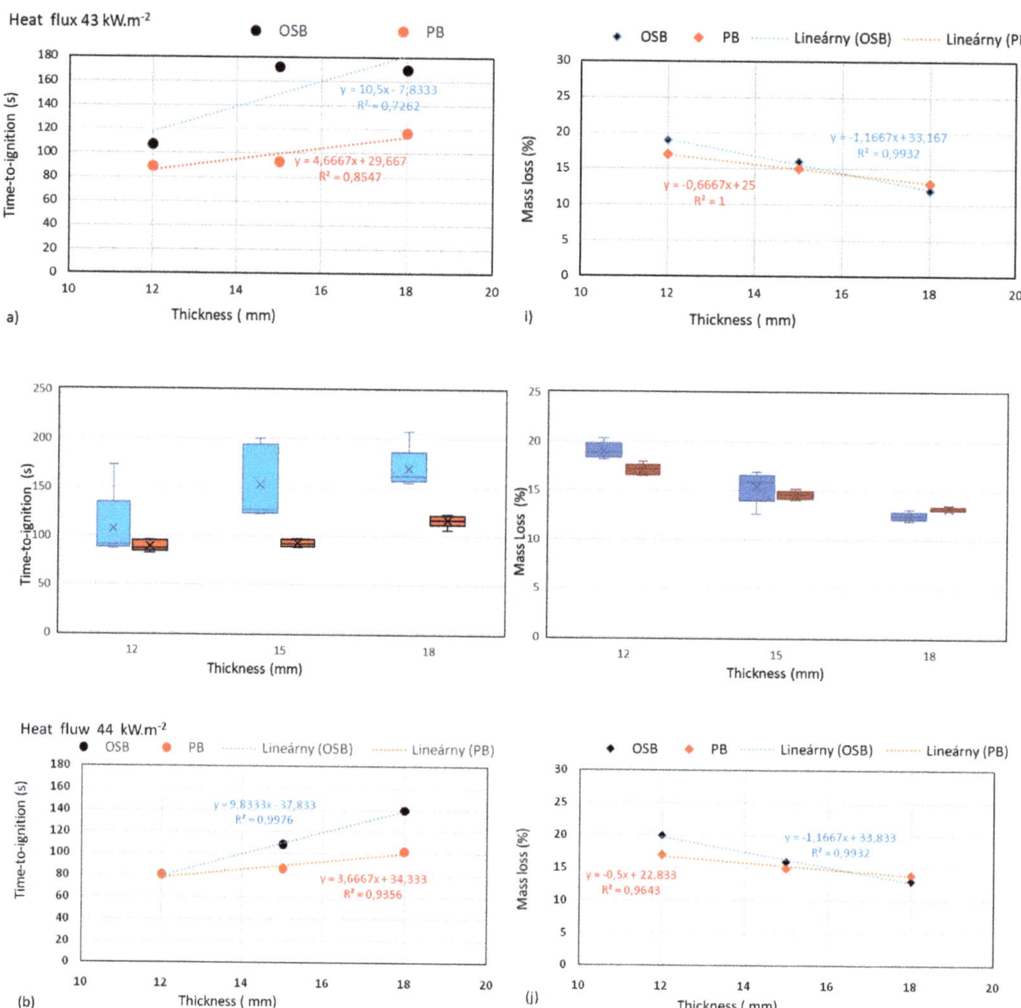

Figure 6. *Cont.*

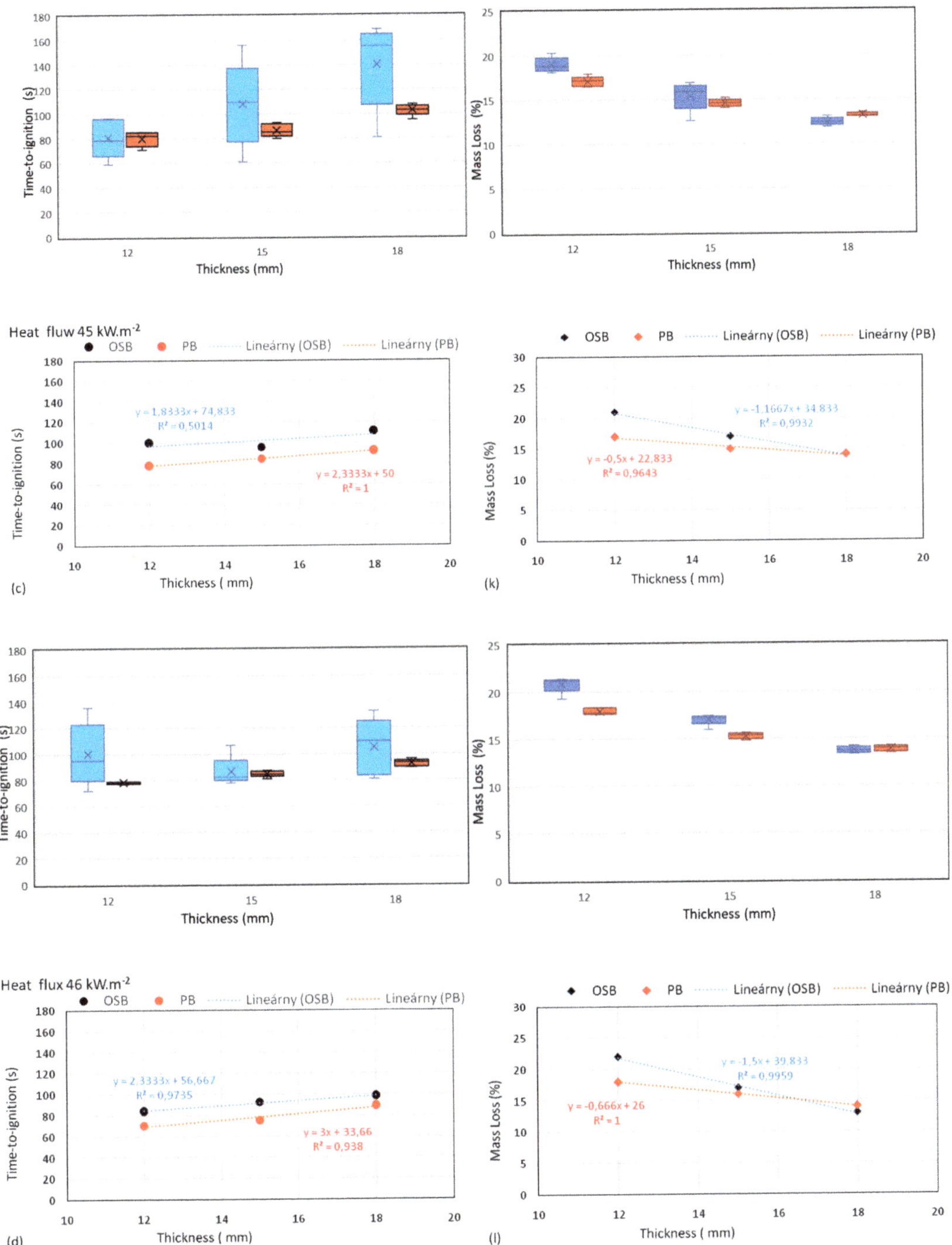

Figure 6. *Cont.*

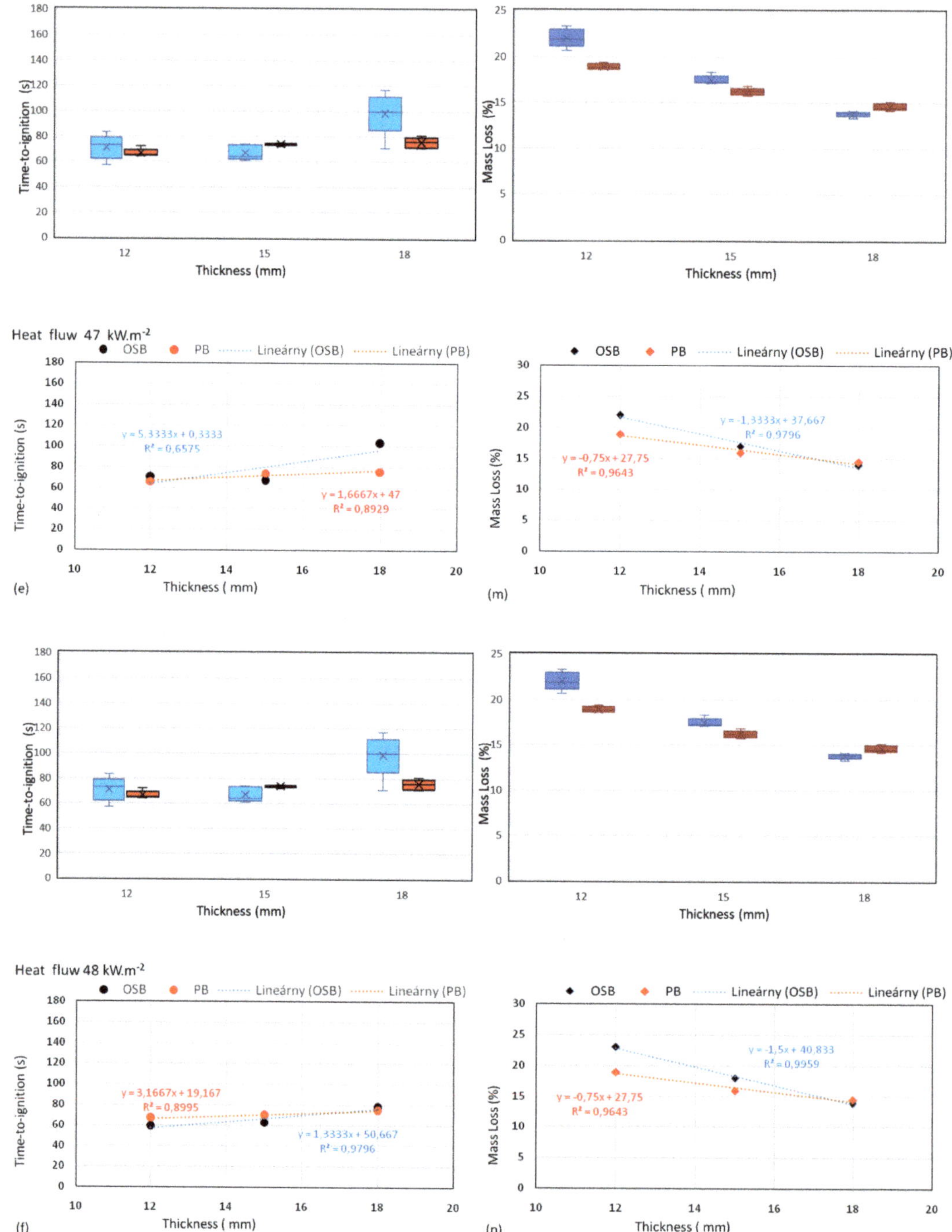

Figure 6. *Cont.*

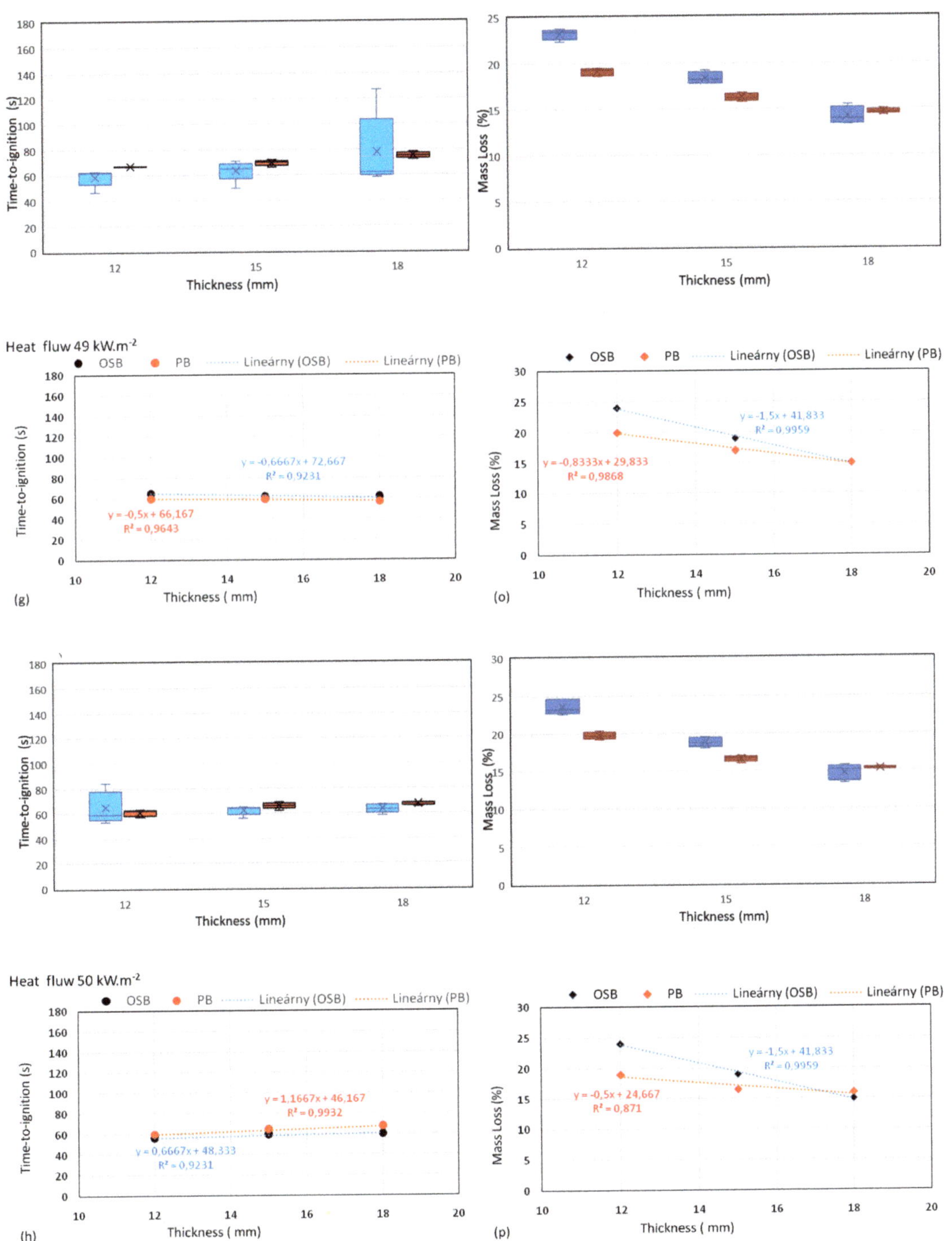

Figure 6. *Cont.*

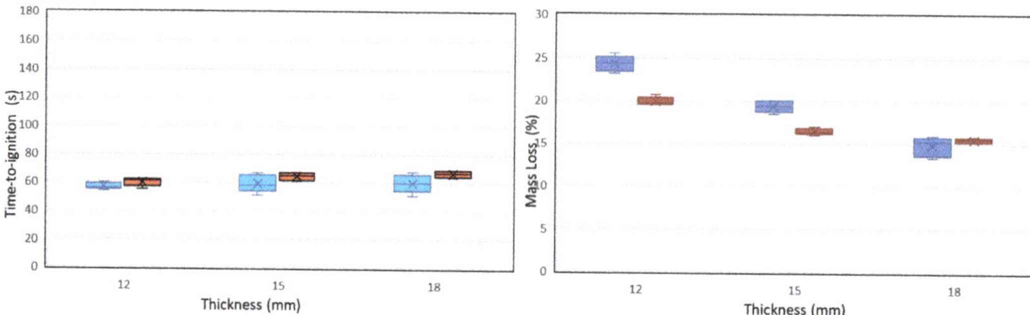

Figure 6. Graphical representation of time-to-ignition and mass loss dependence on board thickness and heat flux with box plots. Legends: blue colour is marked for OSB samples, PB is marked by red, lineárny OSB is linear OSB curve and Lineárny PB is linear curve of PB. Confidential interval 95%.

When comparing the mass loss of particleboards and OSB, lower mass loss values are observed in particleboards of all thicknesses. This difference decreases with increasing sample thickness. Mass loss values of particleboard and OSB samples with a thickness of 18 mm are the same (Figure 5b). A detailed analysis of time-to-ignition and mass loss results for individual sample thicknesses exposed to selected heat flux values is shown in Figure 6. The comparison of time-to-ignition values of particleboards and OSB showed interesting results, apart from the results with the heat flux of 43 kW.m^{-2} (Figure 6a). Figure 6 shows the linear dependences of time-to-ignition increase on the sample thickness. At the same time, the graphs are supplemented with quantitative analysis through box graphs. The presented graphs confirm the description of the behaviour of OSB and PB due to the action of radiant heat. Particleboards record lower time-to-ignition values than OSB up to the heat flux of 47 kW.m^{-2} (Figure 6b–e). Subsequently, the particleboard and OSB time stamps become identical (Figure 6f–h). All linear dependences maintain an increasing tendency (Figure 6a–h), i.e., the time-to-ignition increases with increasing sample thickness. The given increasing tendency was, however, no longer found at heat flux of 49 and 50 kW.m^{-2} (Figure 6g,h).

Naturally, mass loss (Δm) results show the opposite tendency: Δm decreases with increasing sample thickness (Figure 6i–p), while the Δm of OSB is generally greater than the Δm of particleboards. Interesting results can be seen at the heat fluxes of 43 (Figure 6i), 44 (Figure 6j) and 46 (Figure 6l) kW.m^{-2}, where there is a change in Δm occurring in samples with a thickness of 18 mm. These cases show higher Δm values of particleboard samples compared to OSB.

The results confirm relatively similar behaviour of particleboard and OSB samples. OSB have generally higher time-to-ignition values, i.e., they withstand the effect of radiant heat longer than particleboards. On the other hand, OSB have a higher Δm value compared to particleboards during thermal degradation and subsequent combustion.

Our results show that as the thickness of samples increases, the differences in the behaviour of the samples disappear under action radiant heat, which can be seen in Figure 6. Practice should take into account the importance of thickness when applying these materials in building structures or elements.

For the purpose of this analysis, another parameter evaluating the behaviour of solids in the event of a fire was calculated, namely the burning rate of OSB (Figure 7a) and particleboards (Figure 7b). The process of thermal degradation of wood-based materials is associated with the charring of the surface, hence some authors [49] call this parameter the charring rate. Once again, dependence between the increase in the rate of burning and the increase in heat flux was confirmed. The burning rate (g.m^{-2}.s^{-1}) is calculated as the ratio of mass loss Δm to the time of thermal degradation. The results show a decrease in the rate of burning with increasing thickness of the sample (Figure 8), which is also stated by Richter et al. [49]. This fact confirms that particleboards act as thermal insulators.

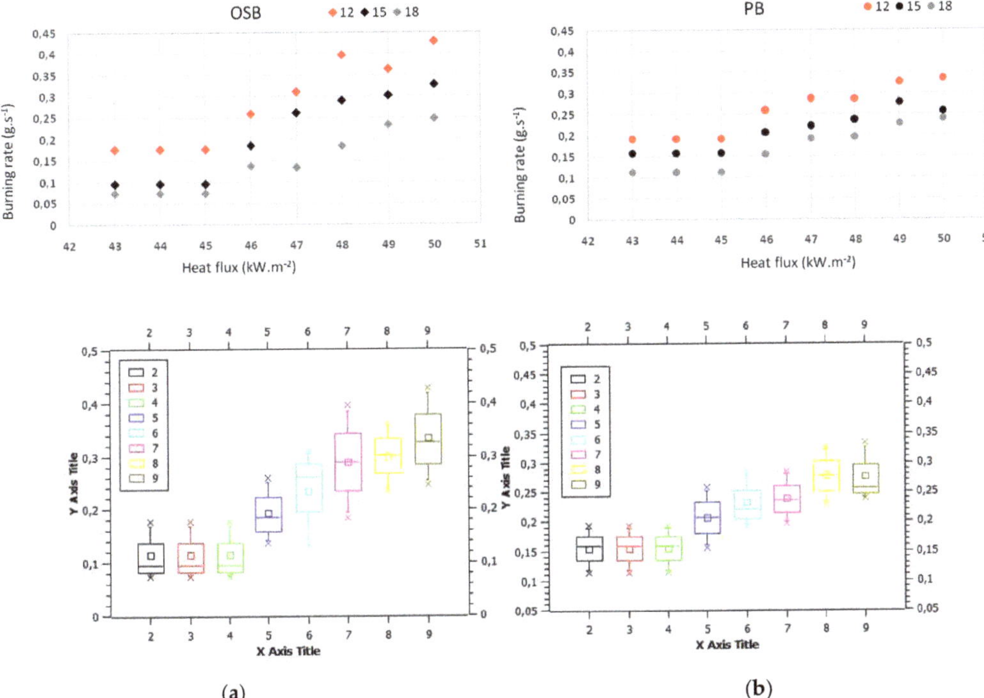

(a)   **(b)**

Figure 7. Graphical dependence of burning rate of OSB and PB samples depending on thermal stress. Legend: 12, 15, 18 are values of thickness. Box plots, have X Axis marks as 2—43 kW.m^{-2}; 3—44 kW.m^{-2}; 4—45 kW.m^{-2}; 5—46 kW.m^{-2}; 6—47 kW.m^{-2}; 7—48 kW.m^{-2}; 8—49 kW.m^{-2}; and 9—50 kW.m^{-2}. Confidential interval 95%.

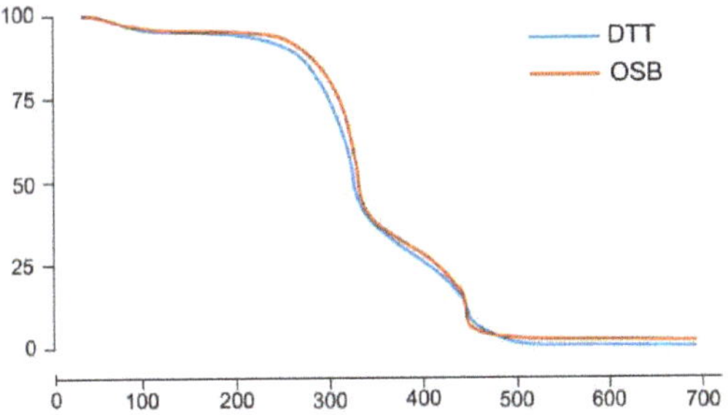

Figure 8. Comparison of thermogravimetric records showing the decomposition of selected board materials at a heating rate of 10 °C.min^{-1} in an atmosphere of air.

The box plots added to Figure 7 show the same tendency for the burning rate to increase. The values of 43,44, 45 and 46 kW.m^{-2} have exactly the same burning rate values, and significant changes occur at heat flows of 48-50 kW.m^{-2}.

Despite the previous linear dependences, it is not possible to draw a clear conclusion. This fact is also confirmed in Figure 7. The results show a relationship between the thickness of the samples and the burning rate, which is again linear, but the lines differ (Figure 7).

Richter et al. [49] addressed the effect of oxygen concentration and heat flux on the ignition and burning of particleboards. The experiments were performed on samples of particleboards with different oxygen concentrations (0%–21%), heat fluxes (10–70 kW.m^{-2}), sample densities (600–800 kg.m^{-2}) and sample thicknesses (6–25 mm). The results of Richter et al. [49] showed the effect of heat flux and oxygen concentration on the rate of burning, ignition time and combustion type (pyrolysis, smouldering, combustion).

Maciulaitis et al. [70] watched, among other things, the influence of 30, 35, 40, 45 and 50 kW.m^{-2} heat flows in accordance with LST ISO 5657: 1999 [65] with 6 mm, 10 mm, 15 mm and 18 mm thick oriented strand boards (OSBs).

Statistical evaluation of measurement data

The assessment of the impact of the kind samples (PB, OSB) and the impact of thickness (12,15 and 18 mm) on time-to-ignition and mass loss was carried out by statistical analysis. We used the multifactor analysis of variance (ANOVA) using LSD (95% level of provability) of the test (software Statistica 10).

Table 4 confirms significant differences for thickness. The OSB 18 mm has the highest time-to-ignition value.

Table 4. The impact of samples (PB, OSB) and impact of thickness (12.15 and 18 mm) on the time-to-ignition through the 1-factor analysis of variance (ANOVA) (α= 0.05).

Samples	Thickness (mm)	Heat Flux (kW.m^{-1})								Average	Hd $_{\alpha0.5}$
		43	44	45	46	47	48	49	50		
OSB	12	107.40	80.00	100.2	84.4	71.0	58.6	65.00	56.8	77.9a	
OSB	15	152.80	108.00	86.40	87.2	67.0	63.4	62.2	59.4	85.8b	
OSB	18	170.00	140.00	105.20	98.8	103.6	77.6	63.2	60.2	102.3c	
PB	12	89.00	80.00	78.2	71.6	66.4	67.0	60.6	59.8	71.6a	
PB	15	92.60	86.00	84.4	76.0	73.8	69.4	66.0	64.4	76.6ab	
PB	18	117.00	102.00	92.2	89.0	75.6	75.0	67.2	66.8	85.6b	
Average		121.4e	99.6d	91.1d	84.5c	75.4b	68.5ab	64.0a	61.2a		4.47

ANOVA–LSD test (α = 0.5): a, b, c, d, e—statistically significant difference.

The mass loss for all samples was 15% of the original weight of the samples. The obtained statistical data did not confirm the significance of the influence of the kind of sample and its thickness on mass loss (Table 5).

Table 5. The impact of samples (PB, OSB) and impact of thickness (12.15 and 18 mm) on the time-to-ignition through the 1-factor analysis of variance (ANOVA) (α = 0.05).

Samples	Heat Flux (kW.m^{-1})								Average	Hd $_{\alpha0.5}$
	43	44	45	46	47	48	49	50		
OSB	15.6	16.5	17.2	17.5	17.7	18.5	19.0	19.5	17.1a	
PB	14.5	15.6	15.7	15.9	16.6	16.7	17.2	17.3	16.1a	
Average	15.2b	16.1a	16.5a	16.8a	17.2a	17.6a	18.1c	18.4c		3.45

ANOVA–LSD test (α = 0.5): a, b, c—statistically significant difference.

Thermal analysis is another method which uses constant heating to analyse the sample. The results confirm thermal decomposition of samples in two stages [49], as is the case with other cellulosic materials (Table 6). Individual stages of thermal decomposition of particleboard and OSB samples were defined with the use of thermogravimetric analysis in an atmosphere of air.

Table 6. Thermogravimetric analysis of OSB and particleboard samples.

Sample	Drying Processes			Thermal Degradation Processes						
				I. Stage			II. Stage			
	Temperature Range (°C)	Tp (°C)	Δm (%)	Temperature Range (°C)	Tp (°C)	Δm (%)	Temperature Range (°C)	Tp (°C)	Δm (%)	C_{rezist} (%)
OSB	42–136	72.3	4.86	179–381	325.7	65.07	381–524	443.0	29.34	0.61
Particleboard	42–136	72.3	5.32	146–378	320.3	64.65	378–525	445.7	29.53	0.64

Thermal decomposition of the OSB sample (Figure 8) took place in two stages. The first stage of thermal decomposition, the main decomposition of the sample, occurred at a temperature of 179 °C. The highest mass loss (65.07%) was recorded at 325.7 °C within the first stage of decomposition, which ranged between the temperature of 179 °C and 381 °C. The second stage of thermal decomposition began at 381 °C. At this stage, the second maximum rate of mass loss was recorded at 443 °C, with a mass loss of 39.34% and a resistant residue of 0.61% after decomposition.

A similar course of thermal degradation was observed in particleboards. The main decomposition of the particleboard sample occurred at a temperature of 146 °C within the temperature range of up to 378 °C. At the same time, the highest mass loss of 64.65% was recorded at the temperature of 320.3 °C. In the second stage of thermal decomposition, which took place at the temperature range of 378 °C to 525 °C, the second maximum rate of mass loss was recorded at 445.7 °C. At this stage, there was a mass loss of 29.53% and the resistant residue after decomposition amounted to 0.64%.

Given values show the behaviour of boards subjected to thermal stress, where the OSB with a thickness of 12 mm begins to thermally degrade at 179 °C and its ignition time is 107 s at a heat flux of 43 kW.m^{-2}.

Particleboard with the thickness of 12 mm begins to degrade at 146 °C and its ignition time is 89 s. The reported results are consistent in all sample thicknesses and heat flux values.

Sinha et al. [76] studied the effect of exposure time on the flexural strength of OSB and plywood at elevated temperatures. They reached a critical temperature of 190 °C at which the strength decreased and thermal degradation occurred. Very interesting research on time-to-ignition on Ancient Wood was conducted by Wang et al. [77].

4. Conclusions

Based on the performed experiments, it is possible to draw the following conclusions:

1. The heat flux and thickness had a significant effect only on time-to-ignition.
2. OSB had a higher time-to-ignition than particleboards and the thermal degradation of OSB started later, i.e., at a higher temperature than that of particleboards. Above 47 kW.m^{-2}, the samples yielded the same results, but OSB had a higher mass loss value than particleboards.
3. Thermal analysis also confirmed a higher thermal decomposition temperature of OSB (179 °C) compared to particleboards (146 °C). The difference in mass loss in both stages did not exceed 1%, and other parameters did not show a significant difference in the behaviour of the samples.
4. Our results show that as the thickness samples increases, the differences in the behaviour of the samples disappear under action radiant heat, which can be seen in Figure 6. Practice should take into account the importance of thickness when applying these materials in building structures or elements.

Author Contributions: Conceptualization, I.T. and M.I.; methodology, I.T.; software, I.M.; validation, I.T., M.I. and J.H.; formal analysis, I.T. and L.M.O.; investigation, I.T. and M.I.; resources, I.T., J.H. and M.I.; data curation, M.I.; writing—original draft preparation, I.T.; writing—review and editing, I.M. and I.T.; project administration, L.M.O.; funding acquisition, I.T. All authors have read and agreed to the published version of the manuscript.

Funding: This article was supported by Institute Grant of University of Žilina No. 12716 and the Cultural and Educational Grant Agency of the Ministry of Education, Science, Research and Sport of the Slovak Republic on the basis of the project KEGA 0014UKF-4/2020 Innovative Learning e-modules for Safety in Dual education.

Informed Consent Statement: Not applicable.

Data Availability Statement: Not applicable for studies not involving humans or animals.

Acknowledgments: This article was supported by the Institute Grant of University of Žilina No. 12716 and Project KEGA 0014UKF-4/2020 Innovative Learning e-modules for Safety in Dual Education.

Conflicts of Interest: The authors declare no conflict of interest.

References

1. Mantanis, G.I.; Athanassiadou, E.T.; Barbu, M.C.; Wijnendaele, K. Adhesive systems used in the European particleboard, MDF and OSB industries. *Wood Mater. Sci. Eng.* **2007**, *13*, 104–116. [CrossRef]
2. Pedzik, M.; Auriga, R.; Rogozinski, T. Physical and Mechanical Properties of Particleboard Produced with Addition of Walnut (*Juglansregia L.*) Wood Residues. *Materials* **2022**, *15*, 1280. [CrossRef] [PubMed]
3. Ligne, L.D.; Van Acker, J.; Baetens, J.M.; Omar, S.; De Baets, B.; Thygesen, L.G.; Van Den Bulcke, J.; Thybring, E.E. Moisture dynamics of wood-based panels and wood fibre insulation materials. *Front. Plant Sci. Sec. Plant Biophys. Model.* **2022**, *13*, 951175. [CrossRef] [PubMed]
4. Seng, H.L.; Lum, W.C.; Boon, J.G.; Kristak, L.; Antov, P.; Pędzik, M.; Rogoziński, T.; Taghiyari, H.R.; Lubis, M.A.R.; Fatriasari, W.; et al. Particleboard from agricultural biomass and recycled wood waste: A review. *J. Mater. Res. Technol.* **2022**, *20*, 4630–4658. [CrossRef]
5. Hellmeister, V.; Barbirato, G.H.A.; Lopes, W.E.; dos Santos, V.; Fiorelli, J. Evaluation of Balsa wood (Ochromapyramidale) waste and OSB panels with castor oil polyurethane resin. *Int. Wood Prod. J.* **2021**, *12*, 267–276. [CrossRef]
6. Bušterová, M. *Effect of Heat Flux on the Ignition of Selected Board Materials*; Faculty of Materials Science and Technology in Trnava, Slovak University of Technology in Bratislava: Bratislava, Slovakia, 2011; p. 149.
7. Makowski, M.; Ohlmeyer, M. Impact of drying temperature and pressing time factor on VOC emissions from OSB made of Scots pine. *Holzforschung* **2006**, *60*, 417–422. [CrossRef]
8. Boruszewski, P.; Borysiuk, P.; Jankowska, A.; Pazik, J. Low-Density Particleboards Modified with Blowing Agents-Characteristic and Properties. *Materials* **2022**, *15*, 4528. [CrossRef]
9. Adamová, T.; Hradecký, J.; Pánek, M. Volatile Organic Compounds (VOCs) from Wood and Wood-Based Panels: Methods for Evaluation, Potential Health Risks, and Mitigation. *Polymers* **2020**, *12*, 2289. [CrossRef] [PubMed]
10. Copak, A.; Jirouš-Rajković, V.; Španić, N.; Miklečić, J. The Impact of Post-Manufacture Treatments on the Surface Characteristics Important for Finishing of OSB and Particleboard. *Forests* **2021**, *12*, 975. [CrossRef]
11. Bekhta, P.; Noshchenko, G.; Reh, R.; Kristak, L.; Antov, P.; Mirski, R. Properties of eco-friendly particleboards bonded with lignosulfonate-urea-formaldehyde adhesives and PMDI as a crosslinker. *Materials* **2021**, *14*, 4875. [CrossRef]
12. Krišťák, Ľ.; Réh, R. Application of Wood Composites. *Appl. Sci.* **2021**, *11*, 3479. [CrossRef]
13. Mirski, R.; Derkowski, A.; Dziurka, D. Dimensional stability of OSB panels subjected to variable relative humidity: Core layer made with fine wood chips. *BioResources* **2013**, *8*, 6448–6459. [CrossRef]
14. Gaff, M.; Kacik, F.; Gasparik, M. The effect of synthetic and natural fire-retardants on burning and chemical characteristics of thermally modified teak (*Tectonagrandis L. f.*) wood. *Constr. Build. Mater.* **2019**, *200*, 551–558. [CrossRef]
15. Wang, S.Q.; Gu, H.M.; Wang, S.G. Layer thickness swell and related properties of commercial OSB products: A comparative study. In Proceedings of the 37th International Wood Composite Materials Symposium Proceedings, Pullman, WA, 7–10 April 2003; pp. 65–76.
16. Georgescu, S.-V.; Șova, D.; Campean, M.; Coșereanu, C. A Sustainable Approach to Build Insulated External Timber Frame Walls for Passive Houses Using Natural and Waste Materials. *Forests* **2022**, *13*, 522. [CrossRef]
17. Salem, M.Z.M.; Böhm, M.; Šedivka, P.; Nasser, R.A.; Ali, H.M.; Abo Elgat, W.A.A. Some physico-mechanical characteristics of uncoated OSB ECO-products made from Scots pine (*Pinussylvestris L.*) and bonded with PMDI resin. *BioResources* **2018**, *13*, 1814–1828. [CrossRef]
18. Paes, J.B.; Maffioletti, F.D.; Lahr, F.A.R. Biological resistance of sandwich particleboard made with sugarcane, thermally-treated Pinus wood and malva fiber. *J. Wood Chem. Technol.* **2022**, *42*, 171–180. [CrossRef]
19. Kup, F.; Vural, C. Determination of physical and mechanical properties of particleboard obtained from cotton and corn stubble with fibreglass plaster net. *J. Environ. Prot. Ecol.* **2022**, *23*, 657–667.
20. Zheng, N.H.; Wu, D.N.; Sun, P.; Liu, H.G.; Luo, B.; Li, L. Mechanical Properties and Fire Resistance of Magnesium-Cemented Poplar Particleboard. *Materials* **2020**, *13*, 115. [CrossRef]
21. Han, G.P.; Wu, Q.L.; Lu, J.Z. Selected properties of wood strand and oriented strandboard from small-diameter southern pine trees. *Wood Fiber Sci.* **2006**, *36*, 621–632.

22. Makovická Osvaldová, L.; Mitrenga, M.; Jančík, J.; Titko, M.; Efhamisisi, D.; Košútová, K. Fire Behaviour of Treated Insulation Fibreboards and Predictions of its Future Development Based on Natural Aging Simulation. *Front. Mater.* **2022**, *9*, 891167. [CrossRef]
23. Efe, F.T. Investigation of some physical and thermal insulation properties of honeycomb-designed panels produced from Calabrian pine bark and cones. *Eur. J. Wood Wood Prod.* **2022**, *80*, 705–718. [CrossRef]
24. Malý, S.; Vavrečková, K.; Malme, K.; Kubás, J.; Hollá, K.; Makovická Osvaldová, L. Occupational Health and Safety of Food Industry Employees with Emphasis on Specific Diseases. In *Innovation Management and Sustainable Economic Development in the Era of Global Pandemic, Proceedings of the 38th International Business Information Management Association Conference, IBIMA Publishing, Seville, Spain, 23–24 November 2021*; IBIMA Publishing: Seville, Spain, 2021; ISBN 978-0-9998551-7-1. ISSN 2767-9640.
25. Ramos, A.; Briga-Sa, A.; Pereira, S.; Correia, M.; Pinto, J.; Bentes, I.; Teixeira, C.A. Thermal performance and life cycle assessment of corn cob particleboards. *Build. Eng.* **2021**, *44*, 102998. [CrossRef]
26. Górski, J.; Podziewski, P.; Borysiuk, P. The Machinability of Flat-Pressed, Single-Layer Wood-Plastic Particleboards while Drilling—Experimental Study of the Impact of the Type of Plastic Used. *Forests* **2022**, *13*, 584. [CrossRef]
27. Langová, N.; Réh, R.; Igaz, R.; Krišťák, Ľ.; Hitka, M.; Joščák, P. Construction of wood-based lamella for increased load on seating furniture. *Forests* **2019**, *10*, 525. [CrossRef]
28. Tabarsi, E.; Kozak, R.; Cohen, D.; Gaston, C. A market assessment of the potential for OSB products in the North American office furniture and door manufacturing industries. *For. Prod. J.* **2003**, *53*, 19–27.
29. Zamarian, E.H.C.; Iwakiri, S.; Trianoski, R.; de Albuquerque, C.E.C. Production of particleboard from discarded furniture. *Rev. Arvore* **2013**, *41*, 410407. [CrossRef]
30. Réh, R.; Krišťák, L.; Sedliačik, J.; Bekhta, P.; Božiková, M.; Kunecová, D.; Vozárová, V.; Tudor, E.M.; Antov, P.; Savov, V. Utilization of Birch Bark as an Eco-Friendly Filler in Urea-Formaldehyde Adhesives for Plywood Manufacturing. *Polymers* **2021**, *13*, 511. [CrossRef]
31. Eickner, H.W. Fire resistance of solid-core wood flush doors. *For. Prod. J.* **1976**, *23*, 38–43.
32. Harada, K. A review on structural fire resistance. In Proceedings of the Fourth Asia-Oceania Symposium on Fire Science & Technology, Tokyo, Japan, 24–26 May 2000; Asia-Oceania Association for Fire Science & Technology/Japan Association for Fire Science & Engineering: Tokyo, Japan, 2000; pp. 155–163.
33. White, R.H. Fire resistance of exposed wood members. In Proceedings of the Wood & Fire Safety: Proceedings, 5th International Scientific Conference, Zvolen, Slovakia., 18–22 April 2004; Tech. University of Zvolen: Zvolen, Slovakia.
34. Ayrilmis, N.; Kartal, S.; Laufenberg, T.; Winandy, J.; White, R. Physical and mechanical properties and fire, decay, and termite resistance of treated oriented strandboard. *For. Prod. J.* **2005**, *55*, 74.
35. Östman, B.A.L.; Mikkola, E. European classes for the reaction to fire performance of wood products. *HolzalsRoh Und Werkst.* **2006**, *64*, 327–337. [CrossRef]
36. Tudor, E.M.; Scheriau, C.; Barbu, M.C.; Réh, R.; Krišťák, Ľ.; Schnabel, T. Enhanced resistance to fire of the bark-based panels bonded with clay. *Appl. Sci.* **2020**, *10*, 5594. [CrossRef]
37. Salek, V.; Cabova, K.; Wald, F.; Jahoda, M. Numerical modelling of fire test with timber fire protection. *J. Struct. Fire Eng.* **2022**, *13*, 99–117. [CrossRef]
38. Rantuch, P. Comparing the Ignition Parameters of Various Polymers. In *Ignition of Polymers. Springer Series on Polymer and Composite Materials*; Springer: Cham, Switzerland, 2021. [CrossRef]
39. Liu, Y.T. Cone Calorimeter Analysis on the Fire-resistant Properties of FRW Fire-retardant Particleboard. *Adv. Mater. Process.* **2011**, *311*, 2142–2145.
40. Turkowski, P.; Węgrzyński, W. Comparison of a Standard Fire-Resistance Test of a Combustible Wall Assembly with Experiments Employing Pre-defined Heat Release Curves. *Fire Technol.* **2022**, *58*, 1767–1787. [CrossRef]
41. Baranovskii, N.V.; Kirienko, V.A. Ignition of Forest Combustible Materials in a High-Temperature Medium. *J. Eng. Phys.* **2020**, *93*, 1266–1271. [CrossRef]
42. Kristak, L.; Ruziak, I.; Tudor, E.M.; Barbu, M.C.; Kain, G.; Reh, R. Thermophysical properties of larch bark composite panels. *Polymers* **2021**, *13*, 2287. [CrossRef]
43. Madyaratri, E.W.; Ridho, M.R.; Aristri, M.A.; Lubis, M.A.R.; Iswanto, A.H.; Nawawi, D.S.; Antov, P.; Kristak, L.; Majlingová, A.; Fatriasari, W. Recent Advances in the Development of Fire-Resistant Biocomposites—A Review. *Polymers* **2022**, *14*, 362. [CrossRef]
44. Rybinski, P.; Syrek, B.; Szwed, M.; Bradlo, D.; Zukowski, W.; Marzec, A.; Sliwka-Kaszynska, M. Influence of Thermal Decomposition of Wood and Wood-Based Materials on the State of the Atmospheric Air. Emissions of Toxic Compounds and Greenhouse Gases. *Energies* **2021**, *14*, 3247. [CrossRef]
45. Gong, J.H.; Zhou, H.G.; Zhu, H.; McCoy, C.G.; Stoliarov, S.I. Development of a pyrolysis model for oriented strand board: Part II-Thermal transport parameterization and bench-scale validation. *J. Fire Sci.* **2021**, *39*, 477–494. [CrossRef]
46. Dietenberger, M.A.; Shalbafan, A.; Welling, J. Cone calorimeter testing of foam core sandwich panels treated with intumescent paper underneath the veneer (FRV). *Fire Mater.* **2018**, *42*, 296–305. [CrossRef]
47. Mikkola, E.; Wichman, I.S. On the thermal ignition of combustible materials. *Fire Mater.* **1989**, *14*, 87–96. [CrossRef]
48. Hao, H.; Chow, C.L.; Lau, D. Effect of heat flux on combustion of different wood species. *Fuel* **2020**, *278*, 118325. [CrossRef]
49. Richter, F.; Jervis, F.X.; Huang, X.Y.; Rein, G. Effect of oxygen on the burning rate of wood. *Combust. Flame* **2021**, *234*, 111591. [CrossRef]

50. Renner, J.S.; Mensah, R.A.; Jiang, L.; Xu, Q.; Das, O.; Berto, F. Fire Behavior of Wood-Based Composite Materials. *Polymers* **2022**, *13*, 4352. [CrossRef] [PubMed]
51. Kuracina, R.; Szabova, Z.; Bachraty, M.; Mynarz, M.; Skvarka, M. A new 365-litre dust explosion chamber: Design and testing. *Powder Technol.* **2021**, *386*, 420–427. [CrossRef]
52. Rantuch, P.; Hrusovsky, I.; Martinka, J.; Balog, K. Calculation of critical heat flux for ignition of oriented strand boards. *Fire Prot. Saf. Secur.* **2017**, *207*, 214–222.
53. *CEN Standard EN ISO 13943: 2018*; Fire safety. Vocabulary. European Committe for Standartion:: Brussels, Belgium, 2018.
54. *ISO 3261:1975*; Fire tests—Vocabulary. International Organization for Standardization: Geneva, Switzerland, 1995.
55. Rantuch, P.; Kacíková, D.; Martinka, J.; Balog, K. The Influence of Heat Flux Density on the Thermal Decomposition of OSB. *ActaFacultatisXylologiaeZvolen res PublicaSlovaca* **2015**, *57*, 125–134.
56. Babrauskas, V. Ignition of wood. A Review of the State of the Art. In *Interflam 2001*; Interscience Communications Ltd.: London, UK, 2001; pp. 71–88.
57. Babrauskas, V. *Ignition Handbook*, 1st ed.; Fire Science Publishers: Issaquah, WA, USA, 2003.
58. Babrauskas, V. Charring rate of wood as a tool for fire investigations. *Fire Saf. J.* **2005**, *40*, 528–554. [CrossRef]
59. Baranovskiy, N.V.; Kirienko, V.A. Mathematical Simulation of Forest Fuel Pyrolysis and Crown Forest Fire Impact for Forest Fire Danger and Risk Assessment. *Processes* **2022**, *10*, 483. [CrossRef]
60. Janssens, M.L. Modeling of the thermal degradation of structural wood members exposed to fire. In *Second International Workshop on "Structures in Fire" (SiF '02)*; Department of Civil Engineering, University of Canterbury: Christchurch, New Zealand, 2002; pp. 211–222.
61. Chen, X.J.; Yang, L.Z.; Ji, J.W.; Deng, Z.H. Mathematical model for prediction of pyrolysis and ignition of wood under external heat flux. *Prog. Nat. Sci. Mater. Int.* **2002**, *12*, 874–877.
62. Reszka, P.; Borowiec, P.; Steinhaus, T.; Torero, J.L. A methodology for the estimation of ignition delay times in forest fire modelling. *Combust. Flame* **2012**, *159*, 3652–3657. [CrossRef]
63. Technical and Safety Data Sheet. *Boards Made of Oriented Flat Triples without Surface Treatment—OSB SUPERFINISH ECO, Type OSB/3*; Bučina DDD: Zvolen, Slovakia, 2019.
64. Safety data sheet Particleboard. In *Raw Un-Sanded*; Bučina DDD: Zvolen, Slovak republic, 2019.
65. *ISO 5657:1997*; Reaction to Fire Tests—Ignitability of Building Products using a Radiant Heat Source. International Organization for Standardization: Geneva, Switzerland, 1997.
66. Božiková, M.; Kotoulek, P.; Bilčík, M.; Kubík, Ľ.; Hlaváčová, Z.; Hlaváč, P. Thermal properties of wood and wood composites made from wood waste. *Int. Agrophys.* **2021**, *35*, 251–256. [CrossRef]
67. *EN 323:1993*; Wood-Based Panels—Determination of Density. European Committe for Standartion: Brussels, Belgium, 1993.
68. Turekova, I.; Markova, I.; Ivanovicova, M.; Harangozo, J. Experimental Study of Oriented Strand Board Ignition by Radiant Heat Fluxes. *Polymers* **2021**, *13*, 709. [CrossRef] [PubMed]
69. Turekova, I.; Ivanovicova, M.; Harangozo, J.; Gaspercova, S.; Markova, I. Experimental Study of the Influence of Selected Factors on the Particle Board Ignition by Radiant Heat Flux. *Polymers* **2022**, *14*, 1648. [CrossRef] [PubMed]
70. Maciulaitis, R.; Praniauskas, V.; Yakovlev, G. Research into the fire properties of wood products most frequently used in construction. *J. Civ. Eng. Manag.* **2013**, *19*, 573–582. [CrossRef]
71. Hakkarainen, T. Rate of heat release and ignitability indices in predicting SBI test results. *J. Fire Sci.* **2001**, *19*, 284–305. [CrossRef]
72. Kucera, P.; Lokaj, A.; Vicek, V. Behavior of the Spruce and Birch Wood from the Fire Safety Point of View. MATERIALS, MECHANICAL AND MANUFACTURING ENGINEERING. *Book Ser. Adv. Mater. Res.* **2014**, *842*, 725. [CrossRef]
73. Fagbemi, O.D.; Andrew, J.E.; Sithole, B. Beneficiation of wood sawdust into cellulose nanocrystals for application as a bio-binder in the manufacture of particleboard. *Biomass Conv. Bioref.* **2021**, 1–12. [CrossRef]
74. Michalovič, R. Assessment of different floors materials from the point of view of fire safety. In Proceedings of the 19th International Scientific Conference Solving Crisis Situations in a Specific Environment, Faculty of Security Egineering University of Žilina, Žilina, Slovakia, 21–22 May 2014; pp. 497–504.
75. Osvald, A.; Tereňová, Ľ.; Štefková, J. The Impact of Radiant Heat on Flexural Strength and Impact Toughness in OSB Panels. *Delta* **2020**, *14*, 26–35. [CrossRef]
76. Sinha, A.; Nairn, J.A.; Gupta, R. Thermal degradation of bending strength of plywood and oriented strand board: A kinetics approach. *Wood Sci. Technol.* **2021**, *45*, 315–330. [CrossRef]
77. Wang, Y.; Wang, W.; Zhou, H.; Qi, F. Burning Characteristics of Ancient Wood from Traditional Buildings in Shanxi Province, China. *Forests* **2022**, *13*, 190. [CrossRef]

Article

Physical and Mechanical Properties of *Paulownia tomentosa x elongata* Sawn Wood from Spanish, Bulgarian and Serbian Plantations

Marius Cătălin Barbu [1,2], **Katharina Buresova** [1], **Eugenia Mariana Tudor** [1,2,*] and **Alexander Petutschnigg** [1,3]

1   Forest Products Technology and Timber Construction Department, Salzburg University of Applied Sciences, Markt 136a, 5431 Kuchl, Austria
2   Faculty of Furniture Design and Wood Engineering, Transilvania University of Brasov, B-dul. Eroilor nr. 29, 500036 Brasov, Romania
3   Institute of Wood Technology and Renewable Materials, University of Natural Resources and Life Sciences (BOKU), Konrad Lorenz-Straße 24, 3340 Tulln an der Donau, Austria
*   Correspondence: eugenia.tudor@fh-salzburg.ac.at

Abstract: The aim of this research is the characterization of physical and mechanical properties of Paulownia sawn wood from three plantation sites in Europe, namely Spain, Bulgaria and Serbia. As a fast-growing wood species, Paulownia has a significant positive forecast for the European markets and a wide range of possible applications that still need to be explored. For this purpose, *Paulownia tomentosa*(Tunb.) *x elongata*(S.Y. Hu) wood species was investigated. Sorption behaviour, Brinell hardness, 3-point bending strength, flexural modulus of elasticity, tensile strength, compressive strength and screw withdrawal resistance were examined in detail. The samples from Spain have the higher average bulk density (266 kg/m^3), 3-point flexural strength (~40 N/mm^2), 3-point flexural modulus of elasticity (~4900 N/mm^2), compressive strength (~23 N/mm^2), tensile strength (~44 N/mm^2) and screw withdrawal resistance (~56 N/mm). The plantation wood from Bulgaria has the highest average of annual ring width (46 mm). Paulownia wood has potential in lightweight applications and can replace successfully expensive tropical species as Balsa.

Keywords: *Paulownia tomentosa x elongata*; plantation wood; lightweight; physical and mechanical properties

Citation: Barbu, M.C.; Buresova, K.; Tudor, E.M.; Petutschnigg, A. Physical and Mechanical Properties of *Paulownia tomentosa x elongata* Sawn Wood from Spanish, Bulgarian and Serbian Plantations. *Forests* **2022**, *13*, 1543. https://doi.org/10.3390/f13101543

Academic Editor: Ian D. Hartley

Received: 2 September 2022
Accepted: 19 September 2022
Published: 21 September 2022

Publisher's Note: MDPI stays neutral with regard to jurisdictional claims in published maps and institutional affiliations.

1. Introduction

Of Asian origin, Paulownia is a fast-growing deciduous tree, with at least nine sub-species [1]. In Europe, in the last decade, is growing interest on Paulownia as regards tree cultivation and agro-forestry plantations for industrial use [2]. Paulownia was also introduced in North America, Australia and Japan [3], and is cultivated worldwide in more than 40 countries [4]. Other objectives of Paulownia plantations are to reduce soil hazards by tree-crop intercropping in farmlands [5], to protect systems against erosion, flooding or wind damage [6], to reduce air pollution and to secure the increasing energy demand [7]. Paulownia trees have exceptional root systems and can adapt easily to various soil conditions [8]. Moreover, Paulownia has high tolerance to drought and salinity, representing an easy solution for sand fixation, also for water and soil conservation [9]. Nevertheless, the optimum soil and climate conditions and the most suitable plantation sites have not been clarified yet, and at present there is no best practice guidance for forest farmers who intent to manage Paulownia plantations [10]. This lightweight tree species gained interest worldwide and is named a miracle tree, empress tree or princess tree [11], due to its high rate carbon absorption and rated as fast-growing energy crop with C4 photosynthesis [12], easy processability, and good fire resistance [13].

Paulownia trees can be harvested in 6–7 years for low-quality lumber and in 15 years for worthwhile timber. The height of an adult Paulownia tree is from 10 to 20 m, its growth rate is up to 3 m in one year (in ideal conditions), allowing exploitation in short rotation periods. The diameter of a 10 year Paulownia is 30–40 cm with a timber volume of 0.3–0.5 m^3 per log [3]. Paulownia wood is semi-ring porous to ring porous, soft, frequently knot free, with an average density under 300 kg/m^3 [14]. The wood is light coloured, soft, lightweight and easy to handle [15]. It dries quickly and is easy to shape; these properties recommend it for industrial applications as furniture, building timber, packaging, plywood, insulation [16], for sculptures and handicrafts [17] and as reinforcing filler for thermoplastic composites [18]. The primary use of Paulownia includes solid wood products, veneer, and pulp as a source for fine papers [19]. Products made from Paulownia wood do not warp, crack, deform or decay easily [18], and are rot resistant [20]. It has a low thermal conductivity and a high ignition point [14,19]. Less valuable stems can be chipped to produce biofuels, biomass, electricity and contribute to air purification [21] or can be used as raw materials for particleboard, OSB and MDF production [22]. Other uses of Paulownia stems were introduced by the Chinese for medical purposes as a component of remedies for infections, e.g., poliovirus [23], due to its antioxidant properties [24].

The variability of Paulownia species depends on plantation site, climatic conditions, irrigations and forestry management [25]. Agro-forestry Paulownia plantations ensure sustainability for small rural communities. The trees represent a source of lumber, firewood, compost and coal and can easily adapt to new places. The Paulownia leaves are rich in nitrogen and can be introduced as feed for livestock [26].

Although Paulownia has a significant value for agroforestry, wood processing industry and ensuring ecological maintenance, this wood species is still under-rated and understudied [9]. *Paulownia tomentosa* is anywise considered invasive species, so its future spread should be attentive managed [21,27].

The objective of this study is to determine the physical and mechanical properties of *Paulownia tomentosa x elongata* sawn wood, grown in Spain, Bulgaria and Serbia and to compare it with similar Paulownia plantation wood from Europe and with other lightweight wood species as Balsa and poplar.

2. Materials and Methods

The Paulownia wood (*Paulownia tomentosa x elongata*) was provided from Glendor Holding GmbH (Kilb, Austria) and originates from plantations from Spain, Bulgaria and Serbia. Mostly juvenile wood samples from 5–7 years old trees were used for the tests. The plantation wood was delivered as rough-sawn lumber of approximately 150 cm length, 20–30 cm width and thickness of 20–25 mm. Prior to testing, the raw material was conditioned to constant weight at 20 °C and 65% relative air humidity for at least 14 days, until constant weight was achieved.

Several tests were carried out to determine the physical and mechanical properties of Paulownia wood from Spain, Bulgaria and Serbia (Table 1).

Table 1. Tests, norms, sample dimension and number of testing specimens for Paulownia wood from Spain, Bulgaria and Serbia.

Test	Norm	n Samples nr.	Sample Dimension [mm]
Swelling and shrinkage	DIN 52184:1979-05	12	20 × 20 × 10
Bulk density (kg/m^3)	ISO 3131:1996	12	20 × 20 × 10
Brinell-hardness (N/mm^2)	EN 1534:2011-01	10	-
3-point modulus of elasticity (MOE) (N/mm^2)	DIN 52186:1978-06	12	20 × 20 × 360
3-point modulus of rupture (MOR) (N/mm^2)	DIN 52186:1978-06	12	20 × 20 × 360
Compressive strength (N/mm^2)	DIN 52185:1976-09	12	20 × 20 × 50
Tensile shear strength (N/mm^2)	DIN 52188:1979-05	15	20 × 6 at predetermined breaking point
Screw withdrawal resistance (N/mm)	EN 320:2011-07	9	50 × 50

Physical and mechanical properties of the Paulownia wood samples were evaluated according to European and German norms, specific for massive wood testing.

The differential swelling and shrinking (in all cutting directions of wood) after 24 h water immersion was determined according to DIN 52184:1979-05 [28]. To measure the density according to ISO 3131:1996 [29], the weight and dimensions (20 mm × 20 mm × 10 mm) of samples were measured (i) after conditioning and (ii) after 24 h kiln drying.

The annual ring width was calculated as mean value from two opposing radii manually. The authors would like to emphasize that the reported values are indicative.

The test for the Brinell-hardness, with respect to EN 1534:2011-01 [30], was performed with the hardness tester Emco Test Automatic (Kuchl, Austria). The metal ball, with a diameter of 10 mm, was pressed with the force ranging from 199–207 N into the three main directions (radial, tangential and axial).

The tests for mechanical properties as bending strength, compressive strength, tensile strength and screw withdrawal resistance were carried out using a universal testing machine Zwick Roell Z 250 (Ulm, Germany). For the bending test, the specimen was placed on two bearings and continuously loaded with vertical force until breakage, according to DIN 52186:1978-06 [31]. During the test, a sensor records the bending stress in N/mm^2 and thus determines the bending strength—modulus of rupture (MOR) and flexural modulus of elasticity (MOE).

According to DIN 52185:1976-09 [32], for the compression test the specimen is loaded in vertical direction with an increasing force until breakage.

The tensile strength was measured in the fibre direction with samples prepared as DIN 52188 [33] requires (Figure 1).

Figure 1. Measuring of the tensile strength of a Paulownia sample from Bulgaria with universal testing machine Zwick Roell Z 250 (in the background the device Emco Test Automatic for testing of Brinell hardness).

The screw withdrawal resistance was tested on samples cut according to EN 320:2011 [34] (Figure 2). The force was applied perpendicular to the screw. The standardized screw with a diameter of 4.2 mm and a nominal length of 38 mm was countersunk into the wood at a right angle to the grain. In this case predrilling was carried out in the pillar drilling machine with a 2.8 mm drill bit. The specimens were fixed to the universal testing machine Zwick Roell Z 250 (Ulm, Germany) in a special fixture with the screw clamped at the top. The screw was pulled out with a specific and constant force.

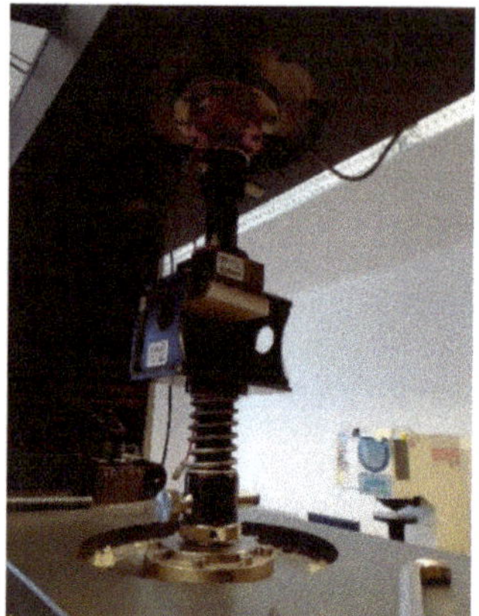

Figure 2. Measuring of the screw withdrawal resistance of a Paulownia sample from Spain with universal testing machine Zwick Roell Z 250.

3. Results and Discussion

The results of this study include density (Table 2), sorption behaviour (Table 3), width of annual rings (Table 4), Brinell hardness (Table 5), 3-point-bending strength (Table 6), flexural 3-point-modulus of elasticity (Table 7), compressive strength (Table 8), tensile strength (Table 9) and screw withdrawal resistance (Table 10). For each physical and mechanical property, Paulownia wood sourced from Spain, Bulgaria and Serbia was compared with the values of Paulownia wood from other European plantations and with Balsa, poplar and spruce from the literature sources.

3.1. Density (ISO 3131:1996)

Physical properties of wood, especially density and water-related properties, are important factors affecting wood quality [35].

Table 2. Basic statistics for bulk density of Spanish, Bulgarian and Serbian Paulownia plantation wood and other Paulownia species from the literature; n (Spain) = 12; n (Bulgaria) = 12; n (Serbia) = 12 (standard deviation in parentheses) (kg/m^3).

Wood Type	Mean Value Bulk Density (kg/m^3)	Min/Max (kg/m^3)	Source
Paulownia *tomentosa x elongata* (Spain)	266 (22)	238/297	Present study
Paulownia *tomentosa x elongata* (Bulgaria)	250 (26)	198/307	Present study
Paulownia *tomentosa x elongata* (Serbia)	259 (31)		Present study
Paulownia *tomentosa* (Hungary)	246		Koman and Feher [1]
Paulownia *tomentosa* (Hungary)	300 (26.59)		Koman and Vityi (2017)
Paulownia *tomentosa* (Türkiye)	272	201/313	Akyildiz and Kol [3]
Paulownia *tomentosa* (Portugal)	460	152/237	Estevez et al. [11]
Paulownia COTE-2 (Spain)	216	262/360	Lachowiz et al. [36]
Paulownia Sp. Siebold and Zucc. (Bulgaria)	220	178/270	Bardarov and Popovska [37]
Balsa	160	179/270	Wiekiping and Doyle [38]
Poplar	440		Grosser [39]
Spruce	430		Grosser [39]

Table 2 shows the values of density for Spanish, Bulgarian and Serbian Paulownia plantation wood measured according to ISO 3131:1996 [29]. Results from the technical literature about Paulownia and other lightweight wood species are listed for comparison.

The average density of wood from all three sites for Paulownia *tomentosa x elongata* is 258 kg/m^3. The Paulownia wood from Spain had the highest average density of 266 kg/m^3, followed by the Serbian wood with 259 kg/m^3, and the lowest average value had the Bulgarian wood with 250 kg/m^3 (Table 2).

In their study, Akyildiz and Kol [3] determined an average density of 272 kg/m^3 for the basic species *Paulownia tomentosa* from Türkiye. This value is lower for Paulownia wood from Hungary at 246 kg/m^3 [1] or from Spain at 215 kg/m^3 [36]. Estevez et al. [11] reported a value of 460 kg/m^3 for Portuguese Paulownia wood, which is even higher than the average density of spruce with 430 kg/m^3 according to [39].

Lachowicz et al. [36] measured the lowest value (Paulownia wood sourced from Spain) with a mean density of 216 kg/m^3. Considering the values reported from [40,41], Balsa wood has a lower density of 160 kg/m^3. The lightweight hardwood species poplar, has a density of 440 kg/m^3 [39,42].

At 12% moisture content, Paulownia wood density varies from 220 to 350 kg/m^3, with an average of 270 kg/m^3 [5]. This variability in density is determined by growth conditions. Higher Paulownia densities, about 400 kg/m^3, were reported for *Paulownia tomentosa* [3,11], and for *Siebold* and *Zucc.* (Bulgaria) [37].

3.2. Sorption Behavior (DIN 52184:1979)

The measurement of the sorption behaviour for Spanish, Bulgarian and Serbian Paulownia wood was carried out according to DIN 52184:1979 [28] and are shown in Table 3. From raw 4 to raw 7 (Table 3) are listed comparative results from the literature.

Table 3. Differential swelling and shrinkage of Spanish, Bulgarian and Serbian Paulownia wood compared to other Paulownia species from the literature; n (Spain) = 12; n (Bulgaria) = 12; n (Serbia) = 12 (standard deviation in parenthesis) (%).

Wood Type	Mean Value Axial (%)	Mean Value Radial (%)	Mean Value Tangential (%)	Source
Paulownia (Spain)	0.375 (0.048)	0.504 (0.077)	1.58 (0.234)	Present study
Paulownia (Bulgaria)	0.157 (0.057)	0.52 (0.17)	0.978 (0.181)	Present study
Paulownia (Serbia)	0.199 (0.050)	0.456 (0.087)	1.266 (0.277)	Present study
Paulownia (Hungary)	0.69	3.2	5	Koman and Feher [1]
Paulownia (Türkiye)		0.07	0.17	Akyildiz and Kol [3]
Paulownia (Spain)	0.172 (0.118)	1.99 (0.44)	5.19 (0.62)	Lachowicz et al. [36]
Paulownia (Croatia)	0.35 (0.332)	2.47 (0.631)	5.3 (0.969)	Sedlar et al. [35]

Spanish Paulownia wood had shrinkage in the axial direction of 0.375%, in the radial direction of 0.50%. In the tangential direction, with 1.58%, Paulownia wood has the highest average values for shrinkage. Bulgarian Paulownia wood has an axial shrinkage of 0.157%, in radial direction 0.52%, and in tangential direction 0.978%. Serbian paulownia wood has the shrinkage in axial direction of 0.199%, a radial shrinkage of 0.456%, and a tangential shrinkage of 1.266%.

It can be observed that the Bulgarian Paulownia wood swells and shrinks the least in all cutting directions. Compared to the Paulownia, Balsa wood has a lower sorption behaviour. In tangential direction, it shrinks and swells between 3.4%–7%. Radial shrinkage is 1.4%–2.1% and volume shrinkage is 5.1%–9.3% [38].

It is important to emphasize the lower ratios of swelling. This behaviour of Paulownia wood can be attributed to narrower core rays. The rays are narrow, occupying a single row up to 0.5 mm, but also multi-seriate rays can occur [5]. Firstly, the core rays control the wood in a radial direction and ensure values of swelling up to 4% [35], such as for most species (at this density). Secondly, the small width of core rays did not influence higher rates of swelling in tangential direction.

3.3. Width of Annual Rings

The widths of annual rings for Spanish, Bulgarian and Serbian Paulownia wood are shown in Table 4.

Table 4. Tree ring width—comparison of Spanish, Bulgarian and Serbian Paulownia wood. n (Spain) = 59; n (Bulgaria) = 7; n (Serbia) = 18 (standard deviation in parentheses) (cm).

Wood Type	Mean Value Annual Ring Width (cm)	Min./Max. (cm)
Paulownia (Spain)	2.8 (1.08)	1.2/7.5
Paulownia (Bulgaria)	4.6 (0.62)	3.7/5.7
Paulownia (Serbia)	1.7 (6.77)	0.6/3.1

The average annual ring width of entire batch of Serbian Paulownia wood was 1.7 cm. Paulownia trees from Spain had larger annual ring width of 2.83 cm, but the largest annual ring width was measured for Bulgarian Paulownia, namely 4.6 cm.

Serbian Paulownia wood had the smallest annual ring width, which is due to soil and climatic conditions. The tree ring width decreases as the height of the tree increases. The diameter of the tree tapers with increasing height. Thus, the diameter decreases, and the annual ring width consequently decreases. As already noted by [16], there are very large fluctuations in tree ring width within the first five years (up to 30%) (from 1 to 3.5 cm). From the beginning of the fifth year, the annual ring width becomes constant and is hardly subject to fluctuations anymore.

3.4. Brinell Hardness (DIN 1534:2022)

The testing of Brinell hardness for Spanish, Bulgarian and Serbian Paulownia wood was carried out according to DIN 1534:2022 (Table 5). In this Table 5, after the third row, are listed comparative results from the literature.

Table 5. Brinell hardness—in axial, radial and tangential directions—comparison of Spanish, Bulgarian and Serbian Paulownia wood with other wood species. n (Spain) = 10; n (Bulgaria) = 10; n (Serbia) = 10 (standard deviation in brackets) (N/mm^2).

Wood Species	Mean Value Brinell Hardness (N/mm^2)			Source
	Axial	Radial	Tangential	
Paulownia (Spain)	20.6 (5.56)	5.6 (1.53)	4.8 (1.19)	Present study
Paulownia (Bulgaria)	18.7 (3.1)	5.6 (1.35)	5.3 (1.35)	Present study
Paulownia (Serbia)	21.22 (7,64)	6.1 (3.23)		Present study
Paulownia (Hungary)	26.74 (3.22)	9.51 (2.17)	5.81 (2.13)	Koman and Vityi [16]
Paulownia (Bulgaria)	20		9.13 (2.16)	Bardanovand Popovska [37]
Paulownia (Türkiye)	19.7 (0.37)	8.23 (0.09)		Akyildiz and Kol [3]
Balsa	7		9.016 (0.23)	Finger [43]
Black poplar	25–33	10–15		Richter and Ehmke [44]
Spruce	32	12		Richter and Ehmke [44]

In the case of Paulownia wood source from Bulgaria, Brinell hardness in axial direction was 18.7 N/mm^2, 5.6 N/mm^2 in radial direction and 5.3 N/mm^2 in tangential direction. For Paulownia wood from Spain was measured the Brinell hardness in axial direction of 21.22 N/mm^2, 6.1 N/mm^2 in radial direction and 5.81 N/mm^2 in tangential direction. For Paulownia wood from Serbia were measured the highest values of Brinell hardness: 21.22 N/mm^2 in axial direction, 6.1 N/mm^2 in radial direction and 5.8 N/mm^2 in tangential direction. The latter values in axial direction are consistent with the results of [37] and [16]. Compared to Balsa wood, with a Brinell hardness of 7 N/mm^2 [45], Paulownia has significantly increased hardness. Other lightweight hardwood species is poplar, with a hardness of 25–33 N/mm^2 [43], in concordance with the results of [16] of 26.74 N/mm^2.

3.5. Modulus of Rupture and Modulus of Elasticity (DIN 52186:1978)

Table 6 shows the measured values of the three-point bending tests (modulus of rupture, MOR) for Paulownia wood sourced from Spain, Bulgaria and Serbia, measured according to DIN 52186:1978. The fourth to eight rows in Table 6 show the comparative results from the literature.

Table 6. 3-point bending strength (MOR)—comparison of Spanish, Bulgarian and Serbian Paulownia wood with other wood species. n (Spain) = 12; n (Bulgaria) = 12; n (Serbia) = 12 (standard deviation in parentheses) (N/mm^2).

Wood Species	Mean Values MOR [N/mm^2]	Min./Max. [N/mm^2]	Source
Paulownia (Spain)	39.77 (6.98)	28.96/50.5	Present study
Paulownia (Bulgaria)	35.53 (5.53)	24.57/43.99	Present study
Paulownia (Serbia)	37.54 (8.54)	24.84/59.48	Present study
Paulownia (Bulgaria)	35		Baranov and Popovska [37]
Paulownia (Türkiye)	43.56 (7.00)	33.36/60.37	Akyildiz and Kol [3]
Paulownia (Hungary)	32.3 (4.68)	28.65/48.65	Koman and Vityi [1,16]
Paulownia (Portugal)	53.5 (6)	-	Esteves et al. [11]
Paulownia (Spain)	38.63	23.89/53.17	Lachowicz et al. [36]
Balsa	16.63 (1.72)		Kotlarewski et al. [45]
Spruce	80	-	Richter and Ehmcke [44]
Oak	95		Richter and Ehmcke [44]
Black poplar	55–65		Richter and Ehmcke [44]

Paulownia wood from Spain achieved the highest flexural modulus of elasticity of 4866.49 N/mm^2 and the highest flexural strength of 39.77 N/mm^2 (Tables 6 and 7). The Bulgarian Paulownia wood had the lowest MOR of 35.53 N/mm^2. Paulownia wood from Serbia is in the middle range with 37.54 N/mm^2 (Table 6).

Jakubowski [5] analysed in a review article the mechanical properties of Paulownia wood and reported a range for static bending strength from 23.98 to 43.56 N/mm^2 [5]. Lachowitcz et al. [36] measured a similar bending strength ranging from 23.89 N/mm^2 to 53.17 N/mm^2 with a mean value of 38.63 N/mm^2 and Esteves et al. [11] found a higher mean value of 53.5 N/mm^2 for Paulownia from Portugal. All these values are at least two-fold higher compared to MOR for Balsa wood, which is about 17 N/mm^2 [44]. The higher value for MOR achieved by the Palownia from Türkiye [3] was at least 20% lower than that the one for black poplar [44]. Spruce has an MOR of 80 N/mm^2 [44] that is at least two fold higher as MOR for Paulownia, which is also the case of oak (95 N/mm^2) [44].

The modulus of elasticity (MOE) for Paulownia ranges from 2651 to 4917 N/mm^2 [5] (Table 7).

Table 7. Flexural modulus of elasticity (MOE)—comparison of Spanish, Bulgarian and Serbian Paulownia wood with other wood species (N/mm^2).

Wood Species	Mean Values MOE [N/mm^2]	Min./Max. [N/mm^2]	Source
Paulownia (Spain)	4866.49 (797.84)	3580/5941	Present study
Paulownia (Bulgaria)	3714.14 (588.51)	2685/4899	Present study
Paulownia (Serbia)	4532.49 (900.92)	2733/6492	Present study
Paulownia (Spain)	1898.75	1167/2690	Lachowicz et al. [36]
Balsa	2900	-	Sell [46]
Black Poplar	8800	-	Grosser [39]
Spruce	11,000	-	Richter and Ehmke [44]
Larch	13,800	-	Grosser [39]
Oak	13,000	-	Grosser [39]

Lachowicz et al. [36] measured the lowest modulus of elasticity of Paulownia wood with 1899 N/mm^2. This mean value is lower than the minimum value of the wood which was tests in this study.

When non-destructive methods were employed, a higher modulus of elasticity has been reported for trees with larger diameters [5].

In comparison, spruce has an MOE of 11,000 N/mm^2 and an MOR of 80 N/mm^2 [44]. Thus, spruce achieves a value twice as high. MOE for larch has even higher values, with a bending MOE of 13,800 N/mm^2 and a MOR of 99 N/mm^2 [39]. This value is almost three times higher than that of Paulownia wood. MOE of Balsa wood is lower, averaging 2900 N/mm^2 [45], but still higher than the values resulted from the study of [36], namely 1900 N/mm^2.

3.6. Compressive Strength (DIN 52185:1976)

The values of the compressive strength tests for Spanish, Bulgarian and Serbian Paulownia wood measured according to DIN 52188:1979 are shown in Table 8, together with other Paulownia species from Hungary, Spain and Türkyie and Balsa, spruce and black poplar.

Table 8. Compressive strength—comparison of Spanish, Bulgarian and Serbian Paulownia wood with other wood species. n (Spain) = 12; n (Bulgaria) = 12; n (Serbia) = 12 (standard deviation in parentheses) (N/mm^2).

Wood Species	Compressive Strength [N/mm^2]	Min./Max. [N/mm^2]	Source
Paulownia (Spain)	22.53 (3.17)	18.7/28.12	Present study
Paulownia (Bulgaria)	18.77 (1.5)	16.25/21.71	Present study
Paulownia (Serbia)	21.41 (4.55)	14.39/32.01	Present study
Paulownia (Hungary)	19.9 (1.78)	19.63/25.24	Koman and Vityi [1,16]
Paulownia (Spain)	14.24 (1.52)		Lachowicz et al. [36]
Paulownia (Türkyie)	35.56 (6.95)		Kaymakci et al. [47]
Paulownia (Türkyie)	25.55 (2.25)	10.45/18.29	Akyildiz and Kol [3]
Balsa	10		Wiekiping and Doyle [38]
Spruce	45	20.35/29.42	Richter and Ehmke [44]
Black poplar	30		Grosser [39]

Paulownia wood from Spain has a compressive strength of 22.53 N/mm^2. The Bulgarian Paulownia wood has a compressive strength of 18.77 N/mm^2 and the Serbian Paulownia wood has a compressive strength of 21.41 N/mm^2. Other values of compressive strength for Paulownia from other plantations are ranging from 25.55 N/mm^2 [3] and

35.56 N/mm^2 [46] for Paulownia from Türkyie and significant lower, of 14.24 N/mm^2, as results from the research of [36].

In comparison, spruce exhibits a compressive strength of 45 N/mm^2, which is twice as high as the determined compressive strength of Paulownia wood [44]. The value is comparatively similar for black poplar, which has a minimum value of 30 N/mm^2 [39]. The Balsa wood has a the lowest mean value for compressive strength of 10 N/mm^2 [38],

3.7. Tensile Strength (DIN 52188:1979-05)

Table 9 shows the measured values of the tensile strength tests for Spanish, Bulgarian and Serbian Paulownia wood according to DIN 52188:1979. The fourth line in Table 9 shows the comparative results from the literature.

Table 9. Tensile strength—comparison of Spanish, Bulgarian and Serbian Paulownia wood with other wood species. n (Spain) = 15; n (Bulgaria) = 15; n (Serbia) = 15 (standard deviation in parentheses) (N/mm^2).

Wood Species	Tensile Strength [N/mm^2]	Min./Max. [N/mm^2]	Source
Paulownia (Spain)	44.12 (10.66)	28.1/64.59	Present study
Paulownia (Bulgaria)	36.17 (6.69)	27.60/51.08	Present study
Paulownia (Serbia)	40.14 (9.11)	25.62/62.27	Present study
Paulownia (Hungary)	33.25 (8.9)		Koman and Vityi [16]
Balsa	14	21.86/52.96	Forest Products Laboratory [48]
Spruce	95		Richter and Ehmcke [44]
Oak	110		Richter and Ehmcke [44]
Black poplar	77		Richter and Ehmcke [44]

The Spanish Paulownia wood has a tensile strength of 44.12 N/mm^2. The tensile strength of Paulownia wood from Bulgaria was 36.17 N/mm^2 and the tensile strength for the wood from Serbia reached a value of 40.14 N/mm^2. For Paulownia sourced from Hungary Koman and Vityi [16] reported a tensile strength of 33.25 N/mm^2, which is consistent the values presented in this study. The tensile strength of Balsa wood is considerable lower with 14 N/mm^2 [47]. The tensile strength of lightweight species as black poplar is about 40% higher than that of Paulownia and for spruce is two-fold higher. In the case of oak, its tensile strength exceeds with at least 60% the values for Paulownia [44].

3.8. Screw Withdrawal Resistance (EN 320:2011)

Table 10 shows the results of screw withdrawal resistance (SWR) measurements according to EN 320:2011-07.

Table 10. Screw withdrawal resistance—comparison of Spanish, Bulgarian and Serbian Paulownia wood with other wood species. n (Spain) = 9; n (Bulgaria) = 10; n (Serbia) = 9 (standard deviation in parentheses) (N/mm).

Wood Species	Screw Withdrawal Resistance (N/mm)	Min./Max. (N/mm)	Source
Paulownia (Spain)	55.56 (6.6)	41.48/63.47	Present study
Paulownia (Bulgaria)	51.95 (13.66)	31.4/87.74	Present study
Paulownia (Serbia)	56.55		Present study
Paulownia (Türkiye)	50.5 (7.87)		Akyildiz [49]
Black pine	152	34.24/91.72	Aytekin [50]
Fir	108		Aytekin [50]
Oak	170		Aytekin [50]

Paulownia wood from Spain measured a screw withdrawal resistance of 55.56 N/mm. The screw pull-out resistance of Bulgarian paulownia wood was 51.95 N/mm. Serbian Paulownia wood reached a value of 56.55 N/mm. The results for SWR for the Paulownia from Bulgaria are consistent with the findings of [48], where plantation wood was extracted from Türkiye, therefore it can be supposed that Paulownia from Black Sea region exhibits similar properties. Compared with SWR of hardwood species, the overall values for Paulownia are at least two or three fold lower [49].

4. Conclusions

The physical and mechanical properties of Paulownia wood have shown that the location of these plantations (Iberian Peninsula and Balkans), the type of soil and the environmental conditions strongly influence the wood properties. The density is directly corelated with the mechanical properties. The low density of all these tested samples ensures that the wood is filled with a lot of air and thus has heat-insulating and light-weight properties.

As expected, Paulownia wood achieved significantly lower values in physical and mechanical properties compared to conventional species such as spruce, oak or poplar. Paulownia wood can be classified very low and low for MOR, MOE and compression strength. Paulownia is not recommended for structural uses, which require high mechanical strength and stiffness.

For further investigations, it is important to pay close attention from which log section was extracted the sample. There are large variations in strength within a log and therefore different mechanical properties, depending on the spot of the log where the test specimen was cut. There are significant differences in the width of annual rings which greatly affects the woods properties.

In view of all the results, the conclusion is that Paulownia has enormous potential for special lightweight application in construction, model making and thermal insulating. Paulownia offers many possibilities in non-load-bearing structures and can successfully replace other tropical wood species, which are more expensive and rarer. Paulownia wood bears resemblance to Balsa wood concerning its lightweight. It is known that Balsa is one of the best core materials embedded in lightweight sandwich structures, with distinctive stiffness-to-weight and strength-to-weight ratios. The comparisons of mechanical properties of these two species demonstrates that it might be suitable to focus on the possibility of using Paulownia wood as a substitute for Balsa wood as core material for composites.

Author Contributions: Conceptualization, M.C.B. and E.M.T.; methodology, E.M.T.; validation, A.P. and E.M.T.; formal analysis, K.B.; investigation, K.B.; resources, K.B. and E.M.T.; data curation, A.P.; writing—original draft preparation, K.B. and E.M.T.; writing—review and editing, E.M.T.; visualization, M.C.B. and A.P.; supervision, A.P. and E.M.T.; project administration, M.C.B. All authors have read and agreed to the published version of the manuscript.

Funding: This research received no external funding.

Institutional Review Board Statement: Not applicable.

Informed Consent Statement: Not applicable.

Data Availability Statement: Not applicable.

Acknowledgments: The authors would like to thank to Arien Crul, for providing all Paulownia lumber from three sources (Spain, Bulgaria, Serbia), FH-Thomas Schnabel (FH Salzburg) for the support with design of experiment, Thomas Wimmer, for the measurements conducted in the facilities of FH Salzburg and Helmut Radauer. for the support with sample preparing.

Conflicts of Interest: The authors declare no conflict of interest.

References

1. Koman, S.; Feher, S. Physical and mechanical properties of Paulownia clone in vitro 112. *Eur. J. Wood Prod.* **2020**, *78*, 421–423. [CrossRef]
2. Jensen, J.B. An Investigation into the Suitability of Paulownia as an Agroforestry Species for UK & NW European Farming Systems. Master's Dissertation, Coventry University, Coventry, UK, 2016.
3. Akyildiz, M.H.; Kol, H.S. Some technological properties and uses of paulownia (Paulownia tomentosa Steud.) wood. *J. Environ. Biol.* **2010**, *31*, 351–355. [PubMed]
4. Gyuleva, V.; Stankova, T.; Zhyanski, M.; Glushkova, M.; Andonova, E. Growth and Development of Paulownia tomentosa and Paulownia elongata x fortunei in Glasshouse Experiment. *Bulg. J. Soil Sci.* **2020**, *5*, 126–142.
5. Jakubowski, M. Cultivation Potential and Uses of Paulownia Wood: A Review. *Forests* **2022**, *13*, 668. [CrossRef]
6. Feng, Y.; Cui, L.; Zhao, Y.; Qiao, J.; Wang, B.; Yang, C.; Zhou, H.; Chang, D. Comprehensive Selection of the Wood Properties of Paulownia Clones Grown in the Hilly Region of Southern China: Feng, Y. et al. *BioResources* **2020**, *15*, 1098–1111. [CrossRef]
7. Abbasi, M.; Pishvaee, M.S.; Bairamzadeh, S. Land suitability assessment for Paulownia cultivation using combined GIS and Z-number DEA: A case study. *Comput. Electron. Agric.* **2020**, *176*, 105666. [CrossRef]
8. Lucas-Borja, M.E.; Wic-Baena, C.; Moreno, J.L.; Dadi, T.; García, C.; Andrés-Abellán, M. Microbial activity in soils under fast-growing Paulownia (Paulownia elongata x fortunei) plantations in Mediterranean areas. *Appl. Soil Ecol.* **2011**, *51*, 42–51. [CrossRef]
9. Cao, Y.; Sun, G.; Zhai, X.; Xu, P.; Ma, L.; Deng, M.; Zhao, Z.; Yang, H.; Dong, Y.; Shang, Z.; et al. Genomic insights into the fast growth of paulownias and the formation of Paulownia witches' broom. *Mol. Plant* **2021**, *14*, 1668–1682. [CrossRef]
10. Tu, J.; Wang, B.; McGrouther, K.; Wang, H.; Ma, T.; Qiao, J.; Wu, L. Soil quality assessment under different Paulownia fortunei plantations in mid-subtropical China. *J. Soils Sediments* **2017**, *17*, 2371–2382. [CrossRef]
11. Esteves, B.; Cruz-Lopes, L.; Viana, H.; Ferreira, J.; Domingos, I.; Nunes, L.J.R. The Influence of Age on the Wood Properties of Paulownia tomentosa (Thunb.) Steud. *Forests* **2022**, *13*, 700. [CrossRef]
12. Świechowski, K.; Stegenta-Dąbrowska, S.; Liszewski, M.; Bąbelewski, P.; Koziel, J.A.; Białowiec, A. Oxytree Pruned Biomass Torrefaction: Process Kinetics. *Materials* **2019**, *12*, 3334. [CrossRef] [PubMed]
13. Yorgun, S.; Yıldız, D.; Şimşek, Y.E. Activated carbon from paulownia wood: Yields of chemical activation stages. *Energy Sources Part A Recovery Util. Environ. Eff.* **2016**, *38*, 2035–2042. [CrossRef]
14. Kalaycioglu, H.; Deniz, I.; Hiziroglu, S. Some of the properties of particleboard made from paulownia. *J. Wood Sci.* **2005**, *51*, 410–414. [CrossRef]
15. Dogu, D.; Tuncer, F.D.; Bakir, D.; Candan, Z. Characterizing Microscopic Changes of Paulownia Wood under Thermal Compression. *BioResources* **2017**, *12*, 5279–5295. [CrossRef]
16. Komán, S.; Vityi, A. Physical and mechanical properties of paulownia tomentosa wood planted in Hungaria. *Wood Res.* **2017**, *62*, 335–340.
17. Yadav, N.K.; Vaidya, B.N.; Henderson, K.; Lee, J.F.; Stewart, W.M.; Dhekney, S.A.; Joshee, N. A Review of Paulownia Biotechnology: A Short Rotation, Fast Growing Multipurpose Bioenergy Tree. *Am. J. Plant Sci.* **2013**, *04*, 2070–2082. [CrossRef]
18. Ayrilmis, N.; Kaymakci, A. Fast growing biomass as reinforcing filler in thermoplastic composites: Paulownia elongata wood. *Ind. Crops Prod.* **2013**, *43*, 457–464. [CrossRef]
19. Bergmann, B.A. Propagation method influences first year field survival and growth of Paulownia. *New For.* **1998**, *16*, 251–264. [CrossRef]
20. Hussain, K.; Nasir, G.M.; Hussain, T. Comparison of wood anatomy of different Paulownia species grown in Pakistan. *Pak. J. For.* **2016**, *66*, 2.
21. Snow, W.A. Ornamental, crop, or invasive? The history of the Empress tree (Paulownia) in the USA. *For Trees Livelihoods* **2015**, *24*, 85–96. [CrossRef]
22. Hua, L.S.; Chen, L.W.; Geng, B.J.; Kristak, L.; Antov, P.; Pędzik, M.; Rogoziński, T.; Taghiyari, H.R.; Rahandi Lubis, M.A.; Fatriasari, W.; et al. Particleboard from Agricultural Biomass and Recycled Wood Waste: A Review. *J. Mater. Res. Technol.* **2022**. [CrossRef]
23. Kang, K.H.; Huh, H.; Kim, B.-K.; Lee, C.-K. An antiviral furanoquinone from Paulownia tomentosa steud. *Phytother. Res.* **1999**, *13*, 624–626. [CrossRef]
24. He, T.; Vaidya, B.; Perry, Z.; Parajuli, P.; Joshee, N. Paulownia as a Medicinal Tree: Traditional Uses and Current Advances. *Eur. J. Med. Plants* **2016**, *14*, 1–15. [CrossRef]
25. Icka, P.; Damo, R.; Icka, E. Paulownia Tomentosa, a Fast Growing Timber. *Ann. "Valahia" Univ. Targoviste-Agric.* **2016**, *10*, 14–19. [CrossRef]
26. El-Showk, N.; El-Showk, S. *The Paulownia Tree, An Alternative for Sustainable Forestry*; The Farm: Wuhan, China, 2003; Available online: https://docslib.org/doc/11683191/the-paulownia-tree-an-alternative-for-sustainable-forestry (accessed on 15 July 2022).
27. Essl, F. From ornamental to detrimental? The incipient invasion of Central Europe by Paulownia tomentosa. *Preslia* **2007**, *79*, 377–389.
28. DIN 52184:1979-05. Testing of Wood; Determination of Swelling and Shrinkage. Deutsches Institut für Normung: Berlin, Germany, 1979.
29. ISO 3131:1996-06-01. Wood-Determination of Density for Physical and Mechanical Tests. European Committee for Standardization: Brussels, Belgium, 1996.

30. EN 1534:2011-01. Wood Flooring-Determination of Resistance to Indentation-Test Method. European Committee for Standardization: Brussels, Belgium, 2011.
31. DIN 52186:1978-06. Testing of Wood; Bending Test. Deutsches Institut für Normung: Berlin, Germany, 1978.
32. DIN 52185:1976-09. Testing of Wood; Compression Test Parallel to Grain. Deutsches Institut für Normung: Berlin, Germany, 1976.
33. DIN 52188:1979-05. Testing of Wood; Determination of Ultimate Tensile Stress Parallel to Grain. Deutsches Institut für Normung: Berlin, Germany, 1979.
34. EN 320:2011. Particleboards and Fibreboards-Determination of Resistance to Axial Withdrawal of Screws. European Committee for Standardization: Brussels, Belgium, 2011.
35. Sedlar, T.; Šefc, B.; Drvodelić, D.; Jambreković, B.; Kučinić, M.; Ištok, I. Physical Properties of Juvenile Wood of Two Paulownia Hybrids. *Drv. Ind.* **2020**, *71*, 179–184. [CrossRef]
36. Lachowicz, H.; Giedrowicz, A. Charakterystyka jakości technicznej drewna paulowni COTE–2. *Sylwan* **2020**, *164*, 414–423. [CrossRef]
37. Bardarov, N.; Popovska, T. Examination of the properties of local origin Paulownia wood. (Paulownia sp. Siebold & Zucc.). *Manag. Sustain. Dev.* **2017**, *2*, 75–78.
38. Wiepking, C.A.; Doyle, D.V. Strength and Related Properties of Balsa and Quipo Woods, Madison, USA. 1960. Available online: https://ir.library.oregonstate.edu/concern/defaults/sx61dq95c?locale=en (accessed on 15 July 2022).
39. Grosser, D. *Die Hölzer Mitteleuropas: Ein Mikrophotographischer Lehratlas, Reprint der 1. Aufl. von 1977*; Remagen: Kessel, The Netherlands, 2007; ISBN 3935638221.
40. Byrne, C.E.; Nagle, D.C. Carbonization of wood for advanced materials applications. *Carbon* **1997**, *35*, 259–266. [CrossRef]
41. Borrega, M.; Gibson, L.J. Mechanics of balsa (Ochroma pyramidale) wood. *Mech. Mater.* **2015**, *84*, 75–90. [CrossRef]
42. Ciftci, A.; Kaya, Z. Genetic diversity and structure of Populus nigra populations in two highly fragmented river ecosystems from Turkey. *Tree Genet. Genomes* **2019**, *15*, 66. [CrossRef]
43. Finger, M. Balsaholz (Ochroma pyramidale). Available online: http://www.holzwurm-page.de/holzarten/holzart/balsaholz.htm. (accessed on 15 July 2022).
44. Richter, K.; Ehmcke, G. Das Holz der Fichte–Eigenschaften und Verwendung. *LWF Wissen* **2017**, *80*, 117–124.
45. Kotlarewski, N.J.; Belleville, B.; Gusamo, B.K.; Ozarska, B. Mechanical properties of Papua New Guinea balsa wood. *Eur. J. Wood Prod.* **2016**, *74*, 83–89. [CrossRef]
46. Sell, J. *Eigenschaften und Kenngrössen von Holzarten, 4*; Bachfach Verlag: Zürich, Switzerland, 1997; ISBN 3855652236.
47. Kaymakci, A.; Bektas, I.; Bal, B.C. Some mechanical properties of Paulownia (Paulownia elongata) wood. In Proceedings of the International Caucasian Symposyum, Artvin, Turkey, 26–27 September 2013; pp. 917–920.
48. Forest Products Laboratory. *Wood Handbook: Wood as an Engineering Material*; USDA Forest Service: Madison, WI, USA, 1999.
49. Akyildiz, M.H. Screw-nail withdrawal and bonding strength of paulownia (Paulownia tomentosa Steud.) wood. *J. Wood Sci.* **2014**, *60*, 201–206. [CrossRef]
50. Aytekin, A. Determination of screw and nail withdrawal resistance of some important wood species. *Int. J. Mol. Sci.* **2008**, *9*, 626–637. [CrossRef] [PubMed]

Review

Environmentally Friendly Starch-Based Adhesives for Bonding High-Performance Wood Composites: A Review

Muhammad Iqbal Maulana [1], Muhammad Adly Rahandi Lubis [1,2,*], Fauzi Febrianto [3], Lee Seng Hua [4], Apri Heri Iswanto [5,6,*], Petar Antov [7,*], Lubos Kristak [8], Efri Mardawati [2,9], Rita Kartika Sari [3], Lukmanul Hakim Zaini [3,10], Wahyu Hidayat [11], Valentina Lo Giudice [12] and Luigi Todaro [12]

1 Research Center for Biomass and Bioproducts, National Research and Innovation Agency, Cibinong 16911, Indonesia
2 Research Collaboration Center for Biomass and Biorefinery between BRIN and Universitas Padjadjaran, Jatinangor 40600, Indonesia
3 Department of Forest Products, Faculty of Forestry and Environment, IPB University, Bogor 16680, Indonesia
4 Laboratory of Biopolymer and Derivatives, Institute of Tropical Forestry and Forest Product, Universiti Putra Malaysia UPM, Serdang 43400, Selangor, Malaysia
5 Department of Forest Product, Faculty of Forestry, Universitas Sumatera Utara, Medan 20155, Indonesia
6 JATI-Sumatran Forestry Analysis Study Center, Universitas Sumatera Utara, Medan 20155, Indonesia
7 Faculty of Forest Industry, University of Forestry, 1797 Sofia, Bulgaria
8 Faculty of Wood Sciences and Technology, Technical University in Zvolen, 96001 Zvolen, Slovakia
9 Department of Agro-Industrial Technology, Universitas Padjadjaran, Jatinangor 40600, Indonesia
10 Institute of Wood Technology and Renewable Materials, Department of Material Sciences and Process Engineering, University of Natural Resources and Life Sciences (BOKU), 1180 Vienna, Austria
11 Department of Forestry, Faculty of Agriculture, University of Lampung, Bandar Lampung 35141, Indonesia
12 School of Agricultural, Forestry, Food and Environmental Sciences, University of Basilicata, Vialedell'AteneoLucano 10, 85100 Potenza, Italy
* Correspondence: marl@biomaterial.lipi.go.id (M.A.R.L.); apri@usu.ac.id (A.H.I.); p.antov@ltu.bg (P.A.)

Citation: Maulana, M.I.; Lubis, M.A.R.; Febrianto, F.; Hua, L.S.; Iswanto, A.H.; Antov, P.; Kristak, L.; Mardawati, E.; Sari, R.K.; Zaini, L.H.; et al. Environmentally Friendly Starch-Based Adhesives for Bonding High-Performance Wood Composites: A Review. *Forests* 2022, 13, 1614. https://doi.org/10.3390/f13101614

Academic Editor: Zeki Candan

Received: 30 August 2022
Accepted: 28 September 2022
Published: 2 October 2022

Publisher's Note: MDPI stays neutral with regard to jurisdictional claims in published maps and institutional affiliations.

Abstract: In recent years, bio-based wood adhesives have gained an increased industrial and research interest as an environmentally friendly and renewable alternative to the commercial petroleum-based synthetic adhesives used in the wood-based industry. Due to its renewability, abundance, relatively low price, and good adhesion properties, starch is a promising natural feedstock for synthesizing bio-based adhesives for wood-based composites. This review aims to summarize the recent advances in developing sustainable starch-based wood adhesives for manufacturing non-toxic, low-emission wood composites with enhanced properties and lower environmental impact. Recent developments in starch modification, physical, and enzymatic treatments applied to improve the performance of starch-based wood adhesives, mainly in terms of improving their water resistance and bonding strength, are also outlined and discussed.

Keywords: amylose; amylopectin; bio-based adhesives; starch; wood adhesives; wood-based panels

1. Introduction

Starch is an abundant natural polymer and the cheapest industrially available carbohydrate. In recent years, it has attracted an increased commercial and research interest for its potential in a wide range of value-added applications, including papermaking, food processing, cosmetics, pharmaceutical products, additives, and industrial adhesives, due to its annual renewability, relatively low price, and good adhesion characteristics [1–4]. Starch is the mixture of two distinct polysaccharide fractions of amylose and amylopectin, which both are made out of glucose of various sizes and shapes [5,6]. The proportions of these components differ according to the starch botanical origin and subsequently affect adhesives properties. The glucan structure of amylose is linear and relatively long, made out of roughly close to 100% $(1{\rightarrow}4)$- α-linkages and some $(1{\rightarrow}6)$- α-linkages [6,7]. The degree of polymerization (DP) of amylose is around 500–5000, with the molecular weight around 105–106 Da. There are about 9–20 branches per molecule in the amylose structure, and each branch chain has 200–700 units of glucose. Meanwhile, amylopectin is a polysaccharide that has numerous branches consisting of (1→4)-α-linkages (95%) and (1→6)-α-linkages (5%–6%). Amylopectin has a DP that varies around 9600–15,900, with a molecular weight of 107–109 Da. Amylopectin molecular chain is shorter compared to amylose, consisting of 18–25 units of glucose (Figure 1) [6–8].

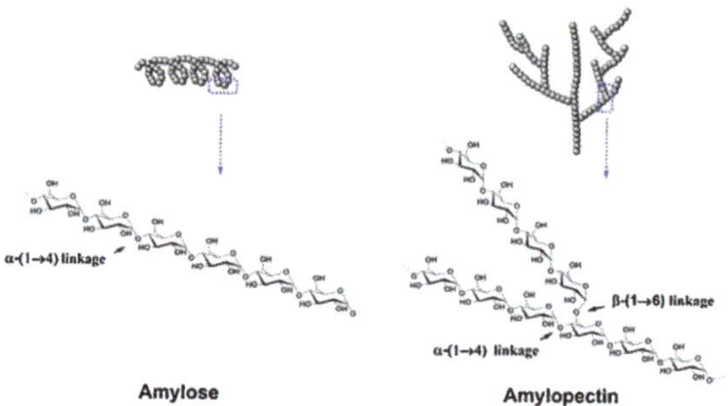

Figure 1. Illustration of amylose and amylopectin structures [8].

Morphologically, the starch granule is a combination of amylose and amylopectin, forming amorphous regions in semi-crystalline regions through hydrogen bonds between molecules (Figure 2) [9]. The crystalline part hinders the water or any chemical components from penetrating the starch structure and causes lower reactivity and a higher gelation temperature [9]. Therefore, some modifications to the starch crystalline part or decreasing the crystalline size has been recommended. Certain modifications have been propounded to reduce the crystallinity of starch, including chemical (esterification, oxidation, cationization, and etherification), physical (heat-moisture treatment, mechanical activation, ultrasonic degradation, and microwave exposure), and enzymatic treatment [2,10–12].

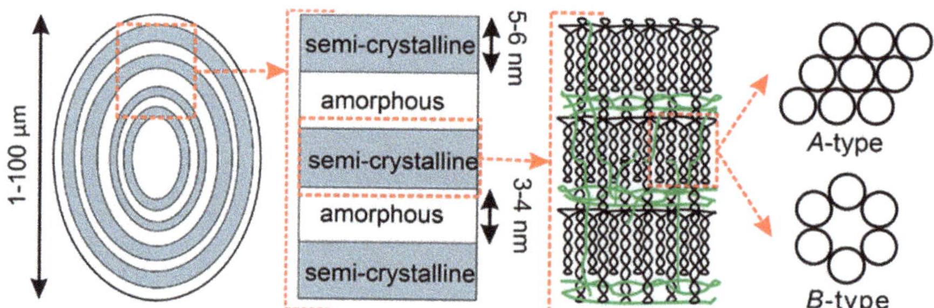

Figure 2. Starch structure illustration form to the unit of glucosyl; black lines represent the branched part and double helices of amylopectin, while green lines represent the single helices of amylose, and the double helices in an A-type or B-type polymorphic crystal from the top view are indicated by circles [9].

The objective of this paper is to present a comprehensive review of the recent advances in the field of starch-based wood adhesives used for manufacturing high-performance, environmentally friendly wood-based composites. The main challenges and future perspectives of using starch as a natural feedstock to develop bio-based wood adhesives have also been outlined.

2. Historical Overview

Starch, a renewable, abundant, and inexpensive biopolymer has always been considered one of the most valuable materials used in many applications [2,4,13–15]. Early starch uses can be seen on Egyptian papyrus strips, dated to the pre-dynastic period (3500–4000 BC) and adhered together using a starch adhesive [16–18]. The historian and philosopher Gaius Plinius Secundus described documents of 130 BC that were created by smoothing the surface of papyrus by sizing it with modified wheat starch. Finely milled wheat flour was used to make the adhesive, then heated with a diluted vinegar solution. Papyrus strips were coated with adhesive and hammered with a mallet, and additional strips were laid over the edges to create a wider sheet. Chinese paper documents, dated about 312 AD, are reported to contain starch [16–18]. Chinese documents were later coated with high viscosity starch to provide resistance to ink absorption and then topped with powdered starch to give more weight and thickness. At that time, a starch made from rice, wheat, and barley was regularly used. Dutch starch was thought to be of great quality throughout the Middle Ages when wheat starch production in the Netherlands became a significant industry. In the early fourteenth century, starch was introduced to stiffen linen in Northern Europe. Coloured and uncoloured starches were used as cosmetics. Uncoloured starch was used particularly as a hair powder. Before Queen Elizabeth banned its use in 1596, blue starch was used by the Puritans. Yellow starch was fashionable until a notorious woman prisoner was publicly executed wearing a bright yellow-starched ruffle. Red starch cosmetics have stayed in fashion for many years [16].

In addition, in 214 BC, Emperor Qin Shihuang of the Qin Dynasty started to use a special concoction to construct the Great Wall of China. The concoction was prepared by boiling a large amount of glutinous rice into a thick soup, mixed with soil, wood, and other materials and spread layer by layer, thus forming an ancient "concrete" structure. It is a super-strong mortar made from sticky rice. Markedly, the archaeological evidence suggests that sticky rice-lime mortar had already reached a mature stage of development by the time of the South-North Dynasty (386–589 AD). Because of its excellent performance, the sticky rice-lime mortar was widely used to construct many significant buildings, including tombs, city walls, and water resource facilities. The resilience of the sticky rice mortar can be attributed to amylopectin, a polysaccharide typically found in rice and other starchy foods [19].

As starch became a more significant industrial commodity, extensive research on its modification was carried out. This included Kirchoff's great discovery in 1811 that the acid-catalyzed hydrolysis of potato starch could produce sugars. Then, the torrefaction methods for producing dextrins, now termed British gums, were accidentally discovered in 1826 after a fire in a Dublin textile factory that used starch as a size. After the blaze was extinguished, a worker noted that some of the heated starch was turned dark and could quickly dissolve in water to create a thick adhesive paste. The new starch was subjected to another roasting, and the resultant product demonstrated the previously noted advantageous characteristics. Thus, heat dextrinization became known and later developed into wider use [20,21].

The first wheat starch plant in America was founded by Gilbert at Utica, New York, in 1807 and later converted into a corn starch-producing factory. In 1842, the shift from wheat to corn starch began with the development of a manufacturing process in which crude corn starch was purified by alkaline treatment [16,22]. The manufacture of potato starch began in 1820 in Hillsborough County, New Hampshire. The utilization of potato starch increased quickly, and by 1895, there were sixty-four factories in operation, forty-four of which were located in Maine. During approximately 3 months of operation, they produce 24 million pounds of starch annually. Most of them are sold to the textile industry. Rice starch manufacture was begun in 1815 using the caustic treatment of rice.

Nevertheless, the production did not rise considerably and most of the later used rice starch was imported [16,20]. By 1890, the number of starch plants in America had risen to eighty; 240 million pounds of starch are produced annually. The National Starch

Company of New Jersey was established in 1902 due to a merger between the United Starch Company and the National Starch Manufacturing Company. The Corn Products Company, which processes 1800 tons per day and accounts for 80% of the corn starch market, was formed by a union of the National Starch Company of New Jersey, the Illinois Sugar Refining Company, the Glucose Sugar Refining Company, and the Charles Pope Glucose Company [16,20–22].

About 77% of the starch utilized globally and 95% in the U.S. comes from corn [22]. In 1995, the United States consumed around 3.6×109 kg of corn starch, with a 2% annual growth rate [7,23]. This amounts to around 3% of the corn produced annually in the United States. During the past decade, corn starch has averaged 0.2–0.3 USD/kg. However, due to 1995's poor weather and the subsequent high export demand, prices were higher in that year. In the United States, potato starch prices are about 0.65 USD/kg and are mostly imported from Europe. There are numerous starches produced that have been engineered by chemical, physical, or genetic methods to suit the needs of various industrial uses better. The price of starch in 2021 was roughly 0.25 to 2.20 USD/kg [24].

3. Sources of Starch

Most plants synthesize starch, a polysaccharide, to store energy [5,6]. It is kept within cells as spherical granules that range in size from 2 to 100 μm [16,20]. Most starches sold in markets come from tubers, such as potatoes, tapioca and cereals, such as corn and wheat. These grains and tubers have a high starch content, often between 60 and 90% of their dry weight [16]. The development of commercial techniques for the recovery of corn starch was naturally prompted by the high starch content of corn, its ability to be stored from one season to another, and its ready availability at steady and comparatively low prices. Beginning with the early 19th century, when the recovery of corn starch was first discovered by crushing soaked grains, the procedure progressively improved into the highly advanced automated procedure used today, which results in a variety of beneficial culinary and industrial products. Much of the early expansion of the corn starch industry was encouraged by mechanical innovation created during full-scale operations [23,25]. Today, rigorous pilot plant assessments, engineering studies, and research are more frequently followed by process and product enhancements.

Starch is mostly derived from corn, wheat, sweet potatoes, cassava, and potatoes, while rice, barley, sorghum, and other grains are minor sources in various regions of the world [26]. About 98%–99% of the dry weight of starch composition includes amylose and amylopectin, and the rest includes a small amount of damaged starch, enzymes, lipids, proteins, ashes, minerals, and phosphorus [6,7,27].

Cassava (*Manihot utilissima* Pohl) is commonly cultivated in tropical climates, while potatoes, wheat, and corn are often grown in temperate climates [23]. Manufacturing plants for cassava starch are located near the root growing areas to minimize transport costs and get the shortest tuber processing time. The roots delivered to the factory are stored in concrete or wooden bunkers. Strict supervision of bunker filling and emptying must be carried out to ensure that the first harvested roots are consumed first. The roots are typically transported to a washing station by a belt conveyor. After washing, the outer skin is removed. The cortex is not eliminated because it has some stable starch in modern processes. Usually, the washer is a U-formed box with paddles that convey the roots to the peeler. The roots are stripped by the abrasion of one against another and against the walls and paddles of the machine. All cell walls must be ruptured to recover the starch. This has been done occasionally by mild fermentation followed by grinding into a pulp and the starch recovered by screening and washing or centrifugation. The fermentation process does not produce good starch yields, and the starch quality is generally inferior [11,14,28–32].

Rice (*Oryza sativa* L.) is the main diet of South, East, and Southeast Asia, wherein 90% of the rice crop in the world is grown and consumed [16,20]. Because brewer's rice is more expensive than other cereals and tubers, its use in the commercial manufacture of starches is limited. In the European Economic Community, only 7000 tons of rice starches are produced annually in Belgium, Germany, The Netherlands, and Italy factories. Factories in Egypt and Syria also produce rice starch, but it has not been produced in the United States since 1943 [16,20]. Since most milled rice protein is an alkali-soluble protein known as glutelin, sodium hydroxide is employed to purify the rice starch [33,34]. Hogan has documented the commercial production of rice starch, and little has changed since the 1960s [16,20]. The step includes soaking rice in 0.3%–0.5% sodium hydroxide solution, wet milling, removing cell walls, extracting the protein with sodium hydroxide solution, washing, and drying. An initial cleaning removes trash and filth. The process of soaking softens the grain and helps extract the protein. The soaking period is typically 24 h, and the temperature ranges from ambient to 50 °C. Because dry milling causes more severe starch degradation and causes more starch to dissolve in alkali, wet grinding is preferred.

Starch and gluten are significant and valuable co-products when wheat flour is wet processed [25,26,35–37]. These products are moving into a new stage of development, mostly due to the abundance of value-added goods being offered because of their sustainability. Large-granule starch, resistant starch, low-moisture starch, cook-up, and pre-gelatinized forms of untreated starch are the products provided on the wheat starch market for usage in food and industry [38,39]. There are six common methods for separating wheat starch and gluten: batter, dough, aqueous dispersion, non-aqueous separation, wet-milling of kernels, and chemical dispersion [16,20,35,36,40,41]. Due to the low quality of the product, high running costs, effluent issues, and ineffectiveness, the latter three methods are not used [20].

The total potato starch production is smaller than the corn starch produced worldwide. Potato starch production was estimated to be only 2.5 million tons per year, whereas corn starch reached 45.8 million tons/year in 2005 [16,20,22,42]. Potatoes should have the most starch possible to produce potato starch effectively. Hence, only specific kinds of industrial potatoes are employed in current potato starch manufacturers in Europe. These potatoes are not eaten as food because they are not particularly palatable due to their high starch content. In Europe, potatoes are collected and processed between August and April; this time frame is known as the starch campaign. Culled food potatoes are typically not used because they have a low content of dry matter, and the starch granules are reasonably small [20], which are more challenging to process. Food potatoes may occasionally be processed in between European starch campaigns when the food potato price is low, owing to excess production. On a limited scale, reclaimed potato starch is processed in Europe and North America from the process waters of other potato processing companies, such as the manufacturers of French fries, chips/crisps, and potato puree [16,20,22,42].

4. Starch-Based Wood Adhesives

Starch is an inexpensive material with good adhesive and film-forming properties and represents a promising candidate for developing bio-based wood adhesives [2,3,36,43–45]. As shown in Figure 3, starch-based adhesives are typically composed of four main constituents [1,17,37]. Although starch can be utilized to make bio-based adhesives, its bonding capacity is based on hydrogen-bonding forces, which are considerably weaker than chemical bonds. Additionally, starch-based adhesives have a low water resistance due to their ease in forming hydrogen bonds with water molecules. Hence, starch modification is required to improve the functionality of starch-based adhesives by enhancing the molecular structure and viscosity of the adhesive [16,33,46–48]. Different starch-based wood adhesives, such as lignin-starch [4,45,49–51], protein-starch [37,52,53], tannin-starch [54,55], starch-polyvinyl alcohol (PVOH) [3,56–58], and starch-isocyanate [59–63], have been explored by several studies. Chemical, physical, and enzymatic treatments are among the most popular strategies to improve the properties of starch-based adhesives [64,65].

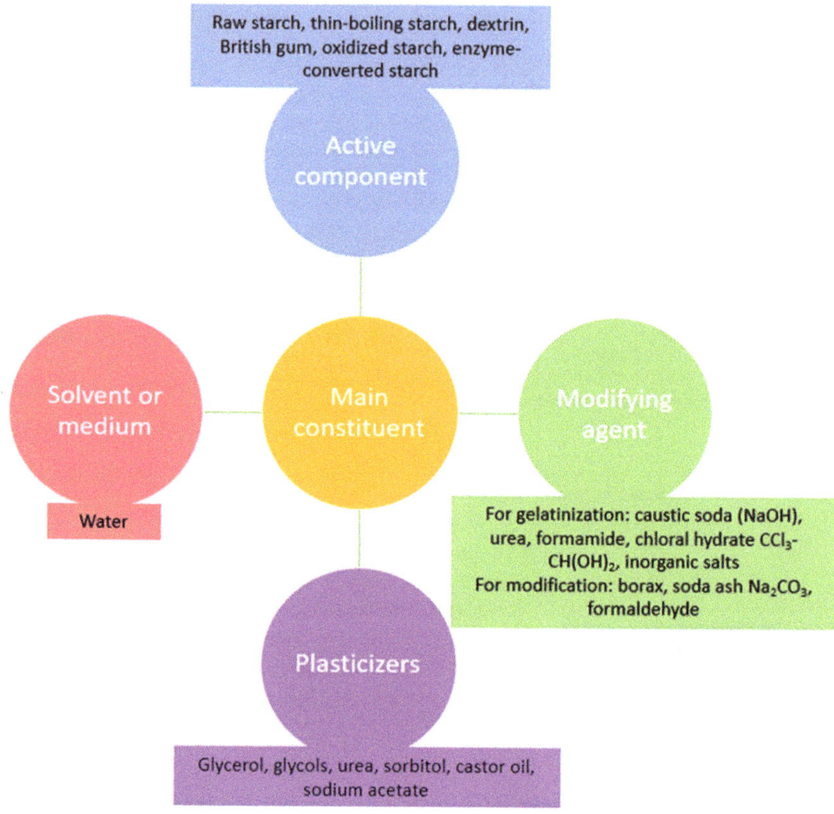

Figure 3. Four main constituents of starch-based adhesives [1,17,37].

4.1. Chemical Treatments

The qualities of the starch-based adhesive have been improved recently using a variety of starch modification techniques, such as acid hydrolysis [66–68]; silane coupling agent [12,69–71]; heat pretreatment [1,16,72]; the addition of nanoparticles, such as nanosilica and nanoclay [73–75]; sodium dodecyl sulfate (SDS) [14,76,77]; and dodecyl succinic anhydride (DDSA) [2,78–80].

As presented in Table 1, acid hydrolysis is among the most prevalent modification methods of starch. Acid hydrolysis significantly influences the amylose component of starch [66,68]. Amylose content greatly affects the structural and functional characteristics of starch. Amylose consists of linear chains that can form strong inter-chain linkages and, therefore, play a vital role in bestowing good water resistance to wood-based composites. In addition, acid hydrolysis could make the modified starch molecules react more readily with grafting monomers by destroying hydrogen bonds and altering starch's crystallinity [67].

Further, acid hydrolysis reduced the viscosity of the wood adhesive from corn starch. Meanwhile, both shear strength in the dry and wet state improved from acid hydrolysis. However, the shear strength decreased beyond 2 h of acid hydrolysis duration [66–68,81–83].

Table 1. Effects of various modification methods on the shear strength of starch-based wood adhesives.

Treatment	Strength (MPa)	Reference
Acid hydrolysis Dissolved in hydrochloric acid (HCl) and stirred at 60 °C (0, 0.5, 1, 1.5, 2, 2.5, and 3 h)	Tensile shear strength Dry state—1.21 MPa (0 h) to 6.65 MPa (2 h) Wet state (23 °C)—0.8 MPa (0 h) to 3.6 MPa (2 h)	[66]
Silane coupling agent γ-Methacryloxypropyltrimethoxysilane (KH570) (0%–10%)	Tensile shear strength Dry state—5.5 MPa (0%) to 6.7 MPa (6%) Wet state (30 °C)—2.2 MPa (0%) to 2.6 MPa (4%)	[84]
Oxidation Hydrogen peroxide (3%–15%) olefin monomer (0%–5%)	Tensile shear strength Dry state—4.43 MPa (3%) to 7.88 MPa (9%) Wet state (30 °C)—0.76 MPa (3%) to 4.09 MPa (9%) Dry state—3.28 MPa (0%) to 7.30 MPa (3%) Wet state (30 °C)—1.40 MPa (0%) to 4.22 MPa (3%)	[85,86]
Heat pretreatment 70, 80, and 90 °C	Tensile shear strength Dry state—8.63 MPa (control) to 10.17 MPa (90 °C)	[87]
Silica nanoparticles (0%–10%)	Tensile shear strength Dry state—3.41 MPa (1%) to 5.12 MPa (10%) Wet state (23 °C)—1.62 MPa (1%) to 2.98 MPa (10%)	[88]
Montmorillonite (MMT, 0%–9%)	Tensile shear strength Dry state—5.60 MPa (0%) to 10.60 MPa (5%) Wet state (23 °C)—1.7 MPa (0%) to 3.9 MPa (3%)	[89]
Anionic surfactant—Sodium dodecyl sulfate (SDS, 0%–2%)	Tensile shear strength Dry state—5.5 MPa (2%) to 6.3 MPa (0%)	[76]
Esterification and polyisocyanate pre-polymer crosslinking(0%–20% prepolymer)	Block shear strength Dry state—2.3 MPa (0%) to ~12.0 MPa (10%) Wet state (30 °C) ~0 MPa (0%) to 4.0 MPa (10%)	[13]
Esterification with dodecenyl succinic anhydride (DDSA, 0%–8%)	Tensile shear strength Dry state—1.51 MPa (0%) to 2.61 MPa (2%) Wet state (63 °C)—0.58 MPa (0%) to 1.0 MPa (6%)	[78]

Another alternative way to improve starch properties is by oxidation. Oxidized starch (OS) can be prepared by treatment with hydrogen peroxide (H_2O_2) [90–92]. Nevertheless, the OS-based adhesive has inferior water resistance and bonding strength. Several additives, such as urea [93], sodium dodecyl sulfate [70], olefin monomer [1], silane coupling agent [84], and isocyanates [59,94], have been introduced to the OS during the preparation of wood adhesive. The prepared OS-based adhesive displayed markedly increased water resistance and bonding strength. The study found that starch-based wood adhesives with superior water resistance and bonding strength can be developed through graft copolymerization of oxidized starch with an olefin monomer and a coupling agent [70,95]. The OS-based wood adhesive's dry and wet shear strength is enhanced with the addition of olefin monomer, which acts as an oxidant. At 3% oxidant, the highest wood adhesive's starch-based shear strength was recorded. However, shear strength reduced as the oxidant content raised from 3% to 5%. On the other hand, coupling agent content also significantly affects wood adhesive's starch-based shear strength. When coupling agent content was expanded from 3 to 9%, the starch-based adhesive's dry and wet shear strength increased considerably but started to drop when more than 9% coupling agent was added [70,95]. A similar study was also made by Chen et al. [84], where the addition of 6% and 4% silane coupling agent, γ-Methacryloxypropyltrimethoxysilane (KH570), recorded maximum dry shear strength and wet shear strength, respectively.

A series of works have been done to enhance the performance of starch-based wood adhesives. Ammonium persulfate was employed as an initiator in the preparation of starch-based wood adhesive using vinyl acetate grafted starch [14,96]. Unfortunately, the produced adhesive has inferior performance due to its poor mobility and storage stability due to starch retrogradation. On that account, a surfactant, such as sodium dodecyl sulfate (SDS), was added to inhibit starch retrogradation [3,58,76]. Adding SDS improved the adhesive's storage stability and mobility (Figure 4). However, the shear strength of the starch-based adhesive was adversely impacted, where the shear strength dropped from 6.3 MPa to 5.5 MPa when 2% SDS was added [76]. Therefore, nanoclay, montmorillonite (MMT), has been added to compensate for the negative effects of SDS on the starch-based wood adhesive [3,89]. The findings proved the addition of MMT promising, as the dry shear strength of starch-based adhesive almost doubled when 5% MMT was added. Owing to the strengthening effects of nanoparticles on the molecular structure of starch adhesive, the incorporation of nanoparticles has been widely adopted to improve the performance of the starch-based wood adhesive. Silica nanoparticles have been used as a reinforcing agent for vinyl acetate (VAc) grafted starch in the production of starch-based wood adhesive [97]. The results revealed that adding 10% silica nanoparticles has increased the dry shear strength and wet shear strength of starch-based wood adhesive by 50% and 84%, respectively.

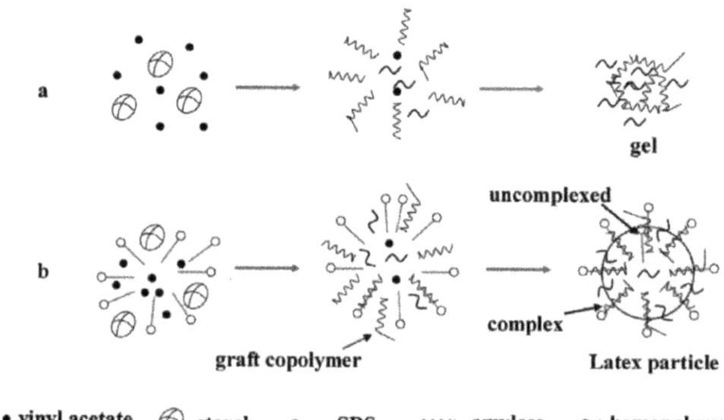

Figure 4. Schematic of the starch-based wood adhesive's amylose-SDS complexes: (**a**) without and (**b**) with SDS [76].

Better water resistance and higher bonding strength of starch-based wood adhesives can be achieved by combining starch with other components, e.g., PVOH [3,56–58], isocyanates [61–63,98], formaldehyde [12,60,99,100], and tannins [54,101,102]. According to several research works, vinyl acetate was grafted onto starch using ammonium persulfate as an initiator to create starch-based adhesives [14,96,97]. The studies demonstrated that graft efficiency significantly affected the bonding performance of the starch-based adhesive. Markedly, a high amylose content starch-based wood adhesive is identified by improved mechanical and water-resistance properties, which are necessary for bonding wood composites in actual applications [103,104].

The hydroxyl groups on C2, C3, and C6 positions in each glucose unit of starch can form hydrogen bonding. The main techniques applied for the chemical modification of starch are presented in Figure 5 [8,83,105,106]. To increase the hydrophobicity of starch, a common chemical modification called esterification transforms hydroxyl groups into esters. The degree of esterification (DS) and the chain length of the esterification agent determine the esterified starch's water absorptivity and solubility. As depicted in Figure 6, maleic anhydride (MA) was used to react maize starch to create esterified corn starch, which was subsequently cross-linked using a poly-isocyanate pre-polymer [13,21]. Synthetic polymer grafting copolymerization onto the starch backbone improved the starch bonding properties. The authors reported that the optimal pre-polymer level was 10 wt%, which produced 12 MPa of dry and 4 MPa of wet shear strength values. Another piece of research described the addition of blocked pMDI (B-pMDI) and an auxiliary chemical to a starch-based adhesive. When the mixture ratio of starch and the blocked isocyanate was 100/25 and 100/20, respectively, the wet and dry bonding strength peaked [59]. In addition, attaching the isocyanate to the starch-based adhesive might make it less viscous. Bentonite could be added to the adhesive to thicken it and increase its water resistance [61].

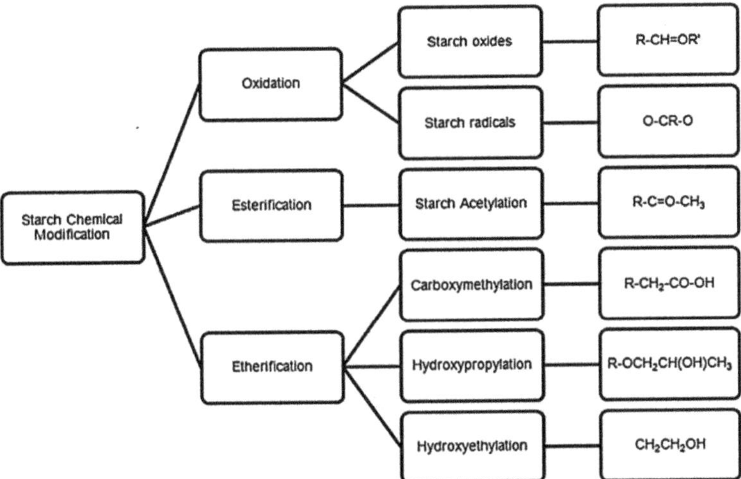

Figure 5. Diagram outlining the classical chemical techniques for starch modification [83]. Reprinted/adapted with permission from Ref. [83]. 2017, Elsevier, License Number 5400551069466.

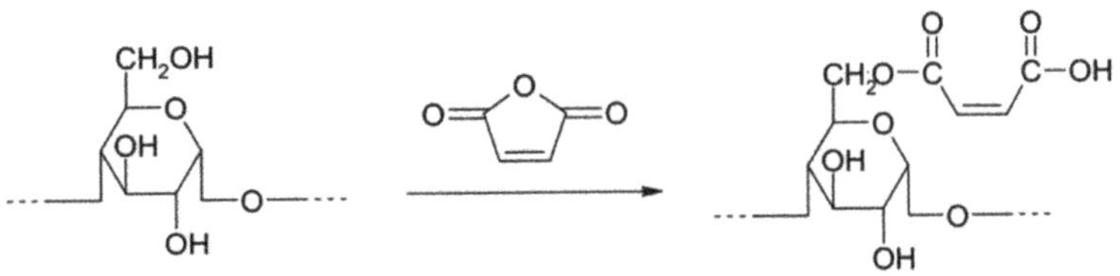

Figure 6. Starch's esterification reaction with maleic anhydride occurs at the C6 position [21].

Another study described the grafting of vinyl acetate monomer onto corn starch and crosslinking polymerization, using N-methylol acrylamide to produce an environmentally friendly starch-based wood adhesive for producing wood-based composites [44,107]. The authors concluded that developing a complex network structure and greater crosslinking density was responsible for the starch-based adhesive's improved performance.

Starch-based adhesives may also be crosslinked using epoxy resin. Combining epoxy resins with polyvinyl acetate grafted starch adhesives has been tested as a method of attaching veneers. According to the authors, epoxy groups can form three-dimensional networks with good shear strength values in dry and humid environments [108–111].

Regarding the chemical modification of starch, an oxidation process forms a more reactive wood adhesive. The peroxide oxidation splits H_2O_2 into two OH radicals by Cu^{2+}, which acts as a catalyst. The radical OH groups and catalyst then oxidized the OH of the starch into aldehyde groups and released H_2O. Further, the remaining H_2O_2 converted the aldehyde groups into carboxyl groups [90–92]. Therefore, greater H_2O_2/starch mole ratios led to a greater DO due to the greater amount of H_2O_2 in the system. B-pMDI and citric acid (CA) have been used as cross-linker to enhance the performance of OS-based adhesives [56,58]. The OS reacted with the isocyanate groups from the B-pMDI to form amide linkages (Figure 7a), while it formed ester linkages by reacting with the CA (Figure 7b).

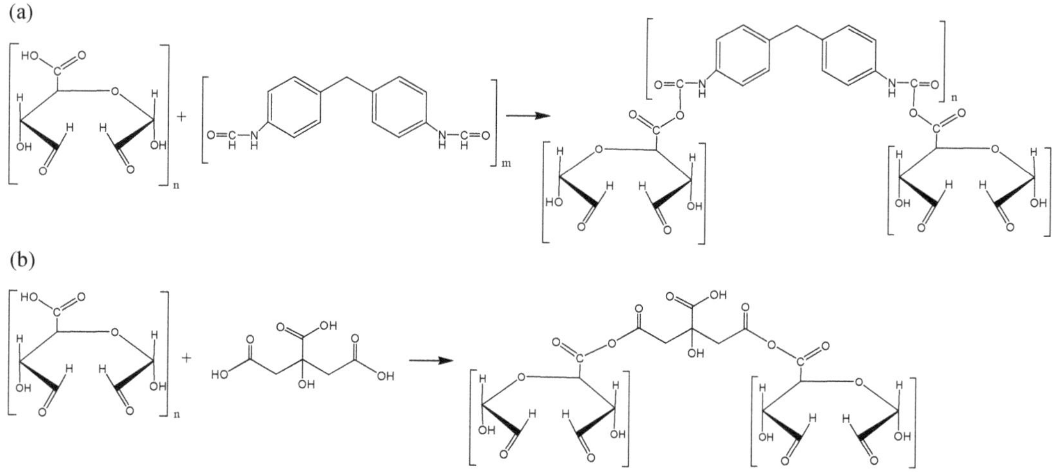

(a)

(b)

Figure 7. Possible cross-linking reactions of (a) B-pMDI/OS and (b) CA/OS [56].

An oxidation-gelatinization technique was used to synthesize a cornstarch adhesive, after which its rheological characteristics were determined [112,113]. The apparent viscosity was determined by shear rate, starch-to-water ratio, and temperature. It was found that apparent viscosity reaches a peak value at 10 °C, and then decreases at higher temperatures, increasing shear rate from 6–60 RPM, causing a slight decrease in viscosity. The starch adhesive has fluid-like, pseudo-plastic characteristics. Oxidized starches typically react responsively to heat, turning yellow or brown when subjected to high temperatures. The aldehyde content has been linked to this drying tendency to turn yellow. The oxidized starch in storage turns yellower with increasing aldehyde content. The yellowing of oxidized starch dispersed in water by cooking or alkali is also related to aldehyde content.

Innovative formulations of starch-based adhesives are made with the addition of a silane cross-linker (CH_2=CH–$Si(OC_2H_5)_3$), hydrogen peroxide, as well as vinyl and butyl acetate [69,70,84]. They reported that when an oxidizing agent was present, the starch hydroxyl groups changed into aldehyde and carboxyl groups. The graft copolymerization enhanced the adhesive's bonding strength, water resistance, and thermal stability. According to the optimization method, adding 3 wt% of oxidant agents and 9 wt% of coupling agent produced a modified starch-based adhesive with wet and dry bonding strengths of around 4.09 and 7.88 MPa, respectively.

Oxidation also affected the solids content, viscosity, and gelation time of OS-based adhesives [56]. OS's solids content and viscosity decreased as the H_2O_2/starch mole ratio increased while gelation time increased (Table 2). This meant that a higher H_2O_2/starch mole ratio decreased the reactivity of the OS adhesive. The excess of H_2O_2 at a higher mole ratio probably remained in the OS, which decreased the solids content and viscosity of OS and eventually increased the gelation time. The lower viscosity of the adhesive generally increased the gelation time due to the adhesive system needing a longer time to evaporate water and solvent. This phenomenon happened to OS adhesives, with a higher degree of H_2O_2/starch mole ratio, which has a greater amount of H_2O_2 in the system. The molecular weight distributions of OS for different H_2O_2/starch mole ratios, such as *Mw*, *Mn*, and PDI, of OS decreased as the H_2O_2/starch mole ratio increased. The *Mw* of native starch has been reported to be around 800,000 g/mole [16,20,47], and the *Mw* decreased by a hundred times after oxidation. It was obvious that the oxidation de-polymerized starch macromolecules into smaller molecules, eventually lowering the molecular weight.

Table 2. Properties of oxidized starch adhesive at different H_2O_2/starch mole ratios [56].

H_2O_2/Starch Mole Ratio	Solids Content (%)	Viscosity (mPa·s)	Gelation Time (s)	*Mw* (g/mole)	*Mn* (g/mole)	Polydispersity Index (PDI)
0.5	48.43	107.7	532	11,882	9881	1.19
1.0	41.82	76.0	547	11,000	9547	1.15
1.5	37.94	60.7	560	9835	8890	1.11
2.0	31.20	45.3	587	8010	7657	1.05

Other researchers have used urea-formaldehyde (UF) resin to create wood-based composite adhesives by mixing it with varying ratios of starch [114], esterified starch [115,116], and oxidized starch [45,93]. It was reported that a new adhesive system made of starch and UF resin had improved water resistance and reduced formaldehyde emission and brittleness. The UF resin adhesive with oxidized starch blending can be utilized in wood adhesive applications because it has strong chemical stability, insulating qualities, temperature resistance, and aging resistance [93,117,118]. Good oil resistance and mildew resistance are also features of oxidized starch-blend UF resin [99].

4.2. Physical Treatments

One of the physical methods of modifying starch that results in its depolymerization is treatment with ultrasounds, using sound waves at or above the range of frequency of 15–20 kHz [119,120]. This procedure is considered more environmentally friendly than chemical processes because it uses fewer chemicals and produces less waste and energy. Without using p-toluene sulphonic acid, ultrasonication reduced the time required to achieve octenyl-succinylated carboxymethylated (OC-CMS) potato starch from as long as 24 h when done traditionally, to as short as a few minutes [8,121]. Numerous reports suggest that ultrasounds physically degrade granules, leaving apparent fissures and pores on the surface, but that the size and shape of the granules remain the same [119,122]. Additionally, the prolonged application of the high-energy ultrasound to polysaccharide solutions results in an apparent decrease in viscosity brought on by the disintegration of macromolecular chains.

Changes in granular structure, swelling power and solubility, gelatinization transition temperature, syneresis, and pasting properties were seen after ultrasonic treatment [119,123]. As depicted in Figure 8, linear amylose was easier to attack by ultrasonic treatment than highly branching amylopectin molecules and it degraded the amorphous portions preferentially [38]. The number of modifications brought about by ultrasonic treatment varied depending on the amylose content and the crystal structure of the starch. It is also demonstrated that physical treatments made it possible to manufacture OS-CMS derivatives, utilizing the microwave or ultrasonic irradiation, reducing the time needed for esterification from the previously described conventional heating method of 24 h to only a few minutes [120].

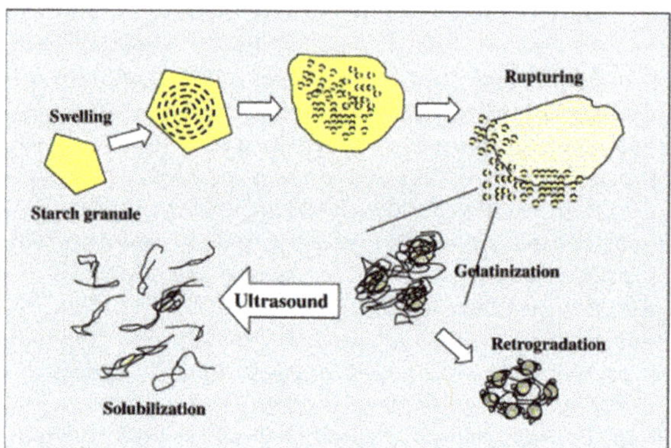

Figure 8. An illustration of the starch granule swelling, gelatinization, and depolymerization process [38]. Reprinted/adapted with permission from Ref. [38]. 2008, Elsevier, License Number 5400560644122.

The associative bonds between the starch in the granule determine the clarity of the starch paste. Waxy or modified starches are typically utilized to avoid opacity if it is an undesirable product attribute. Except for potato starch, native starch pastes exhibit relatively poor clarity [119,120]. For rice, wheat, and corn starches, sonication in ethanol resulted in a slight reduction in paste clarity but had no effect on paste clarity for potato and corn starches. Water modifications greatly impacted the potato starch paste's clarity, increasing it, but they had no discernible impact on the other starches. The disruption of swelling granules during the sonication process increased the clarity of starch pastes. However, there is a lack of knowledge regarding this determination for granulated starches that have been sonicated [119,120,122].

4.3. Enzymatic Treatments

The modification of starch, utilizing the enzyme porcine pancreatic (alpha-amylase), has been investigated, revealing that heat causes the inter- and intra-molecular hydrogen bonds between starch chains to break, which causes water-dissolved starch granules to enlarge and subsequently disintegrate [124]. As gelatinization progresses, more starch chains become available to the digestive enzymes. However, the partially gelatinized starch samples revealed variations in enzymatic action, primarily in the early stages of digestion but less at the full extent. Native and pre-gelatinized starch samples displayed maximum hydrolysis values, comparable but somewhat (by 5%) less than those of gelatinized starch. This suggests that similar levels of enzyme-resistant starch residues existed in native and partially gelatinized samples.

By putting physical restrictions on the accessibility of the enzymes, the remaining crystallites may potentially have an impact on the enzymatic digestibility. Additionally, the crystalline morphology led to variations in the pattern of enzymatic hydrolysis. However, the leftover crystallites' impacts may alter the enzymatic hydrolysis pattern (digestion), particularly in the late and intermediate phases. The formation of the amorphous matrix, which in these studies was exclusively made up of amylopectin, significantly influenced the different digesting behaviors among the partially gelatinized and retrograded waxy rice starch samples, as shown by the hydrolysis patterns [49,124].

The effect of enzymatic hydrolysis by debranching enzymes, such as pullulanase, on the properties of cassava starch-based wood adhesive was investigated by Wang et al. [125]. In contrast to alpha-amylase, which hydrolyzes starch at α-1,4-glucosidic bonds, pullulanase hydrolyzes α-1,6-glucosidic bonds to branch off the starch and produce linear chains of amylose, maltose, or glucose. Markedly, the industrial significance of this debranching enzyme is growing worldwide due to its potential for regulating the existing methods of starch processing [126,127]. The authors reported that moderate enzymatic hydrolysis for 2 h of starch molecule improved the properties of the starch-based wood adhesive, resulting in enhanced bonding strength in both dry and wet states and significantly decreased viscosity of the adhesive, allowing its better workability and penetration in the wood substrate. The higher amylose content produced by this enzyme treatment results in starch with better water resistance and is very useful for several applications, such as adhesives. Starch with pullulanase enzyme pretreatment is more adhesive with lower viscosity than without treatment, resulting in better contact with the substrate. This treatment also provides more grafting sites on the starch, thereby increasing the adhesive formulation's performance and bonding strength. However, it also causes a higher tendency of starch retrogradation [126,127].

In vivo plasma glucose and insulin responses are favorably associated with the degree of gelatinization [124,128]. The amount of native or partially gelatinized starch in foods is particularly significant from a nutritional standpoint because some processed products contain starches that are not fully gelatinized. Because they have health-promoting properties, native or minimally processed cereal products are increasingly being consumed in Korea and other Asian nations in place of traditional meals [16,20]. Enzymatic digestion of starch is substantially slowed down by incomplete gelatinization. The digesting rate decreased as the melting enthalpy rose. The relative melting enthalpy of the retrograded or partially gelatinized starch samples is strongly linked with the percentages of slowly soluble and resistant starch.

5. Starch-Bonded Wood-Based Composites

5.1. Plywood

Wood composites, including plywood, make use of synthetic formaldehyde-based adhesives, such as UF, melamine-formaldehyde (MF), melamine-urea-formaldehyde (MUF), and phenol-formaldehyde (PF), as frequently used resins [129,130]. Environmental and health issues have also prompted efforts to use more friendly adhesives, such as latex [131], soy protein [132], lignin [133,134], and tannin [135,136]. Several studies have also modified conventional adhesives using starch to reduce formaldehyde emissions in plywood products or using starch as a base. Substitution of 20% PF adhesive with cornstarch-quebracho tannin-based adhesive in plywood production can reduce formaldehyde emissions by 26% and improve the water-resistance of the panels [54,137]. The optimal replacement values were 15% cornstarch and 5% quebracho tannin. Markedly, the plywood produced had better mechanical properties and bond quality than 100% PF-bonded plywood. Another study also reported a decrease in plywood formaldehyde emissions, which was proportional to the addition of starch [138]. They reported that adding OS-based adhesives to UF adhesives up to 10% based on resin's solid content could reduce formaldehyde emissions and significantly improve the mechanical properties of plywood.

The dry tensile shear strength (TSS) values of plywood bonded with OS-based adhesives at different H_2O_2/starch mole ratios and different contents and types of cross-linkers are presented in Table 3 [56]. As a control, pure OS without a cross-linker was used to prepare plywood panels, but the TSS value was only approximately 0.61 MPa, which did not meet the minimum plywood requirement of 0.70 MPa. After adding a cross-linker, the plywood's TSS increased with increasing cross-linker content and met the minimum plywood requirement. This result showed that the addition of a cross-linker significantly improved the OS adhesive's performance by increasing the cross-linking density and forming a bigger network than without the cross-linker. However, the results also showed that TSS values decreased as the H_2O_2/starch mole ratio increased. The results suggested that an H_2O_2/starch mole ratio of 0.5 was optimal for B-pMDI/OS, and an H_2O_2/starch mole ratio of 1.0 was selected for CA/OS.

Table 3. Tensile shear strength (MPa) of plywood bonded with OS adhesive at different H_2O_2/starch mole ratios, contents, and cross-linkers [56].

H_2O_2/Starch Mole Ratio	B-pMDI Level (wt%)			CA Level (wt%)		
	5	7.5	10	5	7.5	10
Control		0.61 (0.06)			0.61 (0.06)	
0.5	0.95 (0.08)	1.13 (0.07)	1.35 (0.10)	0.92 (0.07)	0.96 (0.07)	0.98 (0.05)
1.0	0.96 (0.05)	0.97 (0.07)	0.99 (0.06)	1.01 (0.07)	1.05 (0.08)	1.18 (0.07)
1.5	0.94 (0.10)	0.96 (0.04)	0.98 (0.04)	1.00 (0.08)	1.04 (0.09)	1.08 (0.05)
2.0	0.85 (0.12)	0.92 (0.12)	0.96 (0.12)	0.90 (0.11)	0.92 (0.10)	0.94 (0.10)

Furthermore, formaldehyde-free cornstarch-tannin (10:1) adhesive with hexamine as a hardener could produce interior-grade plywood with mechanical properties comparable to PF-bonded plywood [54,137]. This indicates that starch and tannin can be used as interior plywood adhesives. Adding additives to starch-based wood adhesives could improve the water resistance, while a small amount of isocyanate can significantly improve plywood's bonding strength and water resistance [13,59]. The reaction of isocyanates with the hydroxyl groups of wood and starch is the key to this improvement. In addition, esterified starch using blocked isocyanate could increase the strength of plywood. The optimal ratios for plywood dry and wet strength from the adhesive system were 100:20 and 100:25, respectively [61]. The authors also suggested an optimum ratio of additives to starch adhesives varying from 4 to 6%. In addition to isocyanate, dodecyl succinic anhydride (DDSA) can also be used as a modifier of starch adhesive for plywood [76,78]. The crosslinking structure from the polar -NCO groups and the hydroxyl group in starch can prevent water penetration into the adhesive layer, resulting in plywood with better bonding strength and water resistance.

Plywood has also been prepared using OS-based adhesive, modified with pristine-bentonite (P-BNT) and transition metal ion modified-bentonite (TMI-P-BNT) nanoclays to produce a free formaldehyde emission panel [3]. They reported that the modification using 5% TMI-P-BNT nanoclay increased the bonding strength with higher values than the UF-bonded plywood. Recently, cassava starch grafted with glycidyl methacrylate (GMA) and then crosslinked with sodium trimetaphosphate (STMP) has been utilized for the manufacture of plywood [139,140]. As depicted in Figure 9, before the plywood manufacturing process, polyarylpolymethylene isocyanate (PAPI) was mixed as a chain-extending agent [140]. GMA grafting increased the hydrophobicity and shear strength of starch adhesives. The wet shear strength of the grafted starch adhesive increased by 163%, compared to the unmodified starch adhesive. GMA grafted starch particles become smaller, making the resulting adhesive easier to penetrate the plywood bonding interface.

Hellmayr et al. investigated the feasibility of using an aqueous mixture of equal quantities of corn starch and sodium lignosulfonate for bonding beech veneers [4]. The authors reported that the developed adhesive mixture exhibited excellent bonding characteristics comparable with industrial UF adhesives. The presence of sodium lignosulfonate in the adhesive mixture was crucial for its plasticizing, dispersing, and water-retarding properties.

Xi et al. fabricated three-layer plywood panels bonded with chitosan-oxidized starch wood adhesive [141]. They found that utilizing 10% sodium periodate on the weight of the starch to oxidize it led to the best chitosan-oxidized starch adhesive, which was produced by treating a mixture of 8% oxidized starch and chitosan at room temperature. Zhang et al. developed a renewable starch-furanic adhesive with good water resistance using crosslinkers derived from agricultural sources, such as furfural and furfuryl alcohol [142]. Compared to starch, starch-furfural, and phenol-formaldehyde adhesives, the water resistance of the starch-furfural-furfuryl alcohol adhesive was further improved when it was crosslinked with 9% epoxy resin.

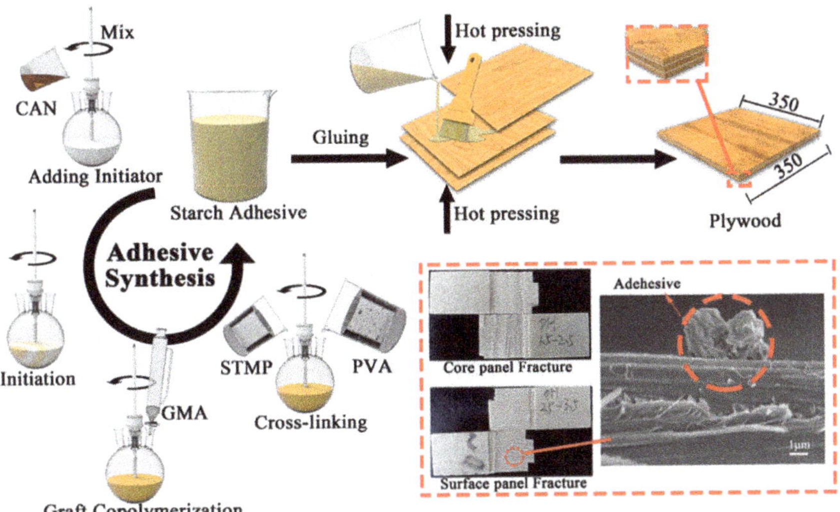

Figure 9. Starch-based adhesive synthesis and plywood preparation [140]. Reprinted/adapted with permission from Ref. [140]. 2022, Elsevier, License Number 5400560927363.

5.2. Particleboard

Conventionally, formaldehyde-based adhesives are commonly used in particleboard production. PF resin is usually used because it provides better water resistance and mechanical properties to the panels [143], while UF adhesives are used for interior-grade particleboards due to their poor resistance in a humid environment [144]. Various bio-based, formaldehyde-free adhesives have also been developed for bonding particleboards, particularly starch-based adhesives. The mechanical characteristics of rubberwood (*Hevea brasiliensis*) particleboards bonded, utilizing oil palm starch, wheat starch, and UF resin were found to meet the relevant Japanese industrial standards (JIS) [145]. Oil palm starch-based particleboards had greater mechanical qualities but inferior dimensional stability than panels bound with wheat starch. Both starch-based adhesive particleboards exhibited lower dimensional stability than the UF-bonded particleboards due to the hydrophilic nature of starch [146].

Sulaiman et al. reported that rubberwood particleboards fabricated using rice starch-based wood adhesive had mechanical properties comparable with the applicable JIS [147]. Modifying rice starch using epichlorohydrin resulted in higher particleboard properties, and further improvement was obtained by adding a small amount of UF resin. Similar results were also reported by epichlorohydrin-modified oil palm starch adhesives [147]. However, a significant drawback of the developed composites was the deteriorated dimensional stability. Selamat et al. made particleboards using carboxymethyl starch adhesive, produced from modified oil palm starch using phosphoryl chloride [148]. The mechanical properties of particleboards bonded using carboxymethyl starch met the JIS for type 8 particleboard except for the modulus of rupture (MOR) value. The addition of 2% UF resin was required to fulfill the standard requirement. Lamaming et al. reported that adding polyvinyl alcohol to carboxymethyl starch adhesive resulted in better mechanical properties and dimensional stability of particleboards than the addition of 2% UF resin reported in previous studies [149]. However, its dimensional stability still did not meet the standard requirements. Furthermore, particleboards bonded with a mixture of oil palm starch, PVA, and nano-silicon dioxide (70:30:3) have better dimensional stability, MOE, and MOR than particleboard bonded with UF [150].

Islam et al. fabricated jute stick-based particleboards bonded with a bio-adhesive, composed of natural rubber latex, combined with starch and formic acid at different blending proportions, i.e., 6:1:1, 2:1:1, and 2:3:3 [15]. The optimal results were obtained using the formula 2:3:3 (natural rubber latex/starch/formic acid). The particleboard's physical and mechanical properties fulfilled the standard's requirements. Markedly, the laboratory-fabricated particleboard panels exhibited satisfactory thermal performance, with thermal decomposition of samples occurring within the range of 265 to 399 °C.

Recently, the development of a fully bio-based wood adhesive for manufacturing three-layer laboratory-scale particleboard panels composed of corn starch, Mimosa tannin, citric acid, and sugar [93,151,152]. The composites were fabricated with the developed bio-based adhesive composition with 20% and 25% resin solid by weight for the surface layers and core layer, respectively. The panels bonded with the bio-based wood adhesive exhibited good physical and mechanical properties, fulfilling the P2-type particleboards (interior grade) requirements according to the EN 312 standard [153]. The same authors developed a green binder formulation for wood-based panel manufacturing that includes oxidized corn starch and urea. The adhesive structure was made stronger by the use of titanium dioxide nanoparticles. The findings demonstrated that the proposed adhesive could be employed in a hybrid adhesive system with MUF resin to produce particleboards with little formaldehyde concentration [55,93].

The feasibility of using agro-forest residues as alternative raw materials for particleboard manufacturing using cassava starch and UF resin as adhesives was evaluated by Mensah et al. [154]. Based on the results obtained for the physical and mechanical properties of the fabricated composites, the authors concluded that the boards could be used in indoor applications for general purposes. Another study investigated the feasibility of manufacturing particleboards from the combinations of insect rearing residue and rice husks bonded with citric acid/tapioca starch-based bio-adhesive [155]. The authors reported that only the laboratory board, composed of 50 wt% rice husk, 20 wt% insect rearing residue, and 30 wt% bio-based adhesives, fulfilled JIS standard requirements for type 8 particleboard.

Chotikhun et al. studied the mechanical characteristics and formaldehyde release of particleboards made from Eastern red cedar (*Juniperus virginiana*) utilizing SiO_2 nanoparticles mixed with modified starch as a bio-based adhesive [156]. The authors fabricated nine different types of boards at three target densities of 600, 700, and 800 kg/m^3 and nanoparticle contents of 0%, 1%, and 3%. The composites were characterized by a very low formaldehyde content of 0.07 ppm.

5.3. Medium-Density Fiberboard (MDF)

The most commonly used synthetic formaldehyde-based resins for manufacturing MDF are UF or PF resins, and these thermosetting wood adhesives have received most of the research and industrial attention due to widespread use of the composites [157,158]. However, their use is associated with serious threats to the environment and human health, related to the number of dust particles generated during processing and the emission of free formaldehyde and other volatile organic compounds, particularly indoors [159,160].

Different bio-based wood adhesives, based on starch [114,161], modified condensed and hydrolyzed tannins [162], soy protein [163], lignin [164], polysaccharides [165,166], and mycelium [167] have been used to partially replace UF or PF resins in the manufacture of MDF to create sustainable, 'green' solutions to formaldehyde-based adhesives. Other alternative examples include binderless fiberboards, which must undergo rigorous chemical and physical processes that result in large volumes of concentrated wastewater [168,169]. An alternative to the above approaches could be the application of thermoplastic starch as a bio-based adhesive. Small polar organic chemicals, such as glycerol, water, urea, formamide, and ethanolamine, can plasticize starch by breaking the internal hydrogen bonds between the anhydro-glucose monomers. The crystalline sections are disrupted, making the structure more amorphous [170]. A study attempted to fabricate MDF from starch using extrusion, where a Prism TSE-24-TC co-rotating twin screw extruder (20 L/D) was used for extrusion together with air swept face-cut pelletizing system and a Prism volumetric feeder [161]. The extruder has five temperature-controlled areas; the first was kept at 80 °C, while the other four were kept at 120 °C.

Recent research reported the modification of starch using oxidation with H_2O_2 and reinforcing it with different levels of B-pMDI [56,58]. The preliminary results showed that MDF could be fabricated with oxidized starch/B-pMDI, and the panels were characterized by a close-to-zero formaldehyde content ($\sim$0.025 mg/L) which probably originated from wood fiber. However, the laboratory-fabricated MDF panels exhibited poor physical and mechanical properties, vastly inferior to the UF-bonded panels. Therefore, further and comprehensive study is needed to understand the modification of starch as an adhesive in MDF manufacturing. Our group attempted to reinforce oxidized starch with 3% PVOH and decrease the level of B-pMDI added to oxidized starch. The MDF bonded with this oxidized starch is expected to overcome the low physical and mechanical properties and compete with MDF bonded with UF resins. The main advantage of using starch as a bio-based wood adhesive is the low formaldehyde release from the panels and their easier recyclability. However, MDF panels bonded with oxidized starch-based adhesives were more susceptible to surface-inhabiting molds than the control UF-bonded panels [57].

5.4. Laminated Veneer Lumber

A type of engineered wood known as laminated veneer lumber (LVL) is produced by stacking multiple wooden layers along the grain direction of wood veneers [171]. Although the production of LVL has increased significantly in recent years, owing to its versatility in many fields, the use of bio-based adhesive has not yet progressed to widespread practical use. Literature reports on applying starch-based adhesive as a binder for LVL manufacturing are scarce. One of the studies reported the application of natural starch modified by sodium hypochlorite (NaOCl) and sulfuric acid (H_2SO_4) as a filler for UF resin, aiming to reduce the formaldehyde emission of LVL [172]. The research indicates that when the modified oxidized starch concentration in the UF resin grew, the strength of the LVL also increased.

Aiming to enhance bio-based adhesive's practicality and utilization rate, Xiong et al. produced LVL, bonded with cornstarch-based adhesive [171]. Despite showing good cohesive and film-forming properties, the cornstarch emulsion prepared by the authors failed to meet the requirements for LVL's water endurance bonding strength and pre-formability. To solve the problem, the authors incorporated a tackifier, reinforcing agent, and filler into the cornstarch emulsion. Wheat flour was utilized as a filler, PVOH as a tackifier, and pMDI pre-polymer as a reinforcing agent. Engineered wood flooring bonded with the modified cornstarch adhesive displayed satisfactory properties, such as high mechanical properties, good adhesion properties, and very low formaldehyde.

Laminates were produced from delignified Norway spruce (*Picea abies*) veneers, bonded with modified corn starch adhesive [173]. The densified wood–starch laminates displayed superior performances, i.e., a relatively low density of 1100 kg/m^3, compared to other matrix-containing composites with comparable mechanical properties. The wood–starch laminates showed better specific tensile properties when compared with jute- or paper-based materials and bio-based flax composites and were very close to that of glass fiber-reinforced epoxy composites. The findings demonstrated a feasible process for producing high-performance, all-bio-based composite laminates, using starch as an adhesive.

Besides plywood, particleboard, fiberboard, and LVL, starch-based adhesives have also been used to produce other types of wood-based panels. Xiong et al. produced a strawboard substrate veneer bonded with a cornstarch-based adhesive, combined with a polyvinyl alcohol solution, flour, and poly-isocyanate pre-polymer [174]. The results showed that the physical and mechanical properties of the mattress board veneer decorative cover surpassed the Chinese national standard for decorative veneer (GB/T 15104-2006).

6. Conclusions

Petrochemical resources have been in short supply since the turn of the century. Society has vigorously advocated for sustainable development, and the global market of wood-based composites has become increasingly strict with formaldehyde emission standards. In this situation, it is vital to use environmentally safe, renewable, and biodegradable resources, such as tannin, starch, lignin, and vegetable protein, to develop wood adhesives. Bio-based adhesives represent a sustainable and eco-friendly alternative to the conventional synthetic adhesive systems widely used in the wood-based panel industry. This review demonstrated that starch, an abundant, renewable, and inexpensive natural raw material, can be efficiently utilized in bio-based adhesives formulation to manufacturing eco-friendly wood-based panels with acceptable properties. This can significantly lower the negative environmental footprint of wood-based panels, stimulate the industry's transition to a low-carbon, circular bio-economy, and reduce its dependence on fossil-derived constituents. However, it should be noted that most of the starch-based wood adhesives presented in this work have only been tested at a laboratory scale and are not commonly adopted in industrial practice. There are still major drawbacks to the wider commercial utilization of starch-based wood adhesives, mainly due to their relatively low water resistance, low bonding strength, and natural variations, which is a result of the growing conditions of starch sources. Future research should be focused on starch modification and optimization of production parameters to develop starch-based wood adhesives with optimal performance.

Author Contributions: Conceptualization, M.A.R.L., A.H.I., F.F., L.S.H., P.A. and L.T.; methodology, M.I.M., L.S.H., L.K. and V.L.G.; validation, M.A.R.L., R.K.S., E.M. and L.H.Z.; resources, M.A.R.L., A.H.I. and R.K.S.; data curation, M.I.M., M.A.R.L. and W.H.; writing—original draft preparation, M.I.M., M.A.R.L., L.S.H., P.A. and L.K.; writing—review and editing, E.M., V.L.G., L.S.H. and L.T.; visualization, M.I.M., M.A.R.L. and L.S.H.; supervision, M.A.R.L., A.H.I., P.A., R.K.S. and L.T.; project administration, M.A.R.L., A.H.I., E.M. and P.A.; funding acquisition, M.A.R.L., A.H.I. and P.A. All authors have read and agreed to the published version of the manuscript.

Funding: This research received no external funding.

Data Availability Statement: Not applicable.

Acknowledgments: This paper was a part of the RIIM-PRN No. 65/II.7/HK/2022 fiscal year 2022–2023 titled "Pengembangan Produk Oriented Strand Board Unggul dari Kayu Ringan dan Cepat Tumbuh Dalam Rangka Pengembangan Produk Biokomposit Prospektif". In addition, this work was also supported by the project "Development, Properties, and Application of Eco-Friendly Wood-Based Composites", No. H C‑‑1145/04.2021, carried out at the University of Forestry, Sofia, Bulgaria, and by the Slovak Research and Development Agency under contract No. APVV-18-0378, APVV-19-0269, APVV-20-0004, and SK-CZ-RD-21-0100.

Conflicts of Interest: The authors declare no conflict of interest.

References

1. Din, Z.U.; Chen, L.; Xiong, H.; Wang, Z.; Ullah, I.; Lei, W.; Shi, D.; Alam, M.; Ullah, H.; Khan, S.A. Starch: An Undisputed Potential Candidate and Sustainable Resource for the Development of Wood Adhesive. *Starch/Staerke* **2020**, *72*, 1900276. [CrossRef]
2. Watcharakitti, J.; Win, E.E.; Nimnuan, J.; Smith, S.M. Modified Starch-Based Adhesives: A Review. *Polymers* **2022**, *14*, 2023. [CrossRef] [PubMed]
3. Lubis, M.A.R.; Yadav, S.M.; Park, B.D. Modification of Oxidized Starch Polymer with Nanoclay for Enhanced Adhesion and Free Formaldehyde Emission of Plywood. *J. Polym. Environ.* **2021**, *29*, 2993–3003. [CrossRef]
4. Hellmayr, R.; Šernek, M.; Myna, R.; Reichenbach, S.; Kromoser, B.; Liebner, F.; Wimmer, R. Heat bonding of wood with starch-lignin mixtures creates new recycling opportunities. *Mater. Today Sustain.* **2022**, *19*, 100194. [CrossRef]
5. Rindlav-Westling, Å.; Gatenholm, P. Surface Composition and Morphology of Starch, Amylose, and Amylopectin Films. *Biomacromolecules* **2003**, *4*, 166–172. [CrossRef]
6. Tester, R.F.; Karkalas, J.; Qi, X. Starch-Composition, fine structure and architecture. *J. Cereal Sci.* **2004**, *39*, 151–165. [CrossRef]
7. Buléon, A.; Colonna, P.; Planchot, V.; Ball, S. Starch granules: Structure and biosynthesis. *Int. J. Biol. Macromol.* **1998**, *23*, 85–112. [CrossRef]
8. Fan, Y.; Picchioni, F. Modification of starch: A review on the application of "green" solvents and controlled functionalization. *Carbohydr. Polym.* **2020**, *241*, 116350. [CrossRef]

9. Dome, K.; Podgorbunskikh, E.; Bychkov, A.; Lomovsky, O. Changes in the crystallinity degree of starch having different types of crystal structure after mechanical pretreatment. *Polymers* **2020**, *12*, 641. [CrossRef]
10. Wang, J. Synthesis and Computer Simulation Analysis of MF Modified Starch Adhesive. *J. Phys. Conf. Ser.* **2021**, *1992*, 022171. [CrossRef]
11. Boonsuk, P.; Sukolrat, A.; Kaewtatip, K.; Chantarak, S.; Kelarakis, A.; Chaibundit, C. Modified cassava starch/poly(vinyl alcohol) blend films plasticized by glycerol: Structure and properties. *J. Appl. Polym. Sci.* **2020**, *137*, 48848. [CrossRef]
12. Wu, L.; Guo, J.; Zhang, Z.; Zhao, S. Influence of Oxidized Starch and Modified Nano-SiO on Performance of Urea-FOrmaldehyde (UF) Resin. *Polymers* **2017**, *41*, 83–89.
13. Qiao, Z.; Gu, J.; Lv, S.; Cao, J.; Tan, H.; Zhang, Y. Preparation and properties of normal temperature cured starch-based wood adhesive. *BioResources* **2016**, *11*, 4839–4849. [CrossRef]
14. Xu, Q.; Wen, J.; Wang, Z. Preparation and properties of cassava starch-based wood adhesives. *BioResources* **2016**, *11*, 6756–6767. [CrossRef]
15. Islam, M.N.; Rahman, F.; Das, A.K.; Hiziroglu, S. An overview of different types and potential of bio-based adhesives used for wood products. *Int. J. Adhes. Adhes.* **2022**, *112*, 102992. [CrossRef]
16. Whistler, R.L.; Bemiller, J.N.; Paschall, E. *Starch: Chemistry and Technology*; Academic Press: London, UK, 1984; ISBN 0127462708.
17. Radley, J.A. Adhesives from Starch and Dextrin. In *Industrial Uses of Starch and Its Derivatives*; Springer: Dordrecht, The Netherlands, 1976; pp. 1–50.
18. Onusseit, H. Starch in industrial adhesives: New developments. *Ind. Crops Prod.* **1993**, *1*, 141–146. [CrossRef]
19. Fuwei, Y.; Bingjian, Z.; Qinglin, M. Study of Sticky Rice-Lime Mortar Technology. *Acc. Chem. Res.* **2010**, *43*, 936–944.
20. Lewandowski, C.M.; Co-investigator, N.; Lewandowski, C.M. *Starch Chemistry and Technology*, 3rd ed.; Elsevier: Amsterdam, The Netherlands, 2015; Volume 1, ISBN 9788578110796.
21. Ačkar, D.; Babić, J.; Jozinović, A.; Miličević, B.; Jokić, S.; Miličević, R.; Rajič, M.; Šubarić, D. Starch modification by organic acids and their derivatives: A review. *Molecules* **2015**, *20*, 19554–19570. [CrossRef]
22. Rutenberg, M.W.; Solarek, D. Starch derivatives: Production and uses. In *Starch: Chemistry and Technology*; Elsevier: Amsterdam, The Netherlands, 1984; pp. 311–388.
23. Shogren, R.L. Starch: Properties and Materials Applications. In *Biopolymers from Renewable Resources*; Springer: Berlin/Heidelberg, Germany, 1998; pp. 30–46. [CrossRef]
24. FAO. *Food Outlook-Biannual Report on Global Food Markets-November 2018*; FAO: Rome, Italy, 2018; ISBN 3906570541.
25. Fetzer, W.R. Methods in the Starch and Dextrose Industry. *Anal. Chem.* **1952**, *24*, 1129–1137. [CrossRef]
26. Horstmann, S.W.; Lynch, K.M.; Arendt, E.K. Starch characteristics linked to gluten-free products. *Foods* **2017**, *6*, 29. [CrossRef]
27. Ellis, R.P.; Cochrane, M.P.; Dale, M.F.B.; Duþus, C.M.; Lynn, A.; Morrison, I.M.; Prentice, R.D.M.; Swanston, J.S.; Tiller, S. A Starch Production and Industrial Use. *J. Sci. Food Agric.* **1998**, *77*, 289–311. [CrossRef]
28. Liu, J.; Guo, L.; Yang, L.; Liu, Z.; He, C. Study on the rheological property of Cassava starch adhesives. *Adv. J. Food Sci. Technol.* **2014**, *6*, 374–377. [CrossRef]
29. Sangseethong, K.; Lertpanit, S. Hypochlorite oxidation of cassava starch. *Starch-Stärke* **2005**, *58*, 53–54.
30. Iamareerat, B.; Singh, M.; Sadiq, M.B.; Anal, A.K. Reinforced cassava starch based edible film incorporated with essential oil and sodium bentonite nanoclay as food packaging material. *J. Food Sci. Technol.* **2018**, *55*, 1953–1959. [CrossRef] [PubMed]
31. Sangseethong, K.; Termvejsayanon, N.; Sriroth, K. Characterization of physicochemical properties of hypochlorite- and peroxide-oxidized cassava starches. *Carbohydr. Polym.* **2010**, *82*, 446–453. [CrossRef]
32. El Halal, S.L.M.; Bruni, G.P.; do Evangelho, J.A.; Biduski, B.; Silva, F.T.; Dias, A.R.G.; da Rosa Zavareze, E.; de Mello Luvielmo, M. The properties of potato and cassava starch films combined with cellulose fibers and/or nanoclay. *Starch-Stärke* **2018**, *70*, 1700115. [CrossRef]
33. Xiao, H.X.; Lin, Q.L.; Liu, G.Q.; Yu, F.X. A comparative study of the characteristics of cross-linked, oxidized and dual-modified rice starches. *Molecules* **2012**, *17*, 10946–10957. [CrossRef]
34. Lian, X.; Wang, C.; Zhang, K.; Li, L. The retrogradation properties of glutinous rice and buckwheat starches as observed with FT-IR, 13C NMR and DSC. *Int. J. Biol. Macromol.* **2014**, *64*, 288–293. [CrossRef]
35. Van Der Borght, A.; Goesaert, H.; Veraverbeke, W.S.; Delcour, J.A. Fractionation of wheat and wheat flour into starch and gluten: Overview of the main processes and the factors involved. *J. Cereal Sci.* **2005**, *41*, 221–237. [CrossRef]
36. He, Z. *Bio-Based Wood Adhesives*; He, Z., Ed.; CRC Press: New York, NY, USA, 2017; ISBN 9781498740746.
37. Glavas, L. Starch and Protein Based Wood Adhesives. Degree Project in Polymer Technology, Nacka, Sweden, 2011. Available online: https://www.diva-portal.org/smash/get/diva2:412056/FULLTEXT02.pdf (accessed on 1 August 2022).
38. Iida, Y.; Tuziuti, T.; Yasui, K.; Towata, A.; Kozuka, T. Control of viscosity in starch and polysaccharide solutions with ultrasound after gelatinization. *Innov. Food Sci. Emerg. Technol.* **2008**, *9*, 140–146. [CrossRef]
39. Wang, S.; Li, C.; Copeland, L.; Niu, Q.; Wang, S. Starch Retrogradation: A Comprehensive Review. *Compr. Rev. Food Sci. Food Saf.* **2015**, *14*, 568–585. [CrossRef]
40. Marshall, J.T. Development of a Simplified Process for Obtaining Starch from Grain Sorghum. Ph.D. Thesis, Texas Technological College, Lubbock, TX, USA, 1969.
41. Radley, J.A. *Industrial Uses of Starch and Its Derivatives*, 1st ed.; Applied Science Publishers Ltd.: London, UK, 1976; ISBN 9789401013314.

42. Pietrzyk, S.; Fortuna, T.; Juszczak, L.; Gałkowska, D.; Baczkowicz, M.; Łabanowska, M.; Kurdziel, M. Influence of amylose content and oxidation level of potato starch on acetylation, granule structure and radicals' formation. *Int. J. Biol. Macromol.* **2018**, *16*, 57–67. [CrossRef] [PubMed]

43. Gadhave, R.V.; Mahanwar, P.A.; Gadekar, P.T. Starch-Based Adhesives for Wood/Wood Composite Bonding: Review. *Open J. Polym. Chem.* **2017**, *7*, 19–32. [CrossRef]

44. Antov, P.; Savov, V.; Neykov, N. Sustainable bio-based adhesives for eco-friendly wood composites a review. *Wood Res.* **2020**, *65*, 51–62. [CrossRef]

45. Zakaria, R.; Bawon, P.; Lee, S.H.; Salim, S.; Lum, W.C.; Al-Edrus, S.S.O.; Ibrahim, Z. Properties of Particleboard from Oil Palm Biomasses Bonded with Citric Acid and Tapioca Starch. *Polymers* **2021**, *13*, 3494. [CrossRef]

46. Shah, U.; Naqash, F.; Gani, A.; Masoodi, F.A. Art and Science behind Modified Starch Edible Films and Coatings: A Review. *Compr. Rev. Food Sci. Food Saf.* **2016**, *15*, 568–580. [CrossRef]

47. Vanier, N.L.; El Halal, S.L.M.; Dias, A.R.G.; da Rosa Zavareze, E. Molecular structure, functionality and applications of oxidized starches: A review. *Food Chem.* **2017**, *221*, 1546–1559. [CrossRef]

48. Yue, L.; Shi, R.; Yi, Z.; Shi, S.Q.; Gao, Q.; Li, J. A high-performance soybean meal-based plywood adhesive prepared via an ultrasonic process and using significantly lower amounts of chemical additives. *J. Clean. Prod.* **2020**, *274*, 123017. [CrossRef]

49. Jimenez Bartolome, M.; Schwaiger, N.; Flicker, R.; Seidl, B.; Kozich, M.; Nyanhongo, G.S.; Guebitz, G.M. Enzymatic synthesis of wet-resistant lignosulfonate-starch adhesives. *New Biotechnol.* **2022**, *69*, 49–54. [CrossRef]

50. Nasiri, A. The Use of Lignin in Pressure Sensitive Adhesives and Starch-Based Adhesives. Ph.D. Thesis, University of Ottawa, Ottawa, ON, Canada, 2019. [CrossRef]

51. Miranda, C.S.; Ferreira, M.S.; Magalhães, M.T.; Bispo, A.P.G.; Oliveira, J.C.; Silva, J.B.A.; José, N.M. Starch-based Films Plasticized with Glycerol and Lignin from Piassava Fiber Reinforced with Nanocrystals from Eucalyptus. *Mater. Today Proc.* **2015**, *2*, 134–140. [CrossRef]

52. Yin, H.; Zheng, P.; Zhang, E.; Rao, J.; Lin, Q.; Fan, M.; Zhu, Z.; Zeng, Q.; Chen, N. Improved wet shear strength in eco-friendly starch-cellulosic adhesives for woody composites. *Carbohydr. Polym.* **2020**, *250*, 116884. [CrossRef] [PubMed]

53. Weakley, F.B.; Carr, M.E.; Mehltretter, C.L. Dialdehyde Starch in Paper Coatings Containing Soy Flour-Isolated Soy Protein Adhesive. *Starch-Stärke* **1972**, *24*, 191–194. [CrossRef]

54. Moubarik, A.; Pizzi, A.; Allal, A.; Charrier, F.; Khoukh, A.; Charrier, B. Cornstarch-mimosa tannin-urea formaldehyde resins as adhesives in the particleboard production. *Starch/Staerke* **2010**, *62*, 131–138. [CrossRef]

55. Oktay, S.; Kızılcan, N.; Bengü, B. Development of bio-based cornstarch-Mimosa tannin-sugar adhesive for interior particleboard production. *Ind. Crops Prod.* **2021**, *170*, 113689. [CrossRef]

56. Lubis, M.A.R.; Park, B.D.; Hong, M.K. Tuning of adhesion and disintegration of oxidized starch adhesives for the recycling of medium density fiberboard. *BioResources* **2020**, *15*, 5156–5178. [CrossRef]

57. Lubis, M.A.R.; Park, B.; Kim, Y.; Yun, J.; Shin, H.C. Visual inspection of surface mold growth on medium-density fiberboard bonded with oxidized starch adhesives. *Wood Mater. Sci. Eng.* **2022**, 1–8. [CrossRef]

58. Lubis, M.A.R.; Park, B.; Hong, M. Tailoring of oxidized starch's adhesion using crosslinker and adhesion promotor for the recycling of fiberboards. *J. Appl. Polym. Sci.* **2019**, *136*, 47966. [CrossRef]

59. Qiao, Z.; Gu, J.; Lv, S.; Cao, J.; Tan, H.; Zhang, Y. Preparation and properties of isocyanate prepolymer/corn starch adhesive. *J. Adhes. Sci. Technol.* **2015**, *29*, 1368–1381. [CrossRef]

60. Lee, S.H.; Md Tahir, P.; Lum, W.C.; Tan, L.P.; Bawon, P.; Park, B.-D.; Osman Al Edrus, S.S.; Abdullah, U.H. A Review on Citric Acid as Green Modifying Agent and Binder for Wood. *Polymers* **2020**, *12*, 1692. [CrossRef]

61. Tan, H.; Zhang, Y.; Weng, X. Preparation of the plywood using starch-based adhesives modified with blocked isocyanates. *Procedia Eng.* **2011**, *15*, 1171–1175. [CrossRef]

62. Valodkar, M.; Thakore, S. Isocyanate crosslinked reactive starch nanoparticles for thermo-responsive conducting applications. *Carbohydr. Res.* **2010**, *345*, 2354–2360. [CrossRef] [PubMed]

63. Fu, L.; Peng, Y. Isocyanate-functionalized starch as biorenewable backbone for the preparation and application of poly(ethylene imine) grafted starch. *Mon. Chem.-Chem. Mon.* **2017**, *148*, 1547–1554. [CrossRef]

64. Ferdosian, F.; Pan, Z.; Gao, G.; Zhao, B. Bio-Based Adhesives and Evaluation for Wood Composites Application. *Polymers* **2017**, *9*, 70. [CrossRef]

65. Li, H.; Qi, Y.; Zhao, Y.; Chi, J.; Cheng, S. Starch and its derivatives for paper coatings: A review. *Prog. Org. Coat.* **2019**, *135*, 213–227. [CrossRef]

66. Wang, Y.; Xiong, H.; Wang, Z.; Chen, L. Effects of different durations of acid hydrolysis on the properties of starch-based wood adhesive. *Int. J. Biol. Macromol.* **2017**, *103*, 819–828. [CrossRef]

67. Yu, H.; Fang, Q.; Cao, Y.; Liu, Z. Effect of HCl on Starch Structure and Properties of Starch-based Wood Adhesives. *BioResources* **2016**, *11*, 1721–1728. [CrossRef]

68. Qiao, Z.; Lv, S.; Gu, J.; Tan, H.; Shi, J.; Zhang, Y. Influence of acid hydrolysis on properties of maize starch adhesive. *Pigment Resin Technol.* **2017**, *46*, 148–155. [CrossRef]

69. Gadhave, R.V.; Sheety, P.; Mahanwar, P.A.; Gadekar, P.T.; Desai, B.J. Silane Modification of Starch-Based Wood Adhesive: Review. *Open J. Polym. Chem.* **2019**, *09*, 53–62. [CrossRef]

70. Zhang, Y.; Ding, L.; Gu, J.; Tan, H.; Zhu, L. Preparation and properties of a starch-based wood adhesive with high bonding strength and water resistance. *Carbohydr. Polym.* **2015**, *115*, 32–57. [CrossRef]

71. Wang, Z.; Zhu, H.; Huang, J.; Ge, Z.; Guo, J.; Feng, X.; Xu, Q. Improvement of the bonding properties of cassava starch-based wood adhesives by using different types of acrylic ester. *Int. J. Biol. Macromol.* **2019**, *126*, 603–611. [CrossRef]

72. Zhang, C.-W.; Li, F.-Y.; Li, J.-F.; Xie, Q.; Xu, J.; Guo, A.-F.; Wang, C.-Z. Effect of Crystal Structure and Hydrogen Bond of Thermoplastic Oxidized Starch on Manufacturing of Starch-Based Biomass Composite. *Int. J. Precis. Eng. Manuf. Technol.* **2018**, *5*, 435–440. [CrossRef]

73. Romero-Bastida, C.A.; Tapia-Blácido, D.R.; Méndez-Montealvo, G.; Bello-Pérez, L.A.; Velázquez, G.; Alvarez-Ramirez, J. Effect of amylose content and nanoclay incorporation order in physicochemical properties of starch/montmorillonite composites. *Carbohydr. Polym.* **2016**, *152*, 351–360. [CrossRef] [PubMed]

74. Kotal, M.; Bhowmick, A.K. Polymer nanocomposites from modified clays: Recent advances and challenges. *Prog. Polym. Sci.* **2015**, *51*, 127–187. [CrossRef]

75. Tudor, E.M.; Scheriau, C.; Barbu, M.C.; Réh, R.; Krišťák, Ľ.; Schnabel, T. Enhanced Resistance to Fire of the Bark-Based Panels Bonded with Clay. *Appl. Sci.* **2020**, *10*, 5594. [CrossRef]

76. Li, Z.; Wang, J.; Cheng, L.; Gu, Z.; Hong, Y.; Kowalczyk, A. Improving the performance of starch-based wood adhesive by using sodium dodecyl sulfate. *Carbohydr. Polym.* **2014**, *99*, 579–583. [CrossRef]

77. Chen, N.; Lin, Q.; Rao, J.; Zeng, Q. Water resistances and bonding strengths of soy-based adhesives containing different carbohydrates. *Ind. Crops Prod.* **2013**, *50*, 44–49. [CrossRef]

78. Sun, Y.; Gu, J.; Tan, H.; Zhang, Y.; Huo, P. Physicochemical properties of starch adhesives enhanced by esterification modification with dodecenyl succinic anhydride. *Int. J. Biol. Macromol.* **2018**, *112*, 1257–1263. [CrossRef]

79. Ait, A.; Boussetta, A.; Kassab, Z.; Nadifiyine, M. Elaboration of carboxylated cellulose nanocrystals filled starch-based adhesives for the manufacturing of eco-friendly particleboards. *Constr. Build. Mater.* **2022**, *348*, 128683. [CrossRef]

80. Li, D.; Zhuang, B.; Wang, X.A.; Wu, Z.; Wei, W.; Aladejana, J.T.; Hou, X.; Yves, K.G.; Xie, Y.; Liu, J. Chitosan used as a specific coupling agent to modify starch in preparation of adhesive film. *J. Clean. Prod.* **2020**, *277*, 123210. [CrossRef]

81. Santoso, M.; Widyorini, R.; Prayitno, T.A.; Sulistyo, J. Bonding performance of maltodextrin and citric acid for particleboard made from nipa fronds. *J. Korean Wood Sci. Technol.* **2017**, *45*, 432–443. [CrossRef]

82. Singh, V.; Ali, S.Z.; Somashekar, R.; Mukherjee, P.S. Nature of crystallinity in native and acid modified starches. *Int. J. Food Prop.* **2006**, *9*, 845–854. [CrossRef]

83. Masina, N.; Choonara, Y.E.; Kumar, P.; du Toit, L.C.; Govender, M.; Indermun, S.; Pillay, V. A review of the chemical modification techniques of starch. *Carbohydr. Polym.* **2017**, *157*, 1226–1236. [CrossRef] [PubMed]

84. Chen, L.; Wang, Y.; Zia-ud-Din; Fei, P.; Jin, W.; Xiong, H.; Wang, Z. Enhancing the performance of starch-based wood adhesive by silane coupling agent(KH570). *Int. J. Biol. Macromol.* **2017**, *104*, 137–144. [CrossRef] [PubMed]

85. Zhang, Y.R.; Wang, X.L.; Zhao, G.M.; Wang, Y.Z. Influence of oxidized starch on the properties of thermoplastic starch. *Carbohydr. Polym.* **2013**, *96*, 358–364. [CrossRef]

86. Zhang, S.D.; Zhang, Y.R.; Wang, X.L.; Wang, Y.Z. High carbonyl content oxidized starch prepared by hydrogen peroxide and its thermoplastic application. *Starch/Staerke* **2009**, *61*, 646–655. [CrossRef]

87. Zheng, X.; Cheng, L.; Gu, Z.; Hong, Y.; Li, Z.; Li, C. Effects of heat pretreatment of starch on graft copolymerization reaction and performance of resulting starch-based wood adhesive. *Int. J. Biol. Macromol.* **2017**, *96*, 11–18. [CrossRef]

88. Wang, Z.; Gu, Z.; Hong, Y.; Cheng, L.; Li, Z. Bonding strength and water resistance of starch-based wood adhesive improved by silica nanoparticles. *Carbohydr. Polym.* **2011**, *86*, 72–76. [CrossRef]

89. Li, Z.; Wang, J.; Li, C.; Gu, Z.; Cheng, L.; Hong, Y. Effects of montmorillonite addition on the performance of starch-based wood adhesive. *Carbohydr. Polym.* **2015**, *115*, 394–400. [CrossRef]

90. Parovuori, P.; Hamunen, A.; Forssell, P.; Autio, K.; Poutanen, K. Oxidation of Potato Starch by Hydrogen Peroxide. *Starch-Stärke* **1995**, *47*, 19–23. [CrossRef]

91. Tolvanen, P.; Sorokin, A.; Mäki-Arvela, P.; Murzin, D.Y.; Salmi, T. Oxidation of starch by H_2O_2 in the presence of iron tetrasulfophthalocyanine catalyst: The Effect of catalyst concentration, pH, solid-liquid ratio, and origin of starch. *Ind. Eng. Chem. Res.* **2013**, *52*, 9351–9358. [CrossRef]

92. Harmon, R.E.; Gupta, S.K.; Johnson, J. Oxidation of Starch by Hydrogen Peroxide in the Presence of UV Light-Part. I. *Starch-Stärke* **1971**, *23*, 347–349. [CrossRef]

93. Oktay, S.; Kızılcan, N.; Bengu, B. Oxidized cornstarch—Urea wood adhesive for interior particleboard production. *Int. J. Adhes. Adhes.* **2021**, *110*, 102947. [CrossRef]

94. Wang, S.; Shi, J.; Xu, W. Synthesis and Characterization of Starch-based Aqueous Polymer Isocyanate Wood Adhesive. *BioResources* **2015**, *10*, 7653–7666. [CrossRef]

95. Meimoun, J.; Wiatz, V.; Saint-Loup, R.; Parcq, J.; Favrelle, A.; Bonnet, F.; Zinck, P. Modification of starch by graft copolymerization. *Starch-Stärke* **2017**, *1600351*, 1600351. [CrossRef]

96. Wang, Z.; Li, Z.; Gu, Z.; Hong, Y.; Cheng, L. Preparation, characterization and properties of starch-based wood adhesive. *Carbohydr. Polym.* **2012**, *88*, 699–706. [CrossRef]

97. Zia-ud-Din; Chen, L.; Ullah, I.; Wang, P.K.; Javaid, A.B.; Hu, C.; Zhang, M.; Ahamd, I.; Xiong, H.; Wang, Z. Synthesis and characterization of starch-g-poly(vinyl acetate-co-butyl acrylate) bio-based adhesive for wood application. *Int. J. Biol. Macromol.* **2018**, *114*, 1186–1193. [CrossRef] [PubMed]

98. Carvalho, A.J.F.; Curvelo, A.A.S.; Gandini, A. Surface chemical modification of thermoplastic starch: Reactions with isocyanates, epoxy functions and stearoyl chloride. *Ind. Crops Prod.* **2005**, *21*, 331–336. [CrossRef]

99. Wang, H.; Liang, J.; Zhang, J.; Zhou, X.; Du, G. Performance of urea-formaldehyde adhesive with oxidized cassava starch. *BioResources* **2017**, *12*, 7590–7600. [CrossRef]

100. Luo, J.; Zhang, J.; Gao, Q.; Mao, A.; Li, J. Toughening and enhancing melamine-urea-formaldehyde resin properties via in situ polymerization of dialdehyde starch and microphase separation. *Polymers* **2019**, *11*, 1167. [CrossRef]

101. Aristri, M.A.; Lubis, M.A.R.; Iswanto, A.H.; Fatriasari, W.; Sari, R.K.; Antov, P.; Gajtanska, M.; Papadopoulos, A.N.; Pizzi, A. Bio-Based Polyurethane Resins Derived from Tannin: Source, Synthesis, Characterisation, and Application. *Forests* **2021**, *12*, 1516. [CrossRef]

102. Aristri, M.A.; Lubis, M.A.R.; Yadav, S.M.; Antov, P.; Papadopoulos, A.N.; Pizzi, A.; Fatriasari, W.; Ismayati, M.; Iswanto, A.H. Recent Developments in Lignin- and Tannin-Based Non-Isocyanate Polyurethane Resins for Wood Adhesives—A Review. *Appl. Sci.* **2021**, *11*, 4242. [CrossRef]

103. Zia-ud-Din; Xiong, H.; Wang, Z.; Fei, P.; Ullah, I.; Javaid, A.B.; Wang, Y.; Jin, W.; Chen, L. Effects of sucrose fatty acid esters on the stability and bonding performance of high amylose starch-based wood adhesive. *Int. J. Biol. Macromol.* **2017**, *104*, 846–853. [CrossRef] [PubMed]

104. Cheng, L.; Guo, H.; Gu, Z.; Li, Z.; Hong, Y. Effects of compound emulsifiers on properties of wood adhesive with high starch content. *Int. J. Adhes. Adhes.* **2017**, *72*, 92–97. [CrossRef]

105. Wilpiszewska, K.; Spychaj, T. Chemical modification of starch with hexamethylene diisocyanate derivatives. *Carbohydr. Polym.* **2007**, *70*, 334–340. [CrossRef]

106. Bayazeed, A.; Farag, S.; Shaarawy, S.; Hebeish, A. Chemical modification of starch via etherification with methyl methacrylate. *Starch/Staerke* **1998**, *50*, 89–93. [CrossRef]

107. Cummings, S.; Zhang, Y.; Smeets, N.; Cunningham, M.; Dubé, M. On the Use of Starch in Emulsion Polymerizations. *Processes* **2019**, *7*, 140. [CrossRef]

108. Duarah, R.; Karak, N. A starch based sustainable tough hyperbranched epoxy thermoset. *RSC Adv.* **2015**, *5*, 64456–64465. [CrossRef]

109. Li, S.; Wang, J.; Qu, W.; Cheng, J.; Lei, Y.; Wang, D.; Zhang, F. Green synthesis and properties of an epoxy-modified oxidized starch-grafted styrene-acrylate emulsion. *Eur. Polym. J.* **2020**, *123*, 109412. [CrossRef]

110. Zhang, S.; Liu, F.; Peng, H.; Peng, X.; Jiang, S.; Wang, J. Preparation of Novel c-6 Position Carboxyl Corn Starch by a Green Method and Its Application in Flame Retardance of Epoxy Resin. *Ind. Eng. Chem. Res.* **2015**, *54*, 11944–11952. [CrossRef]

111. Auvergne, R.; Caillol, S.; David, G.; Boutevin, B.; Pascault, J.P. Biobased thermosetting epoxy: Present and future. *Chem. Rev.* **2014**, *114*, 1082–1115. [CrossRef]

112. Ye, S.; Qiu-hua, W.; Xue-Chun, X.; Wen-yong, J.; Shu-Cai, G.; Hai-Feng, Z. Oxidation of cornstarch using oxygen as oxidant without catalyst. *LWT-Food Sci. Technol.* **2011**, *44*, 139–144. [CrossRef]

113. Yang, L.; Liu, J.; Du, C.; Qiang, Y. Preparation and Properties of Cornstarch Adhesive. *Adv. J. Food Sci. Technol.* **2013**, *5*, 1068–1072. [CrossRef]

114. Jarusombuti, S.; Bauchongkol, P.; Hiziroglu, S.; Fueangvivat, V. Properties of Rubberwood Medium-Density Fiberboard Bonded with Starch and Urea-Formaldehyde. *For. Prod. J.* **2012**, *62*, 58–62. [CrossRef]

115. Menzel, C.; Olsson, E.; Plivelic, T.S.; Andersson, R.; Johansson, C.; Kuktaite, R.; Järnström, L.; Koch, K. Molecular structure of citric acid cross-linked starch films. *Carbohydr. Polym.* **2013**, *96*, 270–276. [CrossRef] [PubMed]

116. Dao, P.H.; Nam, T.T.; Van Phuc, M.; Hiep, N.A.; Van Thanh, T.; Vuong, N.T.; Xuan, D.D. Oxidized Maize Starch: Characterization and Effect of It on the Biodegradable Films.Ii. Infrared Spectroscopy, Solubility of Oxidized Starch and Starch Film Solubility. *Vietnam J. Sci. Technol.* **2017**, *55*, 395–402. [CrossRef]

117. Zhao, X.F.; Peng, L.Q.; Wang, H.L.; Wang, Y.B.; Zhang, H. Environment-friendly urea-oxidized starch adhesive with zero formaldehyde-emission. *Carbohydr. Polym.* **2018**, *181*, 1112–1118. [CrossRef]

118. Wang, H.; Wang, F.; Du, G. Enhanced performance of urea–glyoxal polymer with oxidized cassava starch as wood adhesive. *Iran. Polym. J.* **2019**, *28*, 1015–1021. [CrossRef]

119. Sujka, M. Ultrasonic modification of starch—Impact on granules porosity. *Ultrason. Sonochem.* **2017**, *37*, 424–429. [CrossRef]

120. Čížová, A.; Sroková, I.; Sasinková, V.; Malovíková, A.; Ebringerová, A. Carboxymethyl starch octenylsuccinate: Microwave- and ultrasound-assisted synthesis and properties. *Starch/Staerke* **2008**, *60*, 389–397. [CrossRef]

121. Bi, Y.; Liu, M.; Wu, L.; Cui, D. Synthesis of carboxymethyl potato starch and comparison of optimal reaction conditions from different sources. *Polym. Adv. Technol.* **2008**, *19*, 1185–1192. [CrossRef]

122. Chong, W.T.; Uthumporn, U.; Karim, A.A.; Cheng, L.H. The influence of ultrasound on the degree of oxidation of hypochlorite-oxidized corn starch. *LWT-Food Sci. Technol.* **2013**, *50*, 439–443. [CrossRef]

123. Luo, Z.; Fu, X.; He, X.; Luo, F.; Gao, Q.; Yu, S. Effect of ultrasonic treatment on the physicochemical properties of maize starches differing in amylose content. *Starch/Staerke* **2008**, *60*, 646–653. [CrossRef]

124. Chung, H.J.; Lim, H.S.; Lim, S.T. Effect of partial gelatinization and retrogradation on the enzymatic digestion of waxy rice starch. *J. Cereal Sci.* **2006**, *43*, 353–359. [CrossRef]
125. Wang, Z.; Xing, Z.; Zhang, Q.; Hu, D.; Lv, J.; Wu, C.; Zhou, W.; Zia-ud-Din. Effects of various durations of enzyme hydrolysis on properties of starch-based wood adhesive. *Int. J. Biol. Macromol.* **2022**, *205*, 664–671. [CrossRef] [PubMed]
126. Wang, R.; Liu, P.; Cui, B.; Kang, X.; Yu, B.; Qiu, L.; Sun, C. Effects of pullulanase debranching on the properties of potato starch-lauric acid complex and potato starch-based film. *Int. J. Biol. Macromol.* **2020**, *156*, 1330–1336. [CrossRef] [PubMed]
127. Zheng, Y.; Ou, Y.; Zhang, Y.; Zheng, B.; Zeng, S.; Zeng, H. Effects of pullulanase pretreatment on the structural properties and digestibility of lotus seed starch-glycerin monostearin complexes. *Carbohydr. Polym.* **2020**, *240*, 116324. [CrossRef] [PubMed]
128. Zia-ud-Din; Xiong, H.; Fei, P. Physical and chemical modification of starches: A review. *Crit. Rev. Food Sci. Nutr.* **2017**, *57*, 2691–2705. [CrossRef]
129. Kristak, L.; Antov, P.; Bekhta, P.; Lubis, M.A.R.; Iswanto, A.H.; Reh, R.; Sedliacik, J.; Savov, V.; Taghiyari, H.R.; Papadopoulos, A.N.; et al. Recent progress in ultra-low formaldehyde emitting adhesive systems and formaldehyde scavengers in wood-based panels: A review. *Wood Mater. Sci. Eng.* **2022**, 1–20. [CrossRef]
130. Pizzi, A.; Papadopoulos, A.N.; Policardi, F. Wood composites and their polymer binders. *Polymers* **2020**, *12*, 1115. [CrossRef]
131. Hidayat, W.; Aprilliana, N.; Asmara, S.; Bakri, S.; Hidayati, S.; Banuwa, I.S.; Lubis, M.A.R.; Iswanto, A.H. Performance of eco-friendly particleboard from agro-industrial residues bonded with formaldehyde-free natural rubber latex adhesive for interior applications. *Polym. Compos.* **2022**, *43*, 2222–2233. [CrossRef]
132. Averina, E.; Konnerth, J.; van Herwijnen, H.W.G. Protein-based glyoxal–polyethyleneimine-crosslinked adhesives for wood bonding. *J. Adhes.* **2022**, 1–16. [CrossRef]
133. Lubis, M.A.R.; Labib, A.; Sudarmanto; Akbar, F.; Nuryawan, A.; Antov, P.; Kristak, L.; Papadopoulos, A.N.; Pizzi, A. Influence of Lignin Content and Pressing Time on Plywood Properties Bonded with Cold-Setting Adhesive Based on Poly (Vinyl Alcohol), Lignin, and Hexamine. *Polymers* **2022**, *14*, 2111. [CrossRef] [PubMed]
134. Savov, V.; Valchev, I.; Antov, P.; Yordanov, I.; Popski, Z. Effect of the Adhesive System on the Properties of Fiberboard Panels Bonded with Hydrolysis Lignin and Phenol-Formaldehyde Resin. *Polymers* **2022**, *14*, 1768. [CrossRef] [PubMed]
135. Chen, X.; Pizzi, A.; Fredon, E.; Gerardin, C.; Zhou, X.; Zhang, B.; Du, G. Low curing temperature tannin-based non-isocyanate polyurethane (NIPU) wood adhesives: Preparation and properties evaluation. *Int. J. Adhes. Adhes.* **2022**, *112*, 103001. [CrossRef]
136. Pizzi Tannins: Prospectives and Actual Industrial Applications. *Biomolecules* **2019**, *9*, 344. [CrossRef]
137. Moubarik, A.; Pizzi, A.; Allal, A.; Charrier, F.; Charrier, B. Cornstarch and tannin in phenol-formaldehyde resins for plywood production. *Ind. Crops Prod.* **2009**, *30*, 188–193. [CrossRef]
138. Lubis, M.A.R.; Park, B.-D. Modification of urea-formaldehyde resin adhesives with oxidized starch using blocked pMDI for plywood. *J. Adhes. Sci. Technol.* **2018**, *32*, 2667–2681. [CrossRef]
139. Chen, X.; Pizzi, A.; Zhang, B.; Zhou, X.; Fredon, E.; Gerardin, C.; Du, G. Particleboard bio-adhesive by glyoxalated lignin and oxidized dialdehyde starch crosslinked by urea. *Wood Sci. Technol.* **2022**, *56*, 63–85. [CrossRef]
140. Chen, X.; Sun, C.; Wang, Q.; Tan, H.; Zhang, Y. Preparation of glycidyl methacrylate grafted starch adhesive to apply in high-performance and environment-friendly plywood. *Int. J. Biol. Macromol.* **2022**, *194*, 954–961. [CrossRef]
141. Xi, X.; Pizzi, A.; Lei, H.; Zhang, B.; Chen, X.; Du, G. Environmentally friendly chitosan adhesives for plywood bonding. *Int. J. Adhes. Adhes.* **2022**, *112*, 103027. [CrossRef]
142. Zhang, J.; Liu, B.; Zhou, Y.; Essawy, H.; Chen, Q.; Zhou, X.; Du, G. Preparation of a starch-based adhesive cross-linked with furfural, furfuryl alcohol and epoxy resin. *Int. J. Adhes. Adhes.* **2021**, *110*, 102958. [CrossRef]
143. Iswanto, A.H.; Azhar, I.; Supriyanto; Susilowati, A. Effect of resin type, pressing temperature and time on particleboard properties made from sorghum bagasse. *Agric. For. Fish.* **2014**, *3*, 62. [CrossRef]
144. Que, Z.; Furuno, T.; Katoh, S.; Nishino, Y. Effects of urea–formaldehyde resin mole ratio on the properties of particleboard. *Build. Environ.* **2007**, *42*, 1257–1263. [CrossRef]
145. Salleh, K.M.; Hashim, R.; Sulaiman, O.; Hiziroglu, S.; Wan Nadhari, W.N.A.; Abd Karim, N.; Jumhuri, N.; Ang, L.Z.P. Evaluation of properties of starch-based adhesives and particleboard manufactured from them. *J. Adhes. Sci. Technol.* **2015**, *29*, 319–336. [CrossRef]
146. Lee, S.H.; Lum, W.C.; Boon, J.G.; Kristak, L.; Antov, P.; Pędzik, M.; Rogoziński, T.; Taghiyari, H.R.; Rahandi Lubis, M.A.; Fatriasari, W.; et al. Particleboard from Agricultural Biomass and Recycled Wood Waste: A Review. *J. Mater. Res. Technol.* **2022**, *20*, 4630–4658. [CrossRef]
147. Sulaiman, N.S.; Hashim, R.; Amini, M.H.M.; Sulaiman, O.; Hiziroglu, S. Evaluation of the properties of particleboard made using oil palm starch modified with epichlorohydrin. *BioResources* **2013**, *8*, 283–301. [CrossRef]
148. Selamat, M.E.; Sulaiman, O.; Hashim, R.; Hiziroglu, S.; Nadhari, W.N.A.W.; Sulaiman, N.S.; Razali, M.Z. Measurement of some particleboard properties bonded with modified carboxymethyl starch of oil palm trunk. *Meas. J. Int. Meas. Confed.* **2014**, *53*, 251–259. [CrossRef]
149. Lamaming, J.; Heng, N.B.; Owodunni, A.A.; Lamaming, S.Z.; Khadir, N.K.A.; Hashim, R.; Sulaiman, O.; Mohamad Kassim, M.H.; Hussin, M.H.; Bustami, Y.; et al. Characterization of rubberwood particleboard made using carboxymethyl starch mixed with polyvinyl alcohol as adhesive. *Compos. Part B Eng.* **2020**, *183*, 107731. [CrossRef]

150. Abd Karim, N.; Lamaming, J.; Yusof, M.; Hashim, R.; Sulaiman, O.; Hiziroglu, S.; Wan Nadhari, W.N.A.; Salleh, K.M.; Taiwo, O.F. Properties of native and blended oil palm starch with nano-silicon dioxide as binder for particleboard. *J. Build. Eng.* **2020**, *29*, 101151. [CrossRef]

151. Syahfitri, A.; Hermawan, D.; Kusumah, S.S.; Ismadi; Lubis, M.A.R.; Widyaningrum, B.A.; Ismayati, M.; Amanda, P.; Ningrum, R.S.; Sutiawan, J. Conversion of agro-industrial wastes of sorghum bagasse and molasses into lightweight roof tile composite. *Biomass Conv. Bioref.* **2022**, 1–15. [CrossRef]

152. Sutiawan, J.; Hadi, Y.S.; Nawawi, D.S.; Abdillah, I.B.; Zulfiana, D.; Lubis, M.A.R.; Nugroho, S.; Astuti, D.; Zhao, Z.; Handayani, M.; et al. The properties of particleboard composites made from three sorghum (*Sorghum bicolor*) accessions using maleic acid adhesive. *Chemosphere* **2022**, *290*, 133163. [CrossRef] [PubMed]

153. European Committee for Standardization, EN 312. *Particleboards–Specifications*; European Committee for Standardization: Brussels, Belgium, 2010.

154. Mensah, P.; Owusu, F.W.; Mitchua, S.J.; Kusi, E.; Frimpong-Mensah, K. Some Mechanical Properties of Particleboards Produced from Four Agro-Forest Residues Using Cassava Starch and Urea Formaldehyde as Adhesives. *J. Mater. Sci. Eng. B* **2021**, *11*, 43–58. [CrossRef]

155. Huang, H.-K.; Hsu, C.-H.; Hsu, P.-K.; Cho, Y.-M.; Chou, T.-H.; Cheng, Y.-S. Preparation and evaluation of particleboard from insect rearing residue and rice husks using starch/citric acid mixture as a natural binder. *Biomass Conv. Bioref.* **2022**, *12*, 633–641. [CrossRef]

156. Chotikhun, A.; Hiziroglu, S.; Buser, M.; Frazier, S.; Kard, B. Characterization of nano particle added composite panels manufactured from Eastern red cedar. *J. Compos. Mater.* **2018**, *52*, 1605–1615. [CrossRef]

157. Piekarski, C.M.; de Francisco, A.C.; da Luz, L.M.; Kovaleski, J.L.; Silva, D.A.L. Life cycle assessment of medium-density fiberboard (MDF) manufacturing process in Brazil. *Sci. Total Environ.* **2017**, *575*, 103–111. [CrossRef] [PubMed]

158. McDevitt, J.E.; Grigsby, W.J. Life Cycle Assessment of Bio- and Petro-Chemical Adhesives Used in Fiberboard Production. *J. Polym. Environ.* **2014**, *22*, 537–544. [CrossRef]

159. Lubis, M.A.R.; Hong, M.K.; Park, B.D. Hydrolytic Removal of Cured Urea–Formaldehyde Resins in Medium-Density Fiberboard for Recycling. *J. Wood Chem. Technol.* **2018**, *38*, 1–14. [CrossRef]

160. Lubis, M.A.R.; Hong, M.-K.; Park, B.-D.; Lee, S.-M. Effects of recycled fiber content on the properties of medium density fiberboard. *Eur. J. Wood Wood Prod.* **2018**, *76*, 1515–1526. [CrossRef]

161. Abbott, A.P.; Palazuela Conde, J.; Davis, S.J.; Wise, W.R. Starch as a replacement for urea-formaldehyde in medium density fibreboard. *Green Chem.* **2012**, *14*, 3067. [CrossRef]

162. Roffael, E.; Schneider, T.; Dix, B. Influence of moisture content on the formaldehyde release of particle- and fibreboards bonded with tannin–formaldehyde resins. *Eur. J. Wood Wood Prod.* **2015**, *73*, 597–605. [CrossRef]

163. Yang, I.; Han, G.-S.; Ahn, S.H.; Choi, I.-G.; Kim, Y.-H.; Oh, S.C. Adhesive Properties of Medium-Density Fiberboards Fabricated with Rapeseed Flour-Based Adhesive Resins. *J. Adhes.* **2014**, *90*, 279–295. [CrossRef]

164. Mihajlova, J.; Savov, V.; Simeonov, T. Effect of the content of corn stalk fibres and additional heat treatment on properties of eco-friendly fibreboards bonded with lignosulphonate. *Drewno* **2022**, *65*, 209. [CrossRef]

165. Ji, X.; Li, B.; Yuan, B.; Guo, M. Preparation and characterizations of a chitosan-based medium-density fiberboard adhesive with high bonding strength and water resistance. *Carbohydr. Polym.* **2017**, *176*, 273–280. [CrossRef] [PubMed]

166. Li, X.; Cai, Z.; Horn, E.; Winandy, J.E. Effect of oxalic acid pretreatment of wood chips on manufacturing medium-density fiberboard. *Holzforschung* **2011**, *65*, 737–741. [CrossRef]

167. Sydor, M.; Cofta, G.; Doczekalska, B.; Bonenberg, A. Fungi in Mycelium-Based Composites: Usage and Recommendations. *Materials* **2022**, *15*, 6283. [CrossRef] [PubMed]

168. Hunt, J.F.; Supan, K. Binderless fiberboard: Comparison of fiber from recycled corrugated containers and refined small-diameter whole treetops. *For. Prod. J.* **2006**, *56*, 69–74.

169. Theng, D.; Arbat, G.; Delgado-Aguilar, M.; Vilaseca, F.; Ngo, B.; Mutjé, P. All-lignocellulosic fiberboard from corn biomass and cellulose nanofibers. *Ind. Crops Prod.* **2015**, *76*, 166–173. [CrossRef]

170. Liu, H.; Xie, F.; Yu, L.; Chen, L.; Li, L. Thermal processing of starch-based polymers. *Prog. Polym. Sci.* **2009**, *34*, 1348–1368. [CrossRef]

171. Xian-qing, X.; Ying-ying, Y.; Yi-ting, N.; Liang-ting, Z. Development of a cornstarch adhesive for laminated veneer lumber bonding for use in engineered wood flooring. *Int. J. Adhes. Adhes.* **2020**, *98*, 102534. [CrossRef]

172. Nazerian, M.; Razavi, S.A.; Partovinia, A.; Vatankhah, E.; Razmpour, Z. Prediction of the Bending Strength of a Laminated Veneer Lumber (LVL) Using an Artificial Neural Network. *Mech. Compos. Mater.* **2020**, *56*, 649–664. [CrossRef]

173. Frey, M.; Schneider, L.; Razi, H.; Trachsel, E.; Faude, E.; Koch, S.M.; Masania, K.; Fratzl, P.; Keplinger, T.; Burgert, I. High-Performance All-Bio-Based Laminates Derived from Delignified Wood. *ACS Sustain. Chem. Eng.* **2021**, *9*, 9638–9646. [CrossRef]

174. Xiong, X.Q.; Bao, Y.L.; Guo, W.J.; Fang, L.; Wu, Z.H. Preparation and application of high performance corn starch glue in straw decorative panel. *Wood Fiber Sci.* **2018**, *50*, 88–95. [CrossRef]

MDPI
St. Alban-Anlage 66
4052 Basel
Switzerland
Tel. +41 61 683 77 34
Fax +41 61 302 89 18
www.mdpi.com

Forests Editorial Office
E-mail: forests@mdpi.com
www.mdpi.com/journal/forests